Sustainability in Engineering Design

Sustainability in Engineering Design

An Undergraduate Text

Anthony Johnson **B.Sc., M.I.Mech.E, C.Eng, FHEA**

Andrew Gibson **B.Sc., CIM, MIEx**

AMSTERDAM • BOSTON • HEIDELBERG • LONDON
NEW YORK • OXFORD • PARIS • SAN DIEGO
SAN FRANCISCO • SINGAPORE • SYDNEY • TOKYO
Academic Press is an imprint of Elsevier

Academic Press is an imprint of Elsevier
225 Wyman Street, Waltham, MA 02451, USA
525 B Street, Suite 1800, San Diego, CA 92101-4495, USA
Radarweg 29, PO Box 211, 1000 AE Amsterdam, The Netherlands
The Boulevard, Langford Lane, Kidlington, Oxford, OX5 1GB, UK
32 Jamestown Road, London NW1 7BY, UK

British Library Cataloguing-in-Publication Data
A catalogue record for this book is available from the British Library

Library of Congress Control Number
Johnson, Anthony (Anthony D.), author.
Sustainability in engineering design/Anthony Johnson & Andy Gibson.
pages cm
1. Engineering design–Textbooks. 2. Sustainable engineering–Textbooks. I. Gibson, Andy, author. II. Title.
TA174.J64 2014
620′.00420286–dc23

2013050375

ISBN: 978-0-08-099369-0

For information on all Academic Press publications
visit our website at store.elsevier.com

Printed and bound in United States of America
14 15 16 11 10 9 8 7 6 5 4 3 2 1

Contents

Preface

Anthony Johnson and Andrew Gibson are engineers with extensive experience in several complementary fields.

Anthony started his career as an indentured apprentice with the U.K. Royal Navy. He worked in a commercial capacity as well as a design capacity in several industries, eventually managing a busy design office for a specialist plant manufacturer. Anthony has performed numerous engineering design consultancy projects during his career at the University of Huddersfield, where he has taught mechanics, dynamics, strength of materials, automotive design, computer-assisted design (CAD), and engineering design for 27 years.

Andrew has extensive expertise in customer-driven product development, technical sales, business strategies, export strategies, and legislation related to engineering. After graduating with degrees in engineering and German, Andrew worked in product development and as customer liaison, moving on to export sales and marketing and thence into the management of international subsidiaries of U.K.-based engineering companies. He subsequently developed a consultancy-based business, working with clients to penetrate new markets and focusing latterly on international trade training.

The raison d'être for the book was the result of several influences: a chance meeting with a sustainability entrepreneur, the need to teach sustainability for engineers in a higher education establishment, the embarkation of a post-graduate research program integrating sustainability principles with design methods, and the opportunity to take the message of practical engineering design out into the world.

This book is essentially a text that covers the formulative approach to mechanical engineering design but also combines methods and techniques that accommodate the principles of sustainability for use by practical design engineers performing their day-to-day work. Sustainability is an increasingly important environmental factor that is promoted by environmentalists, national governments, and commercial enterprises alike, and as such it is required by engineering institutions to be woven into the fabric of undergraduate and post-graduate engineering programs.

When new products are created, it makes sense to incorporate the principles of sustainability, but the *only* function in the product creation process that can deliver the principles of sustainability is the design function. This text promotes the total design control approach to sustainability and applies it at a practical level.

The material in the book has been drawn from extensive research as well as from the wide-ranging experience of both authors.

Foreword

Anthony Johnson and Andrew Gibson are both experienced engineers whose combined practical and advanced expertise spans some 86 years. Anthony has brought his extensive practical design experience to the book as well as his 27 years of experience in teaching several analytical subjects, along with engineering design, at the University of Huddersfield in the United Kingdom.

This book expands the normal approach to engineering design to accommodate the approach to modern CAD and analytical techniques, but more important, it incorporates the designer's approach to creating sustainably engineering products. Thus this book is hailed by several sources as a groundbreaking text.

During Anthony's Ph.D. research program, he extensively explored sustainability and its application and discovered that there was a dearth of literature covering practical application of sustainable principles for mechanical engineering designers. Anthony continuously asked the question, "How can practical designers apply the principles of sustainability to their newly designed products?" This book attempts to answer that question.

The book contains a methodical approach to design, taking into consideration Anthony's own practical design experiences and his involvement with various consultancy projects. Sustainability whole-product-life models are put forward, incorporating the six stages of a product's life, from sustainable sourcing through sustainable disposal. The approach to implementing sustainability within the design process has then been engaged within each of the elements.

Andy Gibson is head of a management training company, Segelocum Limited. Andy has strengths in production and process engineering and has particular skills in the fields of marketing-driven product development, and of the impact on companies of financial and legislative processes; these areas have been the focus of Andy's contribution.

The book seeks to address the forces that act on engineering designers and the businesses in which they are involved, including legislation, consumer pressure, and marketing opportunities. It describes in some detail the concepts of open and closed life-cycle loops and argues that the modern design must take into account impacts generated during the full life cycle but concentrate on where the correct selection of material or design modification can make the greatest impact. In that light, the book also looks in brief at some of the modeling methods currently used to measure the sustainability of machines and machine elements across the whole life cycle.

Prof Andrew Ball
Pro-Vice-Chancellor Research
University of Huddersfield
West Yorkshire
United Kingdom

Book Synopsis

Earth's natural resources are becoming increasingly scarce, yet the drive for ever-newer products could not be greater. The strain on our natural resources is therefore increasing, but it is our contention that sustainable engineering design must be part of the solution rather than part of the problem.

Sustainable design incorporates the standard design process with the sustainable sourcing, use, reuse, and disposal of materials ensuring a minimum effect on the planet's resources. The normal design process merely selects materials for manufacture often with no thought as to the source of the material or the method of disposal after the products life has expired, let alone the possibility of renewed or extended life through easy maintenance. Sustainable design incorporates eco-sourcing at the beginning of the design process and eco-dismantling on the end of the design process. It also incorporates an improvement in product life cycle by designing-in ease of repair and renovation.

Central to this theme is the argument that the ultimate responsibility for sustainability in components, products, and systems belongs with the design function; just as Taguchi suggested the same function also controls quality.

If goods and products are to be produced to use resources efficiently and with low cost, then it is important that awareness, knowledge, and ability are generated in our new graduate designers, and this book proposes and details some useful tools and approaches that can be incorporated into a standard designer's armoury to ensure that their output moves the planet toward rather than away from a sustainable future.

Introduction

Traditionally, engineering designers designed and manufacturing engineers manufactured the design. In the late 1950s, engineer and statistician Genichi Taguchi suggested that quality should be placed in the hands of the designer. This revelation prompted a design and manufacturing revolution and was the catalyst that drove production techniques into quality mass production. Though production techniques had been pioneered by inventors and industrialists such as Samuel Colt and Henry Ford, quality of manufacture still largely remained the domain of craftsmen and artisans to produce and create quality goods.

There are a great number of demands placed on the modern design and production environment. Some of these are traditional demands, such as that of reducing cost. Newer demands, however, are becoming prominent, such as those that require reduced environmental impact when a new design is created. This relatively new discipline is *sustainable engineering design.*

Taguchi was one of the first proponents of placing the emphasis on designers and the idea that designers should take control and specify quality. The greater demands and expectations placed on new products effectively demand that designers take a greater role in specifying and controlling the new product, from its inception right through manufacture to packaging and even marketing. This is, effectively, Sustainable Engineering Design. The design function can no longer be compartmentalized, since it is the only function that can oversee and control the entire process of product creation from "cradle to grave."

It is within the designer's gift to apply sustainable design techniques for a long-life product. It is the designer and *only* the designer who has the overview of the whole design and manufacture process. This whole-life process involves the following elements:

- Specifying sustainable materials
- Designing for sustainability
- Designing for sustainable manufacture
- Designing for sustainable use
- Designing for sustainable maintenance
- Designing for sustainable disposal

The designer must therefore take control of the whole-life process. This technique is termed *total design control.*

This book examines traditional design techniques and offers suggestions as to how these techniques can be incorporated into total design control. To achieve this goal, the traditional approach to design compartmentalization must change to a global design approach. Traditionally designers design and manufacturers manufacture, but the new methodology demands that these two major disciplines be combined and joined by several other disciplines. This means that designers' attitudes and approaches have to change to a global, whole-life design strategy. This book shows how to achieve that goal.

Anthony Johnson
University of Huddersfield
West Yorkshire
United Kingdom

Andrew Gibson
Segelocum Ltd
Retford
Nottinghamshire
West Yorkshire

January 2014

CHAPTER

Engineering Design: An Overview

1

1.1 WHAT IS DESIGN?

Design encompasses many fields and can be described as a creative process that brings into being a new product, concept, or process using subjective or qualitative means.

A new design may start in several ways. Often there is a need for an item or a service that leads to a new specification. This specification is followed by a new design and eventually a new product. This approach to a new design may be considered "necessity led." A notable example is the heat shield fitted to the National Aeronautics and Space Administration (NASA) space shuttle. The *need* to reenter Earth's atmosphere and protect the spacecraft from the extreme high temperatures led to a specification (essentially the brief) and the eventual development of a reusable heat shield.

It is rare that a brand-new product is designed without the presence of "need," although there is a notable example in the laser. In 1917 Albert Einstein established the theoretical foundations for the laser, but it was not until Theodore H. Maiman [4] operated the first functioning laser in 1960 that the technology became a reality. This "technology-led" innovation was a system requiring a product. In the intervening years, lasers have become commonplace for optical measurement, CD and DVD players, level sights, and much, much more.

Inventions are often ideas of an individual who is completely unrelated to the topic of the invention in what is sometimes called a *blue-sky moment*, often the result of a previously unperceived need. A great example is when Andrew Speechley was late for a meeting in Halifax, UK, and could not find a parking place. In a blue-sky moment, he imagined the three-dimensional space above the normal car park filled with parked cars. Later, at home he built a model showing the means of parking one car above another. He had just invented the car stacker. Plate 1.1 shows the concrete base version of Speechley's invention.

By whatever means, once there is a need and an idea to fill that need, a product has to be brought into existence. At the first attempt, this product is often very crude and may not work efficiently. Often, to become a sophisticated item, a product requires further development with newer techniques and alternative materials. This further development, building on the original prototype, is usually an evolving process whereby feedback from the user is fed back to the design team. When new information is received, the designer returns to the concept to introduce new elements and further enhance the product concept design.

The bicycle is one of the best examples of design evolution. The first bicycle was originally a wooden frame with two wheels and something by which to steer. See Plate 1.2. Several iterations later, a chain drive was added, and further refinement led to lightweight frames, suspension systems, better brakes, and so on.

A modern concept bicycle is shown in Plate 1.3. This product incorporates lightweight materials streamlining and inbuilt suspension systems and has a mass that is a fraction of the original concept shown in Plate 1.2.

Sustainability in Engineering Design. http://dx.doi.org/10.1016/B978-0-08-099369-0.00001-7

PLATE 1.1

Carstacker Mk2 with Concrete Base [3]

PLATE 1.2

Laufmaschine: One of the First Bicycles [1]

1.2 DEFINITION OF DESIGN AND OVERVIEW OF THE DESIGN PROCESS

The implication of the word *design* is a very familiar concept. It embodies the creation and development of *all* the products, tools, and systems that we use every day. Every product in use has first to be designed, but there is a definitive process all developments go through to reach a conclusion.

The new product first has to be imagined, then communicated, designed and finally developed and produced. The basic design process model is illustrated in Figure 1.1.

PLATE 1.3

Modern Concept Bicycle (fuelyourcreativity.com)

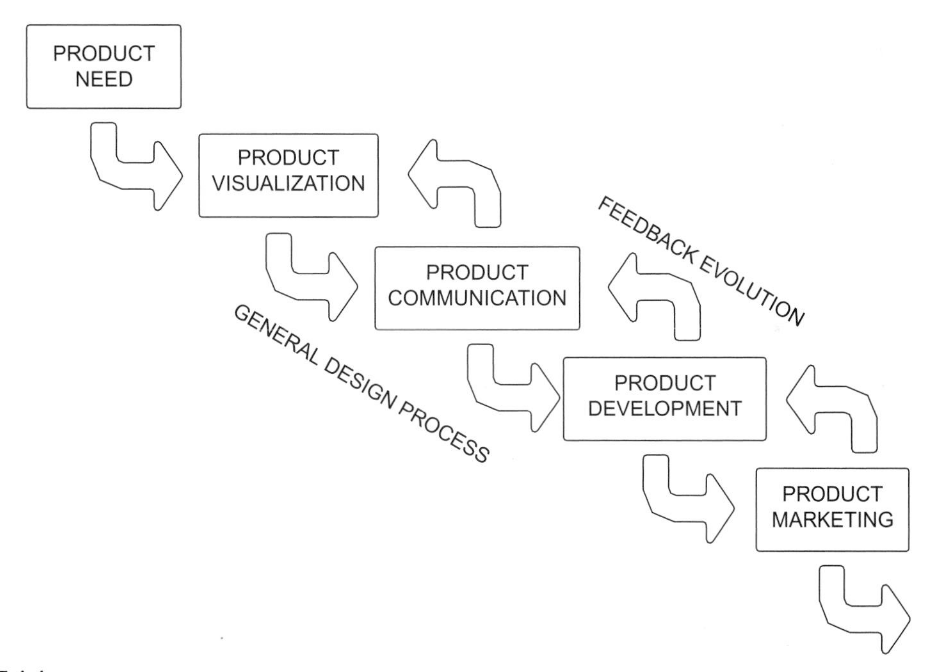

FIGURE 1.1

Basic Design Process Showing Feedback Evolution [3]

Most products are developed because there is a market need, or perhaps a market has been identified. A great example of new market identification is the development of the Apple iPad. This was the first tablet computer on the market and was made possible because of its touch-screen technology. The marketing team at Apple had perceived that there was a need for such a computer by identifying a market gap. Development soon followed.

The Apple marketing team developed a specification for the new tablet computer gleaned from market sources, such as personal interviews, questionnaires, stockist (retailer) responses, and the like. The specifications were then communicated to the design team, each member of which would have had his or her own personal "mind's-eye vision" of the new tablet computer.

Communication of the product specification to each member of the design team—perhaps as sketches or 3-D computer visuals—was then essential. This communication was not limited to the pictorial view of the tablet; there would have been much discussion of how to achieve the technical aspects, such as electronics, touch-screen inclusion, software, and materials. There would also have been much discussion around patents and general management of the design process.

Within this discussion new information would have come to light through new design ideas and research. This new information engaged the design team in evolving certain elements of the concept design. Feedback such as this is essential. Communication of information and ideas is essential for a true concept to be developed.

The design had to pass through concept stage and renewing of the specifications of the finished product, but then it could be manufactured, first as a prototype, then as a preproduction model. Development would also have included specialist tools, production lines, and perhaps even new factories.

The new device then needed to be marketed. The marketing process included marketing strategies, segmentation, marketing communications, packaging, and distribution.

1.3 DESIGN ACTIVITIES

Engineering design combines many disciplines and processes. The design process diagram in Figure 1.2 shows the variety and scale of activities and eventual outputs from the design process.

Engineering design is an enormous subject, stretching from miniature design such as the design of chronometers and watches to the other extreme of large-scale design such as that of oceangoing ships.

The design engineer who is in the process of designing highly technical components is steeped in the discipline of selecting materials, tolerances, surface finishes, seals, and the like so that the components can be combined into perhaps a gearbox or an engine. This is one end of the technical design scale; at the other extreme the graphic designer applies his or her particular visual design skills on areas such as Web sites, advertising media, and business logos to communicate to the market.

Design perception is yet another aspect of design that encompasses the individual view. The word *design* is a familiar word and certainly a familiar concept, since it covers a huge range of creative activity. The householder who has "designed" his kitchen has exercised a form of design but has merely positioned the kitchen appliances and cupboards by moving cardboard cutouts around the sketch of a floor plan. This is really *superficial design,* or "design light."

At the other extreme there is *deep design,* which includes the involvement of many people in creating highly technical and large-scale design developments such as space satellites, atomic power stations, and jet aircraft.

The vastness of scale of the design process is illustrated in Figure 1.2. Designers cannot conceivably become designers across the whole range of design possibilities. Designers must become specialists in a particular field, although that field may be very broad. For instance, the designer whose expertise is the design of mechanical diggers may also apply those skills to other tracked vehicles such as cranes.

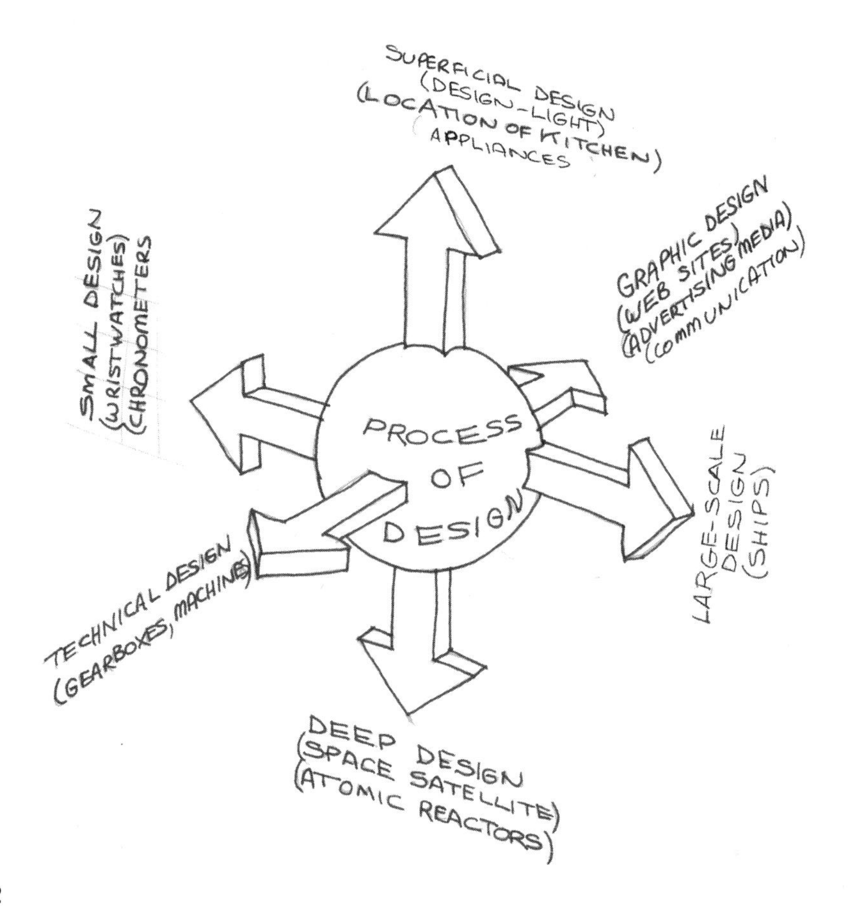

FIGURE 1.2

The Scale of the Design Process [3]

Conversely, atomic power plant designers would not have the specialized skills to design high-speed aircraft.

The scale of design is therefore vast; however, across this enormous subject of multiple disciplines, there are common features in the process of design. The major common features are as follows:

- General approach to design
- Design information
- Design evolution
- Designer skills and attributes
- Design information output
- Attention to detail
- Design within the awareness of the whole picture
- Three basic design objectives related to the triple bottom line

1.4 DESIGN INFORMATION

It is said that "information is the currency of design." Design would not happen without the inward flow and outward flow of information. The designer commences with sometimes a very tenuous brief, which has to be thoroughly investigated to tease out the important information and determine basic requirements. Research is then performed to gain even more information so that concepts can be generated. During this process the designer produces more and more information pertaining to the task in hand. At some point the designer has to communicate this information so that other specialists may add their own input to the project. This information flow is outlined in Figure 1.3, which shows the design information process model.

Information the designer may use can take many different forms. In the early stages of a project, the information may relate to the shape and function of the product. In later stages of the design, the product information will take the form of component specifications or perhaps manufacturing methods. The design of the arm of a digger might depend on the size and specification of a hydraulic ram. Specific ram sizes are required; otherwise the designer cannot complete the envelope within which the ram will fit.

Here is an example of the later stages of the design process: A designer is creating a lifting device that will lift a mass of 250 kg a distance of 2 m. The designer has several options for actuating the lifting device and requires the specific information for the different forms of actuation before the design process can progress. Information on the various actuation systems is as follows:

- Electric motors, sprockets, roller chain
- Electric motors, pulley, steel cable

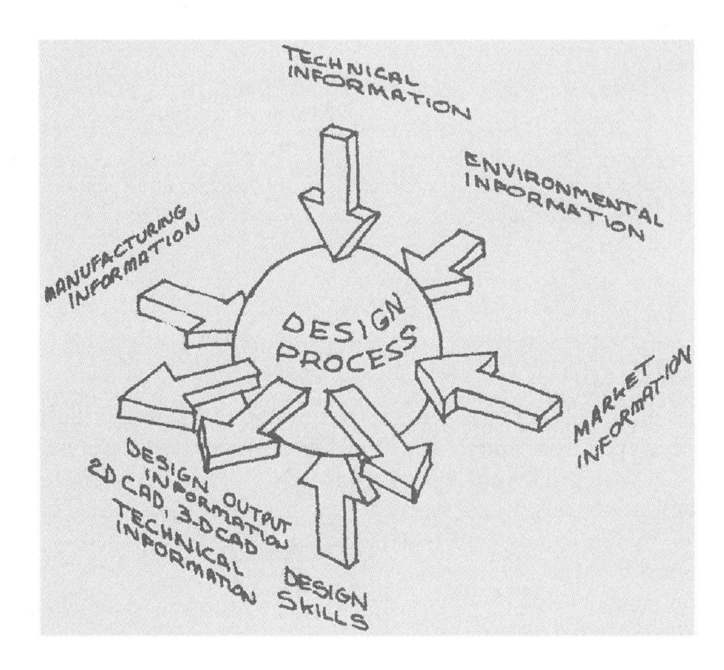

FIGURE 1.3

Design Information Process Model [3]

- Hydraulic power pack, hydraulic rams
- Pneumatic compressor, pneumatic rams

The selection of the actuator depends on many factors. Further information is the key to selecting the appropriate method.

If the device were to be placed in a factory environment, hydraulics or pneumatics could be used. These would be straightforward to maintain, with easy access to specialist personnel, and the noise that each method delivers would normally not be a great problem.

Conversely, if the hydraulic or pneumatic device were to be placed in a domestic environment, the noise levels would not be acceptable. Furthermore, the maintenance and general dirt contamination of the hydraulics device would prohibit its use. Pneumatics may be clean, but the compressor would also be noisy and therefore may not be acceptable in a domestic situation. Here, therefore, further information on the specific environment needs to be input to the concept design.

The electric motor options would be preferable, but the designer's choice has to be between roller chain or cable actuation. The cost of each system would probably be the deciding factor; however, each system would need to be analyzed in depth using specific product information.

It is clear from this example that a designer requires access to information from many different sources in order to complete the design prior to manufacture. Information relating to final destination of the product, usage, maintenance procedures, and so on is necessary for the designer to formulate an efficient design that is suited to the application.

Toward the end of the project when the designer is specifying components, the critical required information could be the size of a hydraulic ram or the physical envelope and power output of a diesel engine. Manufacturers and suppliers are usually very efficient and willing to supply design information that, after all, will help sell their equipment.

For many years, component suppliers have realized the critical nature of design information and its necessary ease of access to the designer. Many suppliers now give free access to a database of components that can be downloaded directly into the designer's 2-D or 3-D computer-aided design (CAD) model. Such standard components as bearings, seals, gearboxes, engines, and many more can be imported in this way. It is human nature that the designer will tend to use components for which the technical details are easy to access and insert. Component suppliers pander to this human tendency.

1.5 DESIGN EVOLUTION

Sophisticated designs are usually the result of *design evolution*. An unsophisticated first design will usually have been developed from initial sources of information, usually from a basic need. During the lifetime of the first product, information in the form of new processes, better materials, and feedback from the product's use will then provide impetus to further develop the design.

This evolutionary process continues as new information is received, both before and after a new product is launched. The product will go through subsequent changes as it is used and as feedback is disseminated. Feedback is important because it tells the designer how to modify the product to elicit improvements.

An excellent example is the development of the passenger vehicle. The motorized carriage shown in Plate 1.4 was originally developed from horse-drawn carriages.

PLATE 1.4

The Benz Patented Motorwagen, Circa 1885 (1.1)

Engines and power transmission devices were first introduced to a driving axle. The result was a horseless vehicle that possessed a top speed of about 12 mph.

Several other engineers were working on gasoline-powered automobiles, but it was Karl Benz who first patented his work and is generally credited as the inventor of the gasoline-powered automobile.

During the 20th century, technological developments took place through design evolution. Engine design was improved along with suspension systems, vehicle structures, tires, brakes, fuel, and a plethora of other devices and systems. The subsequent improvements increased safety, improved passenger comfort, and improved vehicle efficiency. The gas-powered vehicle became a sophisticated people carrier. This happened only because information became available through research, opinion, and trials, which eventually led to the evolution of components and systems.

Sophisticated information-processing systems within computer databases now make it possible to design vehicles that will reach 1,000 mph (Mach 1.42). Information manipulation, especially in terms of analysis, has made it possible to develop designs that would have been unachievable using hand calculations and other manual processing techniques such as graphical synthesis.

By comparison with the Benz motorwagen, modern vehicles are extremely advanced. A modern example of a sophisticated vehicle is the Bloodhound SSC 1,000-mph car, shown in Plate 1.5.

Many elements of this vehicle require state-of-the-art analysis. One of those elements relates to aerodynamics. Much of the prediction of the airflow around the vehicle at speed is due to a technique called *computational fluid dynamics* (CFD). CFD is a sophisticated computational technique that creates tiny imaginary elements around a vehicle. This is termed a *finite element mesh*. In the case of the Bloodhound SSC, 10 million elements were created along the length of the vehicle body.

As fluid flow (air) enters an element and then leaves the element, it transfers the velocity, direction, and pressure to the next element. As the shape of the vehicle develops from a pointed tip at the front of the car to the full body of the car, individual elements record changes to the pressure, velocity, and flow direction. In this way the parameters of the airflow over the surface of the vehicle can be computed and predicted.

The computation of 10 million finite elements is far beyond the capability of human analysis. The Bloodhound team enlisted the assistance of the Swansea University Department of Engineering, which developed a supercomputer to handle the data. Initially the computation spanned four days. Additional sophisticated computing power reduced the time to a little over four hours.

PLATE 1.5

Bloodhound SSC 1,000-mph Car (www.bloodhoundSSC.com)

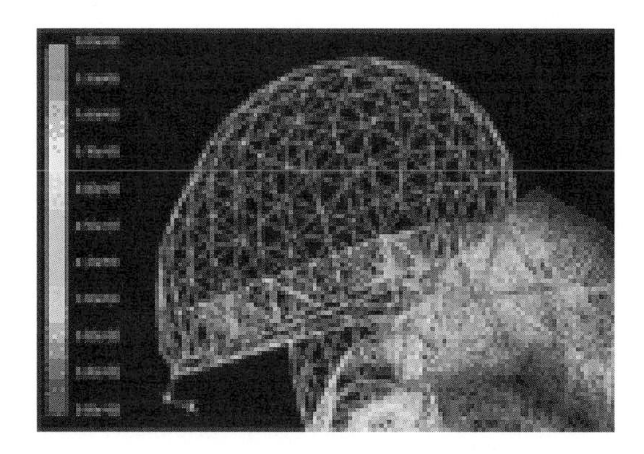

PLATE 1.6

CFD Analysis of Airflow Around a Motorcycle Crash Helmet [3]

CFD output is shown in Plate 1.6, which shows a motorcycle helmet modeled as a finite element mesh. Within the software, fluid of the density of air is blown onto the crash helmet from the left. The discoloration around the rider's neck area shows pressure, velocity, and direction disturbance as the air leaves the helmet. Analysis of this data allows the designer to change the helmet shape and, in so doing, to smooth the fluid flow. This leads to lower "noise" and reduced fluid drag.

Using modern analysis tools, sophisticated designs can be created that could only be dreamed of a few years ago.

1.6 DESIGN INFORMATION OUTPUT

Having processed the information to create a design, the designer needs to communicate that information. This is done in many ways:

- Verbally
- Written technical report
- Two-dimensional drawings
- Three-dimensional models
- Animations and simulations
- Analytical information
- Parts lists
- Bought-out component specifications

Once the information is obtained, applied, and manipulated, the designer requires high-level skills to impart this information to other specialists such as product engineers and marketing specialists. In the modern design and manufacturing environment, in-depth knowledge and use of 2-D CAD and 3-D CAD software are essential to the information dissemination process. CAD software can perform several duties for the designer above and beyond providing 3-D models and technical 2-D drawings.

Packages built into CAD software allow stress analysis, dynamic analysis, deflections, and thermal analysis to be calculated. These sophisticated tools provide high-level information that can be used to disseminate the design information to specialists such as buyers, production engineers, marketing personnel, packaging personnel, and many, many more.

A further development of 3-D modeling has been 3-D printing, in which 3-D models and products are produced by an additive rather than subtractive process (e.g., machining). Layers of product are laid down to build a solid 3-D object in the chosen material.

It can be seen that information in several forms is paramount to any design process and can be summarized as follows:

1. Initial information allows the design to begin.
2. Research information allows the design to unfold
3. Expertise information allows the design to develop.
4. High-level information manipulation techniques allow the design to become sophisticated.
5. Analytical output information allows the design to be performance predictable.
6. Information output methods allow the design to be disseminated.
7. Technical manufacturing information allows the design to be produced.

1.7 QUALITIES OF THE DESIGNER

The engineering designer needs to have a mindset that allows creativity. An open mind and a willingness to investigate solutions from different environments and disciplines are necessary so that simple, effective, and often radical solutions may be designed into a product.

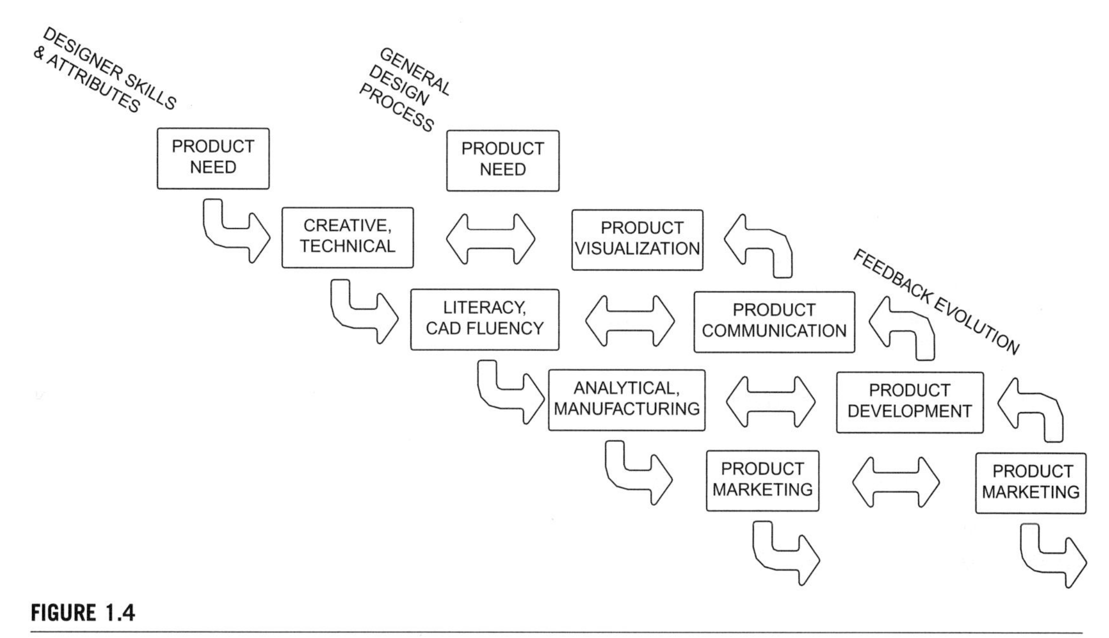

FIGURE 1.4

Designer Skills and Attributes Related to the Design Process Model [3]

Design is a process that brings together many different disciplines. The designer needs to have an awareness of these disciplines and perhaps be a specialist in several of them. His or her skills and attributes are reflected in the general design model and are shown in Figure 1.4.

At the beginning of a project, the designer needs to be creative yet technical, skills that are often in conflict with each other. At the launch of the project, the designer's ideas and concepts have to be unlocked from his mind and disseminated to other team members. This can only be achieved using communication skills in the form of the written word or sketches.

A large element of information dissemination requires the creation of 2-D and 3-D models and, of course, technical drawings. These are not just pictorial representations of the product. They represent a communication tool that describes precisely how the item should be manufactured in terms of shape and size and how it should be precisely assembled. This is absolutely essential for manufacturing to take place, and it is the designer who possesses the skills to create this precise information.

Further along the project, during the manufacturing and development phase, technical details such as material selection, tolerances, stress analysis, and component selection (e.g., bearings or seals) have to be applied. These precise details are usually communicated to the manufacturing function through analytical output, 2-D drawings, or CAD models.

The new design is the brainchild of the designer or the design team. During the design process, the designer must apply her own working knowledge of materials, processes, and manufacturing possibilities. It is the designer who is the vessel for the detail data for bearings, seals, materials, strengths, and all other elements that constitute a full and proper design. It is therefore the designer who is best able to create the drawings and models that will precisely describe the components that make up a product.

True design at a detailed level requires a great deal of skill across many disciplines. The lone designer, however, may not possess the specialist knowledge of all the fields required to create a successful design. He must therefore be aware of his limitations and seek advice from other specialists. In this quest for multiple disciplines, the designer may need to form a multidisciplinary design team that can bring the product design to a successful conclusion.

1.8 ATTENTION TO DETAIL

A major aspect that pervades the whole working ethic of the designer is that of *attention to detail*. Every single minute detail of specification for a design needs to be communicated. This attention to detail may be tedious, but it is of absolutely paramount importance. *The designer must think of everything!*

The designer creates value while in the process of generating new designs. The designer has to think through the manufacturing process and specify parts and components used during manufacture. The detail the designer provides has to be absolutely precise. There must be no allowance for guesswork by the manufacturing team. Guesswork can only lead to wrongly specifying components or costly mistakes.

The cost of a product is based on the parts list the designer provides. Parts that are omitted from the parts list cannot be counted as part of the cost. Furthermore, components such as jigs may be required to aid assembly, and they must also be costed. The minutest detail must be considered. Perhaps a cable tie is temporarily required to hold components together to aid assembly. This cable tie may be cut and discarded once the components are assembled, but it does cost money and it should appear on the parts list.

A single hydraulic fitting worth £3.00 may be omitted from the parts assembly pack for an installation that could be worth £100,000. This could mean that pipe work could not be completed and the installation could not be commissioned. This single fitting could cost hundreds of thousands of pounds in lost revenue simply because its absence delayed commissioning.

The same is true of lubricants. Often lubrication delivery systems or even lubricants are the last item to be considered. Sometimes they are not considered at all, resulting in overheated bearings, seizures, and very expensive downtime. The only person who can specify the type of lubrication delivery system, lubricant, and quantity is the designer. Attention to such detail at the design stage reduces problems in manufacture and further losses of the product in use.

1.9 THE WHOLE PICTURE

Sometimes the design is extremely complex and is impossible to visualize completely. Modern 3-D CAD assists the designer immensely in visualizing the whole picture. These CAD systems are equipped with collision detection, which throws up an error message should one component infiltrate the same space as another component. Components can be fitted within the 3-D computer environment before manufacture. This results in low-cost *prototyping*. The build of full-size prototypes is usually a very expensive process.

A World War II British bomber, the Avro Manchester (see Plate 1.7), could be considered a prototype aircraft and was the forerunner of the Avro Lancaster bomber (see Plate 1.8). The Manchester

PLATE 1.7

Avro Manchester Bomber [6]

PLATE 1.8

Avro Lancaster Bomber [5]

was a twin engine bomber but suffered greatly from underpowered engines and aerodynamic instability.

The practical building and testing of the Manchester enabled the Avro engineers to ascertain its faults and enhance the design by discarding the original Vulture engines, extending the main wingspan, and adding four Rolls-Royce Merlin engines. Further modifications were made to the tailfin arrangement. The result was the Lancaster bomber, which was much more powerful and had a better aerodynamic response. At the time these aircraft were designed, the difficulties in visualizing a complex product were solved by building an expensive full-size prototype.

Even the best modern designers cannot see the whole picture and require feedback from users of the product and the manufacturing team so that elements can be modified to function more efficiently. Feedback from product users is very important. Often the product in service throws up problems that might not have been considered during the design and manufacture phase. Designs may need to be further modified to incorporate this new information.

A classic case of feedback in use and further modification is the London Millennium Footbridge across the River Thames (see Plate 1.9). This is a radical design of a steel suspension bridge.

PLATE 1.9

The London Millennium Footbridge (1.1)

Pedestrians crossing the bridge felt unexpected swaying motion within the first two days of the bridge's opening. The phenomenon that caused the resonance was identified as "synchronous lateral excitation." The natural swaying motion of people walking across a bridge caused small sideways oscillations, which in turn caused people on the bridge to sway in step, thus increasing the amplitude.

Resident vibrational modes in bridges are well understood in bridge design, but in the case of the Millennium Bridge the lateral motion caused by pedestrians was very unusual and was not anticipated by the bridge designers.

In this case, feedback from the bridge users led to the closure of the bridge until modifications could be made. The bridge designers applied much research and testing to the bridge structure, gleaning information that led to appropriate modifications. Thirty-seven viscous dampers were attached to the underside of the structure, reducing the oscillations to a tolerable level.

1.10 CLASSIC ENGINEERING DESIGN-TO-MANUFACTURE MODEL

The classic engineering design-to-manufacture model has been used for thousands of years, and though some elements of sustainability have been practiced, the mindset and pressure on many modern engineers are to design the product so that it can be manufactured at a low cost (see Figure 1.5).

Under normal circumstances, the design engineer would receive a brief and create a target specification, which is often no more than a wish list. This gives the designer or the design team a target toward which the design process should be steered. The designer should fully understand what the brief conveys, perhaps performing some analysis and research before arriving at a concept design.

The concept would then be scrutinized further to provide a concept specification, which is much more detailed than the target specification, since it includes performance feasibility. The concept data provides the design and feasibility information with which to make progress decisions. These are usually financial decisions and a commitment to fund the prototype manufacture. Large products such as

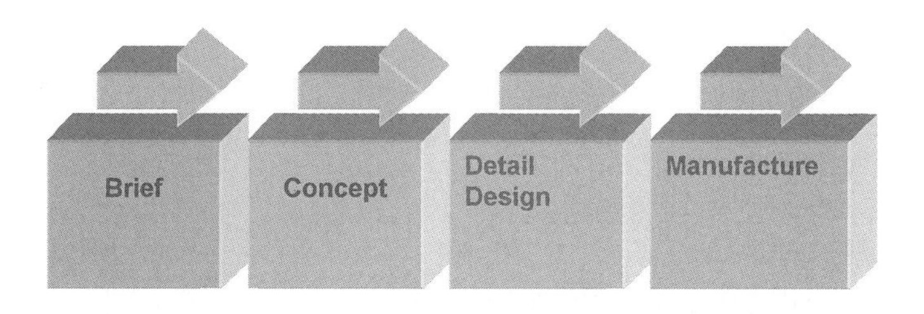

FIGURE 1.5

Classic Design-to-Manufacture Model [3]

ships are basically too large and expensive to simply prototype. The concept design becomes the product in such a case. Smaller products such as automobiles are usually prototyped, then modified before being put into production.

The concept stage is, therefore, a very important phase of the design process, and once accepted, work can begin on converting the feasibility design into a working design.

The working design is achieved by further analysis but also by specifying components such as bearings, seals, motors, hoses, and whatever else is required to complete the product. The designer would draw on the thousands of components that are commercially available and apply them to the design, finally completing either 2-D or 3-D computer models and providing a detailed parts list. The parts list is really a shopping list of raw materials and components and is an important part of the information dissemination. The production team uses the parts list and the drawings in combination in order to manufacture the product.

The design-to-manufacture model shown in Figure 1.5 is the classic approach that has been used by designers for thousands of years. The designer conceives the component, which is then manufactured and sent to the client. In modern times there are increasing pressures on the designer to consider material sourcing and environmental impact through the life of the product right through to its disposal.

1.11 OVERALL DESIGN OBJECTIVES AND THE TRIPLE BOTTOM LINE

1.11.1 Overall design objectives

Let's take a step back from the design and manufacture of a product and ask the question, What is the main objective?

The designer will design and eventually manufacture a product, but subconsciously he is trying to achieve several unspoken objectives:

Objective 1: To create profit for the company (the designer can then be paid a salary).
Objective 2 To create a product that will enhance the lifestyle of other human beings.
Objective 3: To achieve sustainable development of the whole product life cycle.

The first two major objectives have been in every designer's subconscious since the cavemen first used clubs. In modern times there is now a third objective, which is to use as few resources as possible.

These three objectives now form the basis of a relatively new term: *triple bottom line*. A designer should now consider all three objectives when he is confronted with the design challenge.

1.11.2 First bottom line: profit

Any business needs to make money. Some businesses are perhaps nonprofit making, such as charities, but even these businesses need to make a return on their efforts to fund growth and development. Generally, businesses need to make a profit in order to progress and survive.

Early British industrialists, the architects of the Industrial Revolution, were driven by the need for increased output to satisfy the demands of the growing markets within the British Empire. As a result of applying new techniques and materials, they were able to derive a greater financial return on investment in fixed and current assets. Basically, they became wealthy.

This is the reason businesses survive. They have to make money to buy raw materials, to pay workers, to expand, and to make the owners wealthier. Making profit is the primary goal of any business and is the first bottom line: *profit*.

1.11.3 Second bottom line: people and society

During the British Industrial Revolution, some industrialists, such as Titus Salt and George Cadbury, developed a social conscience, seeking to improve the "lot" of their workers by housing them in model villages, educating their children, and even investing money in improving safety in the manufacturing process.

The latter-day social conscience stretches to design and manufacture of products that aid and enhance people's lifestyle. The advent of the car has improved communications and the transport of goods by using mechanical power. The computer has improved communications for business and for personal use. Perhaps the biggest social improvement in recent years has been the development of the cell phone. This personal communicator has many functions that enhance the owner's sociability; in the developing world this device has helped improve local trade, and hence the standard of living, of cell phone owners.

Engineering design, therefore, has played a huge part in enhancing the way people live and work. This has become the second bottom line: *people and society.*

1.11.4 The third bottom line: environmental impact and sustainability

The growing awareness and need to take care of the environment have led some designers to adopt a more sustainable and open approach to their design work. These enlightened designers adopt a *whole-life* approach. This means they have to consider all aspects of the product, from the sourcing of the material right through to its disposal at the end of its life. The following is a list of topics within the whole-life approach that designers need to consider:

- Sustainable material sourcing
- Sustainable design techniques
- Design for sustainable manufacture
- Design for sustainable use
- Design for sustainable maintenance
- Design for sustainable disposal

Sustainable material sourcing has been practiced in some quarters for many years. In the United Kingdom, when purchasing hardwoods for use in, for instance, street furniture, government agencies have required certificates to show that the wood intended for external seating has been sourced sustainably.

Sustainable manufacturing can be achieved through different techniques and methods, but a wider picture needs to be adopted, such as the use of smart factories. These are low energy use, low environmental impact factories. Some forward-thinking companies have incorporated these low-impact factories within their manufacturing profiles.

Examples of smart processing include the use of waste materials from the manufacturing process to generate heat and power for the process (e.g., cocoa processing). European Union (EU) funding has been made available under the *Factories of the Future* Public Private Partnership (PPP) to improve the control systems of "smart" factories, where energy management and efficiency are key considerations in both factory layout and the manufacturing process. Examples of these smart factories include the Brandex apparel factory in Sri Lanka, which uses efficient light-emitting diode (LED) lights for workstations, recycles water and harvests rainwater, and recycles biogas from the factory waste to power the kitchens. The new Renault factory in Tangiers Plate 1.10 has reduced CO_2 emissions by 98% through measures such as energy conservation systems, the use of renewable energy resources, and the recycling and harvesting of natural rainwater.

The use of a product often has a greater environmental impact than its manufacture or its disposal. A car or a truck is just such a product; its day-to-day use—burning fossil fuels and the incumbent emissions—creates an environmental impact much greater than that produced during its manufacture or its disposal.

Vehicle designers, however, have for years been striving to reduce the mass of vehicles, thus reducing the fuel burned during the vehicles' productive lifetimes. In the early years of this process, the goal was always to reduce the running cost, but this also improves the sustainable impact, since a reduction in fuel use is also a reduction in environmental impact.

Electrically powered vehicles have been predicted to be the transport of the future. Though they have an environmental impact during manufacture and disposal, the electric motor drive does not directly burn fossil fuels. Though some fossil fuels are used to generate electricity on which the vehicles will run, developments in fossil-fuel-free power generation are already in place or are in development. The advent of the electric vehicle can be considered a major pointer to true sustainable use and reducing environmental impact.

PLATE 1.10

Renault Smart Factory [2]

PLATE 1.11

Motorcycle 70 cc [3]

Maintenance has always been essential to prolong the life of products and has largely been necessary to reduce the cost of purchasing new products. Maintenance of a product can now be considered a major sustainability tool, since prolonged life means lower environmental impact by avoiding the procurement of new items.

Although designs for industrial plant, equipment, and machinery have included ease of maintenance in their design briefs, most domestic equipment has become increasingly difficult to maintain as labor rates have tended to make manual maintenance uncompetitive. It requires another shift in design thinking to embrace the possibilities presented by modern materials and methods into designing ease of maintenance back into products.

The motorcycle shown in Plate 1.11 is an excellent example of design for sustainable maintenance. The motorcycle can be taken apart very easily to allow components to be replaced. One major aspect of this design is that it is a simple design that requires only basic tools to enable dismantling.

Though this is an excellent example of design for sustainable maintenance, it is possible that this was not the designers' primary goal. The market requirement was really for low-cost, single-person transport. Cost was therefore the driving factor for this simple design. Comfort can be taken, however, from the fact that sustainable design often reduces the cost of a product.

Sustainable disposal is often driven by cost. The designer must now consider disposal aspects of products at the concept stage of the design. The design should therefore accommodate easy removal of components and appropriate reuse, recycling, and refurbishment as appropriate.

The vehicle manufacturing industry is a useful example of this practice. Vehicle manufacturers aim for 95% recycling in new vehicles, and of course there is a healthy reuse market for vehicle components supplied by vehicle dismantlers.

The incorporation into the design process of the whole-life model and its consideration at each stage of the design process incorporates the third bottom line: *environmental impact and sustainability*.

1.11.5 Triple bottom line and design objectives comparison

Though designers are normally working on a practical level in creating products, high-level goals also need to be achieved, as follows:

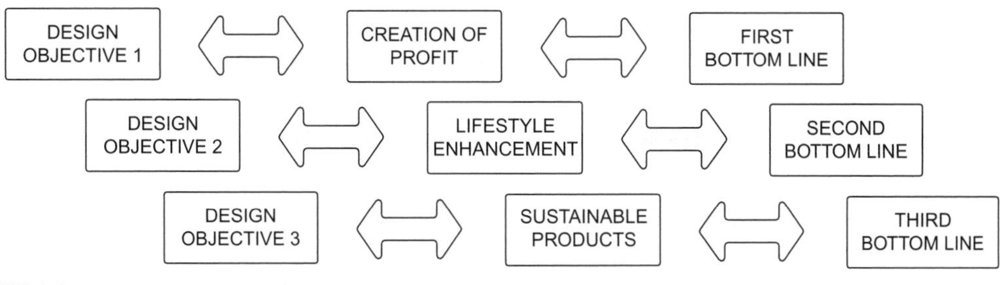

FIGURE 1.6

Major Design Objectives Linked to the Triple Bottom Line [3]

Objective 1: To create profit for the company
Objective 2 To create a product that will enhance the lifestyle of other human beings
Objective 3: To achieve sustainable design and manufacture

These three high-level design objectives fit well with the requirement to provide a triple bottom line. As was stated above the goals of the triple bottom line are as follows:

First bottom line: Profit
Second bottom line: People and society
Third bottom line: Environmental impact and sustainability

Figure 1.6 shows how the design objectives are combined with business objectives leading to elements of the triple bottom line.

Many companies, especially international companies, are combining sustainability into their design and manufacturing programs. In some notable cases the introduction of sustainable themes has improved the environmental impact of products as well as reducing costs for the company. Sustainability built into a product life cycle also gives marketing potential and provides a feel-good factor for the consumer.

References

1. Open source free from copyright restrictions.
2. Open Internet source.
3. Author provision.
4. T.H. Maiman, Stimulated optical radiation in ruby. Nature 187 (4736) (1960) 493–494, http://dx.doi.org/10.1038/187493a0, Bibcode 1960Natur.187.493 M. DOI:10.1038%2F187493a0.
5. www.secondworldwar.org.uk.
6. www.turbosquid.com (2011).secondworldwar.org.uksecondworldwar.org.uk.

CHAPTER

Design Approach, Philosophy, and Normal Approach Design Model

2

2.1 THAT "EUREKA" MOMENT

It is probably built into the human condition that from time to time there is a "eureka" moment, sometimes called a "light-bulb moment." In a moment of clarity, a person is struck by a bright idea. This phenomenon is part of the inventive process and has its place in the development of new products and services. Indeed, a whole industry is devoted to processing and developing inventions.

It has been said that inventors are "2% technical and 98% enthusiasm." This view can be explained by understanding that inventors are often free-thinking individuals who are not tied by the bounds of engineering discipline. Inventors often believe that their "invention" is something that the world cannot do without, and it is going to make them millions. The drawback is that inventors often fail to perform the proper marketing and patents research. They believe so much in their invention that they may spend a huge amount of money in development only to find that someone has developed exactly the same product or perhaps that it has already been patented. After spending a lot of money, the "inventor" may even find that her intervention does not work.

In most cases, an "invention" is merely conceptual and requires proper engineering development and marketing to make it a viable product. The odds of an invention making it to market are massively stacked against success. It is estimated that only one in every 5,000 ideas [7] ever reaches the prototype stage, and even then many prototypes are never developed into marketable products, simply due to the cost of development or lack of design expertise. The development of even the smallest product often swallows up an alarming amount of money.

Almost by definition, engineers and designers who are disciplined in their field are rarely called inventors. A distinction needs to be drawn between inventors and engineering designers. The latter normally approach the design of a new product in a disciplined fashion, having also performed due process in terms of market research and patent searches.

2.2 AN HISTORICAL APPROACH TO DESIGN

Ever since early humans created the first tools, an element of design has existed. Those early attempts were often by trial and error, which these days we would call iteration, and the design concept was in the mind of the person who was doing the manufacturing. This was a general method, but most things were made by hand unless some form of planning was required, usually with large buildings. The pyramids would have to have been designed and planned in order to achieve straight lines, geometric accuracy, and specific alignment to the heavens.

Those people who manufactured were called *artisans*. They were masters of their crafts and could manufacture a chair, a cooking vessel, or a cartwheel without recourse to drawings. The expertise was

Sustainability in Engineering Design. http://dx.doi.org/10.1016/B978-0-08-099369-0.00002-9

in the mind and the skill of the artisan. This design approach, with a few notable exceptions, was still prevalent until the advent of the Industrial Revolution in Britain. Although the actual manufacturing and production processes were still very much manual, requiring hundreds of workers, the machines that assisted the manufacturing process were complicated and themselves required designing prior to being built. The concepts were usually drawn by a visionary, such as Richard Trevithick, James Watt, Isambard Kingdom Brunel, and other notaries, and then given to the artisans to manufacture and fit together.

Toward the end of the 19th century the design process gradually became more sophisticated, and more accurate design drawings were made so that there was increasingly less reliance on the skills of the artisans. A huge step toward more sophisticated design occurred when Samuel Colt, who founded Colt's Manufacturing Company in Connecticut, made mass production of the revolver commercially viable for the first time. Colt's manufacturing methods were directed at beating his competition and were at the forefront of the U.S. Industrial Revolution. He was the first industrialist to successfully employ the assembly line and to use interchangeable parts. This fact alone meant that production and manufacturing had to be planned and "designated." The "artisans" on the production line would have been less artisans and more machine tool operators. The design function had arrived. The use of interchangeable parts meant that tolerances and surface finish were employed, and these would have been specified by the designers.

Though the design function in the Colt factory was in its infancy, it was nevertheless a glimpse of the future of the design element in an engineering context.

Around the same time as Colt in Connecticut, Joseph Whitworth in Manchester, England, proposed a standardized thread system called the British Standard Whitworth (BSW). This became British Standard thread and was designated BS 84 and, later, BS 84:1956. In 1830 a skilled mechanic could be expected to work to an accuracy of 1/16 of an inch (1.5 mm). By 1840 an accuracy of 1/10,000 of an inch (0.00254 mm) was practical. This was largely due to Joseph Whitworth devising a measurement instrument, a bench micrometer, that could measure one millionth of an inch. He also introduced methods of tolerancing* and a standard for surface finish.

As with the Samuel Colt factory in Connecticut, Whitworth's endeavors in England represented a breakthrough for designers because they could specify sizes, tolerances, and finishes and have confidence that the artisans on the production lines could produce components to those instructions. The shift from manufacturing to design was enabled by these advances in measurement and specification.

During the 20th century, design became a major function in the manufacturing industry. Some observers have said that World Wars I and II were as much a contest among engineering designers as among opposing armies. Certainly design and manufacturing advanced dramatically during those eras.

Figure 2.1 shows a drawing, circa 1928. It is a coachbuilders' drawing of the chassis for a Leyland model PH 4, a truck that was a workhorse for the British Army in World War I and was still being used in the late 1920s and early 1930s. This drawing was issued to coach-building artisans so that they could build a wooden body on the chassis. It should be noted that it is typical of drawings of the time and is almost an artistic piece of work.

Light shines on the vehicle from the top-left corner, putting the outlines on the bottom-right corner in shadow. These are double-thickness lines and give the impression of shadow. Though this is essentially

*A tolerance is a specified and allowed variation in a dimension. Physical dimensions cannot be machined perfectly so normally a "tolerance" is applied, e.g. 25 +/− 0.001 mm. This means the machinist can create the dimension to be between: 24.999 mm and 25.001 mm.

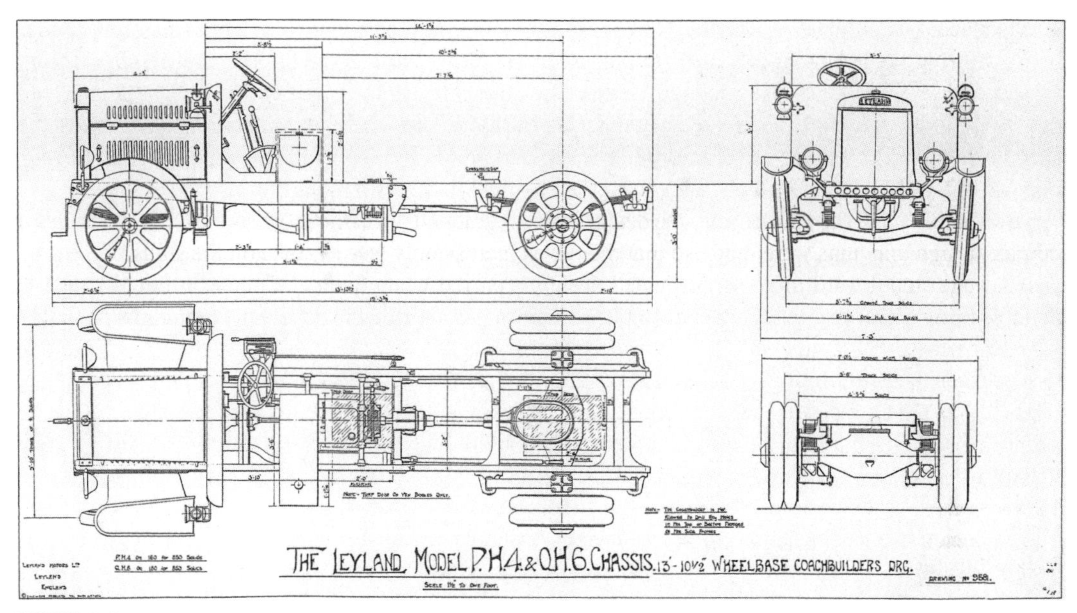

FIGURE 2.1

Coachbuilders' Drawing of the Leyland Model PH 4 Truck Chassis [1]

an overview drawing, note that there are no tolerances. The coachbuilders may have had their own drawings, but they would have relied on their own personal skill to produce accurate coach bodies.

Throughout the 20th century, design grew in importance as accuracy improved, both in manufacture and in drafting methods. The accuracy of information also improved, with more reliable analysis and the use of standards.

Genichi Taguchi, a Japanese manufacturing and statistical engineer whose most significant work was publicized in the 1950s and 1960s, expounded theories relating to the quality of manufactured goods. He was aware that the quality of manufactured goods was usually left to the manufacturing craftsmen to achieve. Taguchi suggested that quality could be designed in before components were manufactured. He also recognized that to achieve designed-in quality, the designers' mindset had to be changed.

As designers were eventually made more aware of the "new" design method, manufactured goods became more available at lower cost and higher quality. Engineering designers created better components by introducing standards, tolerances, and precision machine tools, which led to large-quantity production methods. The design function had now passed from the lone artisan through to a complete design department, with methods and tools to assist very sophisticated design work.

Modern designers are equipped with tools that a few years ago were mere dreams. Tools such as 3-D CAD modeling software are able to reproduce lifelike images of complex components on screen, without having to building them as prototypes. This software is equipped with stress analysis, dynamics, and thermal analysis tools (and many more) that allow the designer to perform complex analytical calculations that often are impossible to perform by hand. Very sophisticated designs are now possible, and accurate modeling and drawing production permits manufacturing production to be much more automated. Indeed, once the image is digitized into computer memory, that data can be moved to other functions such as purchasing, manufacturing, or marketing and easily shared. This capability assists greatly in improving a product's time to market. Designers no longer need to be in the same factory as production. A designer in the United Kingdom or the United States can create designs and e-mail them to a production facility in the Pacific Rim, say, China. The information can now be downloaded directly to computer numerical control (CNC) machine tools, where technical details are fed to the machine electronically and do not appear on paper drawings.

The results of CAD modeling and simulation software are so good that often prototypes do not have to be built. As designers move into a more and more sustainable role, this feature is of great benefit, since it reduces environmental impact. Furthermore, it is possible to print a 3-D image that gives a 3-D physical form. This image may be used to prove analysis, perhaps in a wind tunnel, or even used as an end component.

The possibility of printing a component is very exciting, since it uses a material-build process rather than a material-removal process, therefore giving zero waste. Another plus is that components can be created close to the assembly point rather than being manufactured overseas, thereby sparing a great deal of fossil fueled transport. The 3-D printing process has great potential of being a very sustainable feature of the constantly expanding design technologies.

The design function, whether for a lone designer or a full design department, is now in control of the design process and much of the manufacturing process. The designer creates value and disseminates information so that the product can be manufactured.

2.3 THE DESIGN APPROACH

Design is often performed by nontechnical people, and most of the time this is adequate. They will create something that works for them. Unfortunately, this approach may be expensive, since to make things feel safe people tend to overdesign, which costs money. Rule of thumb and perception of strength always tend to make things bigger than necessary. It is a human failing to overdesign in the belief that "it is better to be safe than sorry."

Before the advent of detailed stress analysis techniques, engineers had only one recourse, and that was to build things so that they *looked* safe. They generally worked to the principle of "if it looks right, it is right." Often buildings were built to very heavy specifications because materials were cheap but also because things were "built to last." The United Kingdom has an enormous number of buildings that are hundreds of years old and that were essentially overdesigned.

Imagine building an aircraft to this regimen. The aircraft would probably be so heavy that it would never fly. The lesson here is to look at the outcome of the product. A multistory car park design may be optimized to give a perception of strength to the users, and indeed it may require some overdesign to achieve that goal. An aircraft, however, will need to be optimized for light weight. In general it can be considered that the lighter the aircraft, the greater the payload and the higher the efficiency. Land vehicle designers have also optimized their vehicles for reduced weight in order to increase fuel efficiency and increase payload.

Generally, designers need to optimize design toward cost, which is the one single parameter that drives almost all designs. A second major optimization parameter that should influence designs is that of sustainability. The common misconception is that sustainability is expensive to incorporate, but true sustainability reduces environmental impact and often gives reduced manufacturing and running costs.

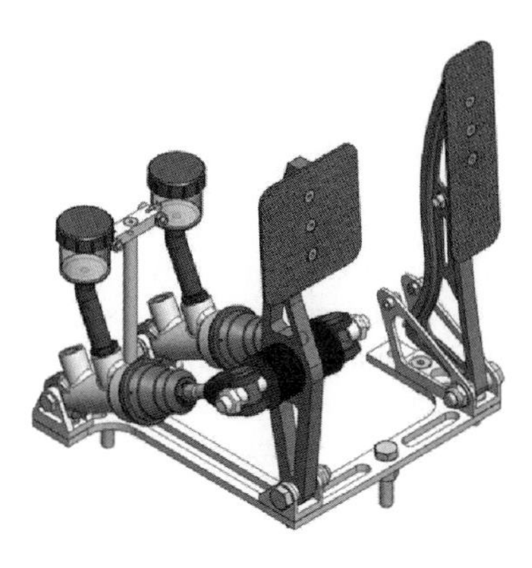

FIGURE 2.2

Final Concept Design for a High-Performance Vehicle [3]

Though designs can be optimized to achieve certain goals, the designer really needs to be led by the product requirements and tools, such as stress analysis, dynamic analysis, thermal analysis, and the like. Using these tools, designers can refine a design to give maximum performance for reduced cost and reduced weight and still have that design be sustainable.

The final concept design for a foot-pedal arrangement for a high-performance, lightweight vehicle is shown in Figure 2.2.

Various criteria were involved in the design of the pedal bracket, one of which was the necessity for reduced weight. Often a reduction in weight also means a reduction in strength. The design of the foot pedals optimized the reduction in weight and further optimized the high strength, even though these two parameters were in conflict.

Figures 2.3, 2.4, and 2.5 show how the step-by-step design process was influenced by stress analysis.

After cursory analysis it became clear that the initial concept did not possess the required strength. A plate version without weight saving was therefore developed (see Figure 2.6).

In Figure 2.6, material has been removed to reduce the mass of the pedal. Stress analysis was necessary to ensure that strength of the new shape was not compromised by the reduction in weight.

Optimization is a very necessary part of the design process, often juggling conflicting parameters. In our example, weight reduction was important, but just as important was the inherent strength of the pedal bracket. Using analysis tools it was possible to obtain an optimized design combining conflicting parameters of low weight and high strength.

2.4 ATTRIBUTES OF A SUCCESSFUL DESIGN ENGINEER

To be a successful design engineer, it is necessary that the designer possesses knowledge and is disciplined across the whole range of engineering skills. Design is a process that brings together and combines a whole range of engineering processes and topics. Some of these are as follows:

- The engineering design process
- Analytical skills

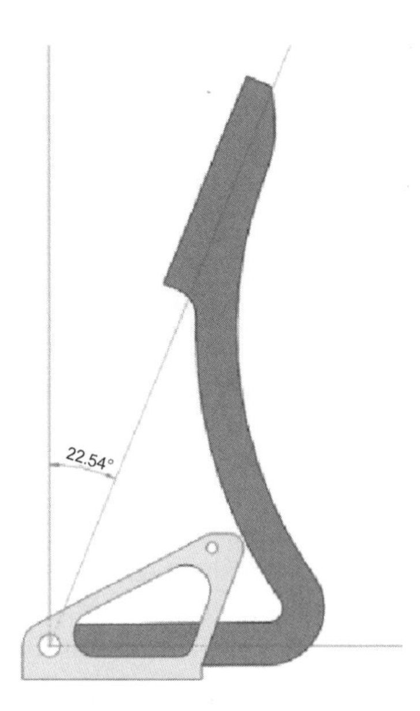

FIGURE 2.3

Side View of the Accelerator Pedal; Initial Concept Design [3]

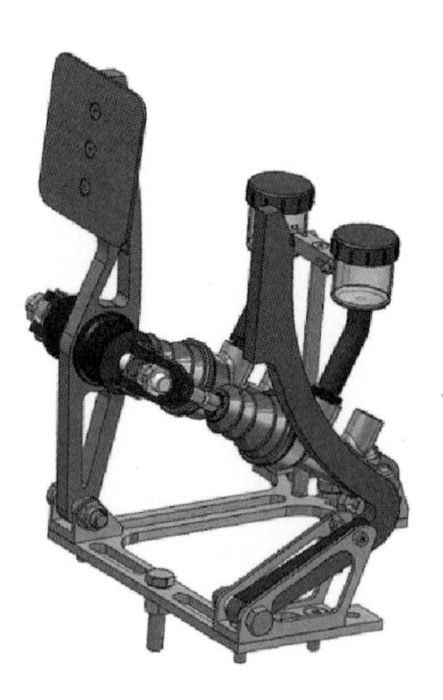

FIGURE 2.4

Initial Concept Design of the Accelerator Pedal Shown *In Situ* in the Pedal Bracket [3]

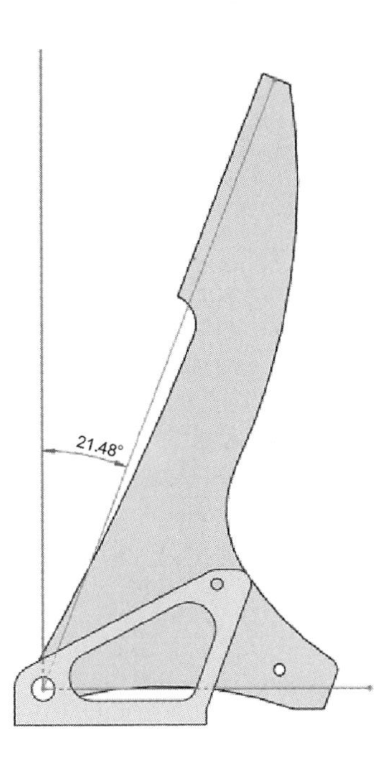

FIGURE 2.5

Second Concept Design Without Weight Saving [3]

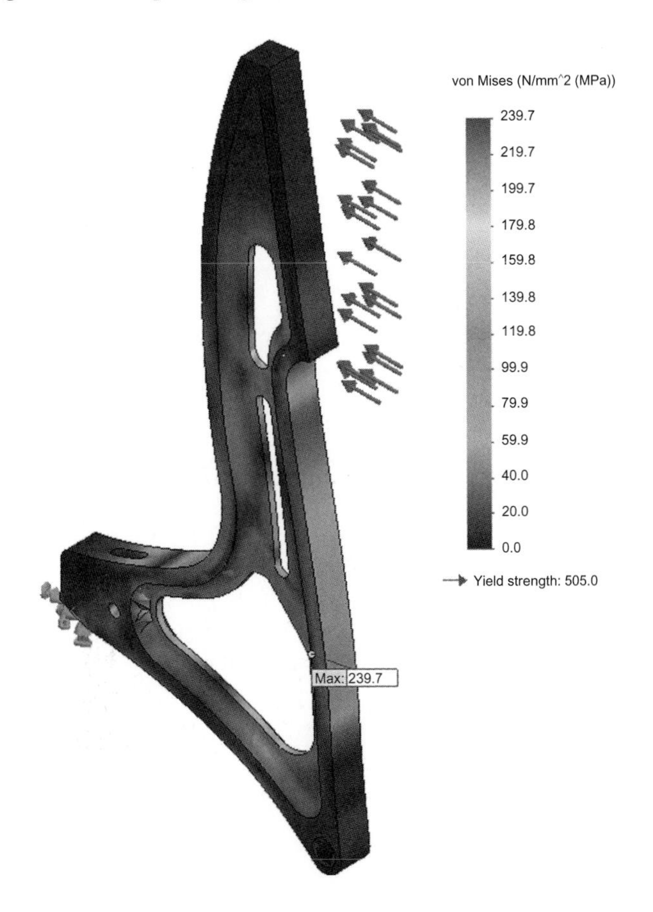

FIGURE 2.6

Finite Element Stress Analysis Model of Concept 2 [3]

- Management of design
- Communication methods and skills
- Communication devices (CAD)
- Manufacturing, including casting machining, forming, etc.
- Protection of intellectual property rights
- Cost analysis processes
- Sustainability in engineering design
- Design for production
- Knowledge of standards
- Diplomacy

This list is not exhaustive, but it outlines the vast knowledge and skills required by a successful engineering designer.

The successful design engineer is someone who understands the needs of his client. A client may only *think* he knows what he wants and will try to persuade the design engineer that the design should be achieved in a particular way. The designer must take a step back and ask himself what the real requirements are. He must see through what the client thinks he needs and employ his skills and experience to provide the client with an efficient solution.

The normal technique when confronted by a new design challenge would be to develop several potentially successful solutions. Depending on the design challenge, the solutions may not necessarily be developed simultaneously; indeed, one solution might appear to be very favorable but on receipt of further information may be rendered completely unusable.

For example, the designer of a flywheel kinetic energy storage system developed a flywheel rotor that rotated at 18,000 revs per minute. According to the information supplied by one particular manufacturer, the rolling element bearings that would support the load could only be used at a maximum speed of 6,600 revs per minute.

The solution to this problem was to suggest magnetic levitation bearings, which held all the advantages the design required to support the flywheel rotor. At this point two pieces of new information negated the use of magnetic levitation bearings:

1. During the investigation and sourcing of magnetic levitation bearings, it was found that they were far too expensive for the device to be manufactured economically.
2. Stress analysis on the rotor showed that the speed of 18,000 revs per minute would overstress the material and burst the rotor.

These two facts meant that the speed of the rotor had to be dramatically reduced to 4,000 revs/minute. At this speed, not only would the rotor stay intact, but the use of off-the-shelf hybrid rolling element bearings was now possible.

This example highlights the case where possible excellent solutions have to be ruled out on receipt of particular pertinent information.

2.4.1 A creative attitude

The designer must possess a creative attitude and have the self-confidence to search for a solution. This is demonstrated as follows:

- A willingness to proceed in the face of incomplete and often contradicting data and an incomplete understanding of the problem
- The recognition of the necessity for developing and using engineering judgment
- A questioning attitude toward every piece of information, specification, method, and result
- Recognition that experimentation is the ultimate method of finding correct information
- A willingness to assume final responsibility

With experience, the designer will develop the confidence to improve on or create solutions based on his or her skill, experience, and abilities. All designers will have at some stage created good designs and poor designs. It is the analysis of these previous designs that builds on experience and later leads to the confidence to create improved designs and new designs. The analysis of previous work often leads to improved designs and new design concepts.

2.4.2 Constructive discontent

Satisfaction with the status quo is often the enemy of discovery and creation. The designer needs to develop attitude and interest that deliberately looks at other designs, shortcomings in designs, and unfulfilled needs.

2.4.3 A positive outlook

The designer needs to nurture the belief that with cultivation, several different solutions can be developed. The designer will then be able to select and develop the best solutions. As the solutions are developed further, novel iterations may come to light.

2.4.4 An open mind

When first considering the brief for a particular design, the designer first needs to look at the requirements and constraints. With an open mind, many possible solutions across many technologies may present themselves. The designer then needs to use her experience and expertise to reduce these possibilities to practical options.

2.4.5 Design courage

The courage to create workable and efficient designs comes to the designer gradually. This courage comes from experience. At the beginning of a designer's career, there is little experience, and some designs may fail or perhaps be inefficient. Learn to overcome the obstacles with persistence and the flexible approach. Courage and experience will grow at the same pace.

Design courage comes in many forms. One notable example was Barnes Wallis, the engineer and inventor of the bouncing bomb. He had courage that his device would work, even in the face of much ridicule. Another example of design courage is that of the NASA design team who, in the 1960s, had the courage to design and build the Saturn V moon rocket. The task was huge, the rocket was huge, and no one had accomplished this trip before. The design could not be allowed to fail, since lives and national pride were at stake. A glimpse of the enormity of the design can be seen in Plate 2.1, which shows the huge size of the five rocket nozzles that lifted the spacecraft off the launch pad.

PLATE 2.1

The Five Rocket Nozzles of the Main Stage, Saturn V Rocket [1]

2.5 THE CLASSIC DESIGN APPROACH

Engineering design lies at the heart of engineering. It is the one discipline that combines all others, including stress analysis, thermal analysis, dynamic analysis, manufacturing, and materials processing.

Design is the process that starts as a visualization in the designer's mind. The designer teases out the specification, applying engineering principles and finally technically communicating his brainchild so that the object or product can be manufactured.

Engineering design is not an exact science but rather an art that uses scientific principles and concepts as its framework. It is at the design stage that the work of many specialists is brought together, and it is the function of the designer to find solutions to design challenges using scientific principles to define design parameters.

The discipline of design is a rational process of decision making, considering each challenge as it arises and searching for solutions. The designer should attempt to create several differing solutions utilizing different methods or components. The next step of the decision-making process should be to select a particular concept for further development or perhaps combine several concepts into one final specification.

Though scientific principle is the cornerstone of good design practice, design challenges may be arrived at by several different approaches:

- Calculation and analysis
- Experience of similar problems
- Intuition
- Application of the results of research

The standard approach to design can be expanded further and can be seen in the Classic Design Approach flowchart in Figure 2.7. As the chart shows, that design journey has three major phases:

1. *Preliminary investigation,* leading to the product design specification
2. *Investigation of possible solutions*, generating the concept design specification
3. *Detail design,* leading to a final design specification

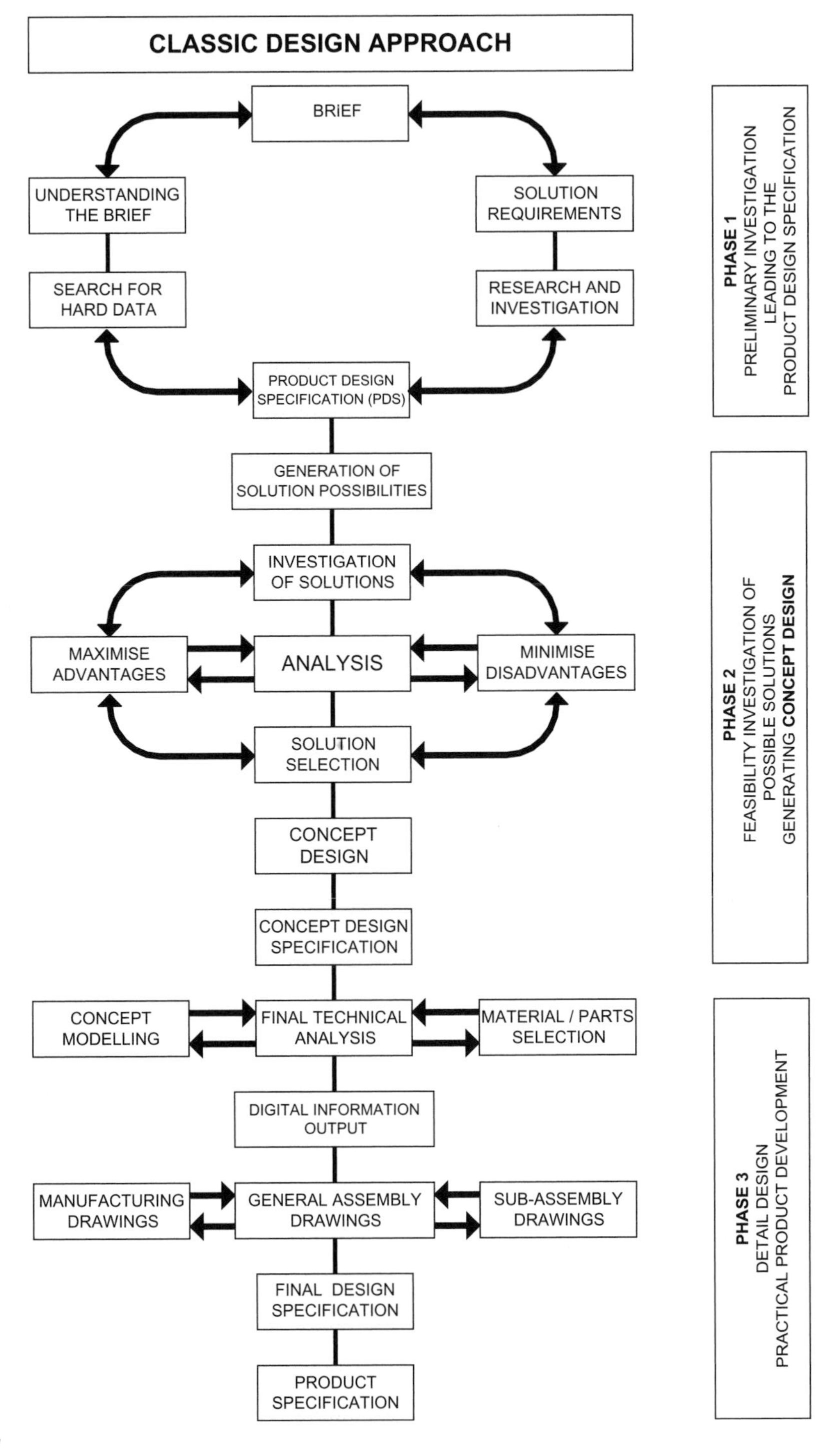

FIGURE 2.7

Classic Design Flowchart [1]

Each of the three design phases finds its completion in a particular specification. The specification element is an important way-marker on the design journey, since it should be compared against the needs of the product. If a specification does not match the needs of the product as set out in the brief, the design process has to be returned to the previous stage.

2.6 DESIGN SPECIFICATIONS

Throughout the design process, specifications are developed and modified as new information and new designs unfold. This re-evaluation is necessary to keep the design on track and to ensure that the original requirements are met. New specifications are really checkpoints along the design journey.

There are normally four levels of specifications derived from the classic design approach shown in Figure 2.7 during the design of a product. These levels are:

1. Product design specification (target specification)
2. Concept design specification
3. Final design specification
4. Product specification

These four specifications represent stage completion on the route to the product's final design. Each specification lists a set of parameters that are either things that the product should be able to do or things that the product can do. In the early stages of a design, the specification outlines parameters that the yet undesigned product should be able to achieve. In the later stages of a design, the specification should define the parameters that the product can achieve.

The designer should be clear on the content and purpose of each of the specifications, since they can be used as a measure of progress throughout the design process.

The *product design specification* (PDS) is a target specification and a statement of what a not-yet-designed product is intended to do. The aim of the PDS is to ensure that the subsequent design and development of a product meet users' needs.

The *concept design specification* (CDS) takes place after a feasibility design and a conceptual design have been formulated. This specification predicts what the concept should be able to achieve and is a checklist against the original need for the product.

The *final design specification* (FDS) must include all necessary technical specifications and should be a complete set of instructions on how to build, operate, and maintain the product.

The *product specification* (PS) is the final specification that should be attached or enclosed with the product. This is formulated after the product has been built and tested and lists specifically what the product can achieve and what the product requires for it to work efficiently.

2.6.1 Product design specification (target specification) (PDS)

The brief makes the first request for a design solution. As the designer teases, investigates, and understands the brief, a target specification or PDS is created. This is a statement, or at least a list of requirements and constraints, of the performance of the new product. Its aim is to ensure that the design and development of the new product meet users' needs.

The PDS is formulated as a consequence of interrogating the brief. The brief may set out what is required, but it is not until the designer fully understands the brief that the PDS can be formulated. The PDS will carry requirements or needs as well as constraints. In the design for a 50-meter-wide car park, for example, the designer may have a requirement to park 25 cars side by side. The constraint would be that each parking place can be a minimum of 2 m.

The proposed specification combines the information that is currently known and suggests a wish list of requirements and constraints in the final design solution. For example, consider planning a trip from Edinburgh to London passing through Carlisle, Liverpool, and Birmingham. As the journey is undertaken, information may come to light that causes us to change direction. For instance, road works in Liverpool have changed the route to go through Leeds instead.

The same is true of an engineering design. A conveyor system that is powered by electric drive motors works well in a dry atmosphere, but when the conveyor is placed in a very wet environment, the electric motors would fail. Hydraulic motors would be an excellent replacement because they would not be affected by the wet environment. Here new information came to light related to the working environment, and that information changed the direction of the design.

2.6.2 Functional parameters and metrics

We previously explained that the PDS contains requirements and constraints. Many of these elements, such as volume, force, velocity, and mass, will be measurable values. The brief will set out requirements and sometimes will give values, but very often the designer has to develop the design to a stage where he can predict the design parameters as a set of numeric values.

Metrics are a measurement element which defines in precise terms the physical attributes of the new product, how the new product should perform and what it should use as consumables to achieve those physical attributes and that performance. These metrics should form a substantial part of the product design specification and match the requirements originally set out in the brief, even if the brief does not specifically state.

2.6.2.1 Example: kinetic energy storage device project (flywheel)

A very wide brief was given to create a flywheel system that would store mechanical kinetic energy in a rotating mass. The intention was to store electrical energy as kinetic energy, then feed it back into the national grid to smooth out high demand for electricity. When the designer requested the physical parameters such as diameter, mass, and speed, the only parameter given to the designer was that it should store 20 KWh of kinetic energy. This parameter was the only metric given in the brief, and it was derived from the requirements for selling the energy back into the national grid. The only other major requirement built into the brief was that the flywheel should be built to comply with sustainable engineering values.

The designer was faced with a very broad field of possibilities that needed to be simplified. Some of the main questions were these:

- Large heavy flywheel?
- Several small flywheels?
- Speed: high or low?

Many flywheels in operation made use of high speed, given the kinetic energy equation:

$$KE = \frac{1 \text{ x } I \text{ x } \omega^2}{2} \text{ Joules} \tag{2.1}$$

where

I = Moment of inertia (Kgm^2)
ω = Angular velocity (rad/sec)

It can be seen that if the angular velocity ω is increased, the kinetic energy is squared. Many designers and researchers used an increase in speed to create a greater kinetic energy capacity. Stresses within the rotor suggested that the angular velocity should be reduced substantially.

The reduction in speed brought enormous benefits to the cost and the sustainable manufacture of the flywheel system. It enabled use of standard bearings, normal machining practices, easily obtainable materials, shorter development time, and low-cost manufacture with a high sustainability factor.

The product design specification was formulated and included the following metrics:

- Kinetic energy storage: 20.1 kwh
- Power available: 5.6 kW
- Angular velocity: 4,000 rpm
- Angular velocity: 419 rad/second
- Diameter: 2 m
- Mass: 1,650 kg
- Bearings: Standard hybrid ball bearings

To formulate the product design specification and meet the targets of 20 kwh kinetic energy storage, a good deal of analysis was conducted to create an accurate target specification. Many of these parameters were taken through to the concept design specification after verification through the selection of a solutions process, which followed in Stage 2.

The PDS would therefore contain essential metrics to assist in Stage 2 Concept Design, where design solutions need to be formulated.

2.6.3 Concept design specification (CDS)

The concept design specification is formulated after much work has been done in researching the possibilities for a solution to the design challenge. As the PDS is investigated, several solutions will emerge. It is suggested that between three and five solutions be formulated, all of which will fulfill the design challenge to varying degrees of efficiency. It is necessary for the designer to evaluate each solution against various constraints such as cost, size, power, environmental impact, and so on.

The outcome will be a concept solution that can also be termed a *feasibility solution*. From this solution, the designer will be able to propose a specification describing the performance of the conceptual product. This is the second specification in the design route and should be checked carefully against the original needs of the design challenge.

2.6.4 Final design specification (FDS)

An FDS must include all necessary drawings, dimensions, environmental factors, ergonomic factors, aesthetic factors, costs, requirements, quality, safety documentation, and disposal methods. This specification is the culmination of all the design work that has led to the concept and the detailed design work that has converted the concept into a practical design. The FDS should be a complete set of instructions on how to build and use the product.

For simple products such as an electric alarm clock, the FDS may be a one-page document that outlines power consumption and instructions for use. In the case of a ship, for example, the document may run into many thousands of pages outlining the working requirements, maintenance requirements, and all necessary information relating to the millions of components that contribute to the efficient working of the ship.

2.6.5 The product specification (PS)

The product specification should be attached to or enclosed with the product. It is the specification the customer will see and is normally derived from the FDS. The product specification is formulated after the product has been built and tested; it lists specifically what the product can achieve and what the product requires for it to work efficiently.

The product specification may be so simple that it is a straightforward copy of the FDS. The complexity of the product, such as a ship, may mean that the FDS becomes the product specification in its entirety.

2.7 DESIGN PHASES

Regardless of the type of product, the design process will follow a broadly similar path involving back-and-forth iteration. There are generally three phases to any design challenge. Phase 1 is primarily an investigation phase that interrogates the brief and looks at similar methods and processes; the culmination of this phase is to create a PDS.

Once the brief is understood and information is gleaned through initial research, the creative design process can commence. This phase involves the creation of several solutions to the design challenge and an analysis of feasibility for each of the solutions. The outcome of the second phase is to create a concept design and a concept design specification.

The third phase converts the concept design into a practical product by selecting materials, components, manufacturing methods, and the like. This phase culminates in a final design specification that describes what the product will achieve and what the product requires in order to function.

Each of the design phases can be subdivided into a progression of activities as follows:

Phase 1: Preliminary Investigation Leading to the Product Design Specification

- The brief
- Exploring the brief
- Defining solution requirements and constraints
- **Product design specification**

Phase 2: Investigation of Possible Solutions and Generation of Concept Design
- Generation of solution possibilities
- Analyzing several solutions for feasibility
- Selecting an appropriate solution (concept)
- **Concept design specification**

Phase 3: Detail Design and Conversion of the Concept into a Practical Product
- Detail development: modeling, selection of parts
- Manufacturing data, drawings, models, instructions
- **Final design specification**
- **Product specification**

2.8 PHASE 1

2.8.1 Preliminary investigation of the design challenge

The preliminary investigation of the brief is arguably the most important part of the design challenge. If the brief is not investigated properly and thoroughly, the solution to the design challenge may be completely inappropriate. It is important that the brief is interrogated thoroughly so that the designer knows exactly what is required and what constraints should be included.

This investigation extends beyond the brief in that other research should be conducted. This research could be in the form of competition or perhaps patent searches. This research is necessary to build a complete picture of the design challenge and formulate the PDS.

2.8.2 The Brief

The creation of the design will naturally start with some form of brief. This brief could be of the following types:

- *Written.* Written briefs are usually much more formal than others and have a commercial content. They should include a list of requirements and constraints. Such a brief may well be accompanied by a formal contract.
- *Verbal.* Verbal briefs tend to be less formal, and since they might not have requirements attached, they may well be less accurate. This kind of brief should be reviewed carefully since verbal misunderstandings may well result in arguments between principals, which will then lead to lost time and perhaps loss of money.
- *Perceived ("I have a good idea").* Perceived briefs are very informal and are often the result of conversations between friends or perhaps individuals who have "good ideas."

2.8.3 Exploration of the brief

The brief is the very first information relating to requirements of the new product. It possesses the key information that will specify the design requirements. The designer must understand the brief completely if the design is to comply with the requirements set out in the brief. He or she must analyze what is required, perceive the final product, and provide a target specification, sometimes called a product design specification (PDS). To do this the designer must "climb inside the brief" and understand every nuance.

An inexperienced designer may look at the design challenge and think it is simple, but an experienced designer will look at the same design challenge and remember the adage, "If you think the challenge is simple, you don't understand the challenge."

If the designer is asked to design a chair, she must understand the material of manufacture, the manufacturing process, and what the chair has to go through when it is in use. He/she must understand that the joints will creak and move, the seat will sag, and the structure will groan.

To take an artist's is point of view (see Figure 2.8):

"Climb inside" the chair
Understand the stresses being applied to its crossbars and legs
Feel the stresses that the chair feels
Live the feelings of the chair
Understand its creaks and groans as it is sat upon
Become the chair!

By understanding the brief in this way, the designer can better understand the design challenge and fully apply design experience to develop a full solution.

The process of understanding the brief is really to ask the correct questions. In a class session the author suggested to the class that he needed to get a car between two towns about 30 miles apart. He invited the class to ask appropriate questions to better understand the challenge.

Some of the questions were as follows:

Does the car have an engine? Answer: No
Is the car able to be steered? Answer: No
Is the car in pieces? Answer: No

FIGURE 2.8

Understanding the Feelings of a Chair [1]

Can it be pushed? Answer: Yes
Does it matter if it rains? Answer: No
Can it be carried on a trailer? Answer: Yes

Many of these questions assumed that the car was a full-size vehicle. It was only when one student asked "How big is the car?" that the true answer was given. The car was a children's toy, and it was carried to the next town in a carrier bag on public transport.

In his quest to understand the brief, the designer should create a wish list of requirements the design should fulfill. Equally, there may be some constraints that may be the same as the requirements. For example, a *requirement* for a vehicle might be that it should carry four passengers. Conversely, a similar *constraint* may be that it should carry no more than four passengers. Armed with the list of requirements and constraints, the designer may then proceed to the research phase.

Though constraints and requirements may be similar, the general approach is that constraints will create an envelope in which the design may be accomplished. Requirements are really "need to know" items and describe the performance of the design. Sometimes requirements and constraints cannot be given in the brief, but they may become evident as a result of research.

2.8.4 Exploration and research outside the brief

To understand the whole design challenge, it is necessary for the designer to immerse himself in the extent of the challenge and all its nuances. If the designer is new to the industry, research may involve understanding working practices and methods within that industry. Should the designer be a specialist within the industry, he may well already be aware of industry practice. Nevertheless, to gain a full understanding of the design challenge, research is always necessary. Depending on the product, research could involve many different areas, from competition research through high-level scientific research.

2.8.4.1 Competition research

Probably the first step in any research would be to "see what is out there" in terms of the competition. This may be very enlightening since the competition's products will already be serving the market and will have already been designed and manufactured. Analyzing the competitors' products will give insight into new techniques, materials, and manufacturing processes and perhaps give insight into improvements.

2.8.4.2 Patents research

Most new product development will require some form of patents search. This search will inform the designer if the design he is proposing has previously been considered. If a product is designed, manufactured, and marketed and closely resembles a patented product, a heavy lawsuit could be applied by the original patent owner. A patents search will also reveal technical details of closely related products. Furthermore, the "claims" within a patent will inform the designer of the exact capabilities of the device. This finding may or may not allow the designer to "break" the patent. Breaking patents by avoiding the patented claims is completely legal, as opposed to illegal patent breaking, where the product is directly copied and marketed. Example: One of the claims for a charcoal ignition system may be that it should be used a s a free standing device. This means that to break this part of the patent it could be incorporated in to a larger device and therefore used in a different way. There may, however be other claims which preclude this. The patent breaker needs to take care not to infringe other claims.

2.8.4.3 Scientific research

Scientific research can be expensive. Much of the development of highly technical components and products such as aircraft, vehicles, silicon chips, solar panels, new materials, and so on has been achieved through enormous injections of money into specific research.

The BAE Systems A380 Airbus, shown in Plate 2.2, has a maximum take-off weight of 650 tonnes. To achieve this mass, a great deal of design effort was injected into the project. This effort included not only the design of aircraft components but also the research and development of new techniques and materials. To reduce the mass yet keep the strength, the fuselage required the development of a large, very specific composite structure and was the result of some high-level research. The total cost of development of the A380 Airbus was estimated at $18 billion.

2.8.4.4 Component research

Many specialist manufacturers make standard components that can be fitted and combined directly in a design. These components may include bearings, seals, screws and fasteners, electric motors, gearboxes, engines, and many, many more off-the-shelf items.

The manufacturers of a stone tumbler (see Plate 2.3) would normally manufacture the special components such as the drum and the chassis. Off-the-shelf standard components were such elements as wheels, bearings, and drive unit. These parts, shown in Plates 2.3, 2.4, and 2.5, would be purchased from component suppliers.

2.8.5 Cost and sustainability

During preliminary investigation, the cost and sustainability of the product are not a large element of consideration. However, cost is always a limiting factor and can be considered a major constraint. It does not matter how well a product might function; if it is too expensive, it will not sell. Cost, and more recently, sustainability, has therefore to be major considerations in the back of the designer's mind at all levels of the design process.

PLATE 2.2

Airbus A380 [2]

PLATE 2.3

Stone Tumbler [1]

PLATE 2.4

Tumbler Bought-Out Parts, Comprising Plumber Block Bearing and Vehicle Wheel [1]

A high-level industrialist recently observed, “Everything costs money.” This doctrine should now be expanded to include sustainability: “Everything costs money and everything has an environmental impact.”

This adage should be adopted by every designer and should be considered at every stage during the design process. As the design progresses, cost and sustainability will grow in importance since they will steer the design toward a low-cost and sustainable product.

PLATE 2.5

Tumbler Bought-Out Part: Drive and Reduction Unit [1]

2.8.6 Factors to consider for the product design specification

The variety of designs in engineering precludes creating a rigid list of parameters that can be applied to developing a design or specification. However, many factors could be considered and are useful as a checklist, as discussed in the following sections.

2.8.6.1 Duty factors

1. Required form of action (what are the working principles?)
2. Mechanical loading (type and magnitude-static/dynamic/shock/constant/variable/cyclic)
3. Speeds (given/determined by the chosen design configuration; constant/variable)
4. Cycle time (strokes per minute or hour/hours per week)
5. Power sources available
6. Required life (total life/life between servicing in years/number of cycles, etc.)
7. Degree of reliability required (operating life without attention, likelihood of premature failure)

2.8.6.2 Environmental factors

1. Temperature extremes (effect on material properties; problems caused by differential expansion)
2. Humidity (problems of corrosion or of drying out)
3. Chemical or dirt contamination (from surrounding sources or contaminating the surroundings
4. Vibration and noise, shock loading
5. Tampering (could the operator or passerby interfere with the machine?)
6. Maintenance (will it be easy or impossible? will it be conscientiously applied?)

2.8.6.3 Sustainability factors (environmental impact)

1. Where will the raw materials be sourced? (transport impact)
2. Will manufacture have a low environmental impact (smart factories' use of high-tech models to reduce build impact)

3. Environmental impact in use (has use been made of low-impact materials and usage methods, e.g., the use of a lean burn diesel engine in an item of plant)
4. Designed for long life (larger bearings ability to maintain components, refurbishment)
5. Design for disposal (4R approach: reuse, recycle, reduce, refurbish)

2.8.6.4 Economic and manufacturing considerations

1. The position in the "cost-quality" spectrum (standard of accuracy required?)
2. Quantities required, delivery rate and date
3. Capital cost
4. Operating cost (energy cost/efficiency; cost of replacing failed or worn parts)
5. Material availability (scrap utilization)
6. Manufacturing facilities available (small firm or large firm? is the manufacturing equipment fully committed on other work?)
7. Use of existing and bought-out components
8. Interchangeability (can it be done? components, subassemblies, whole units)
9. Adaptability (suitable for wider uses? any advantages?)

2.8.6.5 General factors

1. Size (any limitations? often uncertain dimensions; compactness is often a good selling feature)
2. Weight (is it important?)
3. Rigidity/flexibility (not the same as strength)
4. Appearance (another selling feature; create a product with good function and good aesthetic form)
5. Ergonomics (put operating controls in convenient positions; operator seated? foot control?)
6. Safety (guards, etc., for operator safety; overload devices to protect machine.)
7. Safety factors (possible failure modes; consequences of failure; is it fail-safe? reliability of materials and manufacture, accuracy of calculations)
8. Ease of handling and transportation
9. Standards (British Standards, International Organization for Standardization [BSI/ISO], American National Standards Institute [ANSI])

The relative importance of these various factors depends on the application of priorities that must be established by clearly visualizing the design challenge. Not all the factors will be usable in any single design challenge, but the list is an excellent memory jogger when PDS's are formulated. Furthermore, the list is an excellent guide when developing several different solutions.

2.8.7 Metrics

Part of the PDS must include measurable elements of the design. These elements, termed metrics, are indicators of performance or consumption. Metrics for a passenger vehicle might include a mass of 750 kg and a fuel consumption of 75 mpg. These can often be taken from the list of requirements and constraints, but sometimes the brief is extremely vague and it is not until feasibility calculations have been performed that some of these metrics can be formulated. However, they should be included in the product design specification, since these measurable elements will allow the designer to target the design formulation more precisely.

2.8.8 Preliminary investigation overview

During the investigation and research process, a greater understanding of the requirements and constraints will be gained. The investigation and search for hard data, which is then compared against the requirements of the brief, lead to the PDS (target specification). This is really the first list of performance comparators gleaned from the brief and will be used in future design development as a check to ensure that the design is heading in the correct direction.

2.9 PHASE 2

2.9.1 Generation of solution possibilities

The completion of the PDS generally marks the end of Phase 1. It should be noted that the quest for new information and ongoing research will not cease throughout the life of the project.

More important, the PDS marks the end of Phase 1 and the start of Phase 2. At this point the designer or design team should be completely knowledgeable as to the requirements of the brief. Furthermore, the PDS provides a design target structure that will guide the designer.

2.9.2 Strategic thinking

The beginning of Phase 2 instigates the process for the generation of solutions to the design challenge. At this point it is easy to fall into the "inventor syndrome" by inventing a single solution and pursuing that solution at the cost of exploring other avenues. This path must be avoided by all means. Premature commitments can be dangerous since the designer might become attached to a bad concept, or at best the outcome will be an inefficient design.

It is normal practice to adopt a strategy of *least commitment*. Only after all other avenues have been evaluated and exhausted does the designer commit to the best of the solutions. It is good to remember the adage, "Never marry your first design."

Another important strategy in design thinking is that of *decomposition*. This is a strategy of breaking down a much larger product into smaller subchallenges. A complex item such as a passenger vehicle cannot possibly be designed as a whole unit. The design is broken down into a smaller set of design challenges—for instance:

- Engine
- Body
- Suspension
- Braking
- Seating
- Chassis
- Wheels
- Tires

This list contains just a few of the major subcomponents of a vehicle. These may be split down even further—for instance, the engine may be split into the crankshaft, pistons, engine carcass, alternator, camshafts, and so on. When large components are reduced to individual systems and components, the whole design process can be made easier. This technique also allows specialists to work on their particular component or system.

A complex item such as a passenger vehicle will be split down into hundreds of smaller challenges, all of which, when finished, will come together as the complete passenger vehicle.

2.9.3 Solution generation

It is recommended that between three and five or more solutions are generated for each design challenge or subchallenge. Information thus created should be shared with other members of the design team in the form of sketches, models, and any other means that will allow efficient explanation of the various solutions. The explanations should be full and complete so that there is no ambiguity. The solutions should be accompanied by a list of valid and well-reasoned advantages and disadvantages that may need further investigation.

Note: Any statements should be qualified and verified by a particular source or by appropriate research and tests. One cannot say merely that "This cannot be used because it cannot stand the temperature." A qualified statement would be: "This cannot be used because the plastic components will melt at 150° before the operating temperature of 400° is reached."

2.9.4 The tools of creativity

Generating new solutions is a creative process that relies heavily on the designer's personal experiences. Designers will often suffer "mental blocks" relating to a design challenge, but there are tools available to help unlock the logjam. Solution generation is treated in a later chapter; for now, some of the major solution generation tools are described in the following subsections.

2.9.4.1 Heuristic redefinition

Heuristic redefinition is valuable for the individual designer who is working alone, unsupported by a creative team. It is a method of looking at a whole system in which a problem exists. Each major part of the system is itemized and visualized as to its contribution. At each stage the question is asked, "How can we ensure that . . . ?" Answering this question allows a redefinition of each of the elements and a possible solution to the original challenge.

2.9.4.2 Classic brainstorming

Classic brainstorming is a very popular method of generating solutions by a team. The tool allows team members to pool their knowledge creatively and in an open and uncritical environment. There is only one rule, and that is not to be critical of others' ideas. Sharing a seemingly outlandish and obviously unworkable idea may well set another member of the team on a path toward a practical solution.

2.9.4.3 Imaginary brainstorming

Imaginary brainstorming is a variation on classic brainstorming except that it uses a "trick" that tends to shift people's thinking "outside the box." An example is to start with the question, "How can we improve communication between departments?" The trick here is to change the word departments to planets. This shift in thinking is quite effective in generating new solutions.

2.9.4.4 Word-picture associations and analogies

Associations and analogies are often used to generate alternative solutions. Associations are really mental connections triggered by an idea or a memory, picture, or event. For example, a piece of apple pie might stimulate memory of your grandmother's home.

An *analogy,* on the other hand, is usually a more direct comparison of principle, action, or behavior. When R. J. Mitchell, the designer of the Spitfire, first designed the Schneider Trophy aircraft, he used the analogy of birds in flight to assist the design of the aircraft wings.

Analogies and associations can be generated through random words, pictures, biotechniques, and any other memory-provoking stimulant.

2.9.4.5 TILMAG

TILMAG is an acronym for transformation idealer lösungselemente durch matrizen der assoziations und gemeinsamkeitenbildung, which, loosely translated, means "the transformation of ideal solution elements in an association matrix." This technique was developed by Dr. Helmut Schlicksupp. The basic principle is that a matrix of ideal solutions is created and cross-matched to deliver alternative solutions. This method is discussed in depth in a later chapter. It is particularly ideal for generating solutions to stubborn design challenges.

2.9.4.6 The morphological box

The morphological box is a structured method for systematically looking at key characteristics or parameters of a solution and the realistic options for each parameter. It helps the team to identify practical solutions to a problem by first defining the essential characteristics of possible solutions. For each parameter multiple options are defined thus when combined within a matrix a multitude of unique solutions are yielded.

2.9.4.7 Graphical synthesis

Graphical synthesis is a well-tried method whereby a single designer may generate his own solutions by using a large sheet of paper and developing sketched ideas. A large sheet of paper, probably A1 in size, is spread on a flat surface. The designer would have a list of requirements and constraints at hand and sketch a solution in the top-left corner. The development of this sketch could be horizontal across the top of the paper. Another development could be vertically down the left side of the paper. Solutions may develop from combinations of the horizontal and vertical sketches until eventually the whole sheet is full of combinations and progressions of ideas. As solutions develop toward the bottom-right corner of the paper, a unique solution should emerge. The general approach is illustrated in Figure 2.9.

2.9.4.8 Synectics

Synectics is a formalized creative design method that uses analogical thinking as a main process. Various forms of analogy may be used as described here:

- *Direct analogies.* Using a biological example such as in the development of Velcro when the hooks on cocklebures were used as the analogy.
- *Personal analogies.* Here team members imagine what it would be like to use oneself as part of the system.
- *Symbolic analogies.* Poetic metaphors and similes are used to relate elements of one thing to aspects of another, such as the head and claw of a hammer.

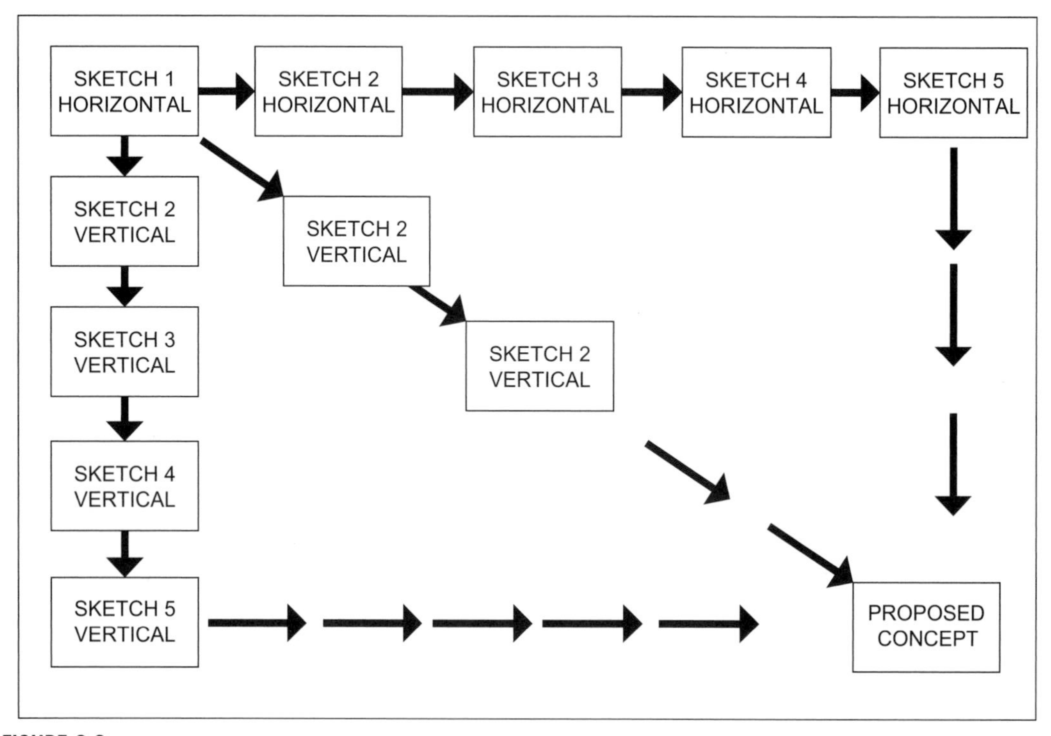

FIGURE 2.9

Graphical Synthesis and Development of Possible Solutions by Sketching on a Large Sheet of Paper [1]

- *Fantasy analogies.* These are impossible wishes for things to be achieved in some magical way. For instance, one may wish to lift a 20,000-tonne ship out of the water using a force beam.

We've identified some of the more popular methods of solution generation that are usually used when there is a tough challenge to solve. Experienced designers may be able to generate ideas without recourse to more formal techniques.

2.9.5 Solution evaluation and selection

Thus far the design process has been dominated by research and creativity, but solution evaluation involves the convergence of the design process toward a single concept. Solution evaluation is often an iterative process whereby solutions are explored and perhaps found to be lacking in some technicality. The process returns then to basic principles to modify or scrap the particular solution.

It is usual that between three and five solutions are developed. This creates a choice of different solutions for the designers to evaluate for their efficiency and effectiveness.

Evaluation involves a large set of possible solutions being reduced to a smaller set, but through the process of iteration these solutions may be subsequently combined and improved to increase the number of solutions available for selection. Figure 2.10 illustrates the successive narrowing and temporary widening of a set of options under consideration during the evaluation process.

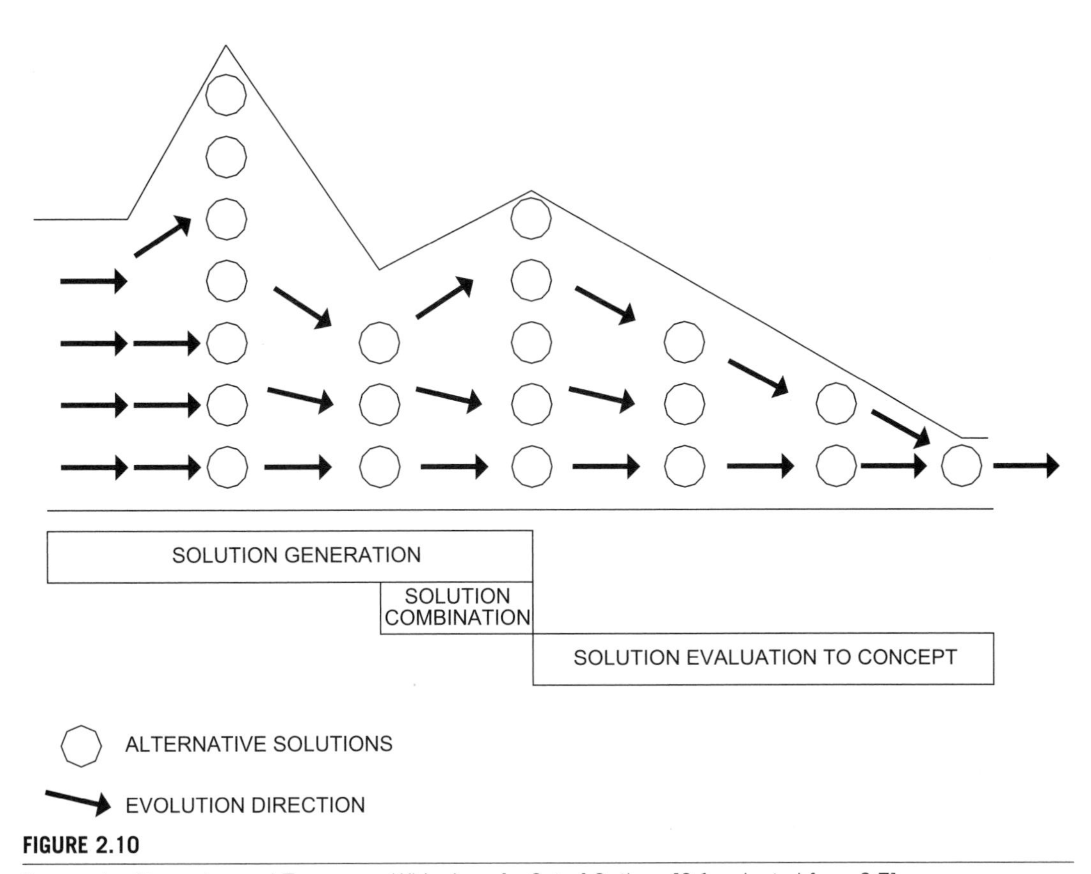

FIGURE 2.10

Successive Narrowing and Temporary Widening of a Set of Options [2.1, adapted from 2.7]

No matter what the product or system or whether the selection process is prescribed or very free, designers and design teams will use methods to evaluate and reduce the concepts to a single workable solution.

Methods vary in their effectiveness and efficiency and include the following:

- *External decisions.* Various solutions are related to the client or another external body for their selection.
- *Product champion.* Often in a team design situation there is a dominant person. The chief designer may be a responsible member of the team who personally makes the final selection. Or there is a member of the team who is very opinionated and tends to speak the loudest. The decision may be based on his opinions. If a product champion is appointed, he may lead the team and may make the final decision.
- *Intuition.* The solution is selected by its "feel." Explicit criteria or trade-offs are not used. The concept just "seems" better. This is rather a dangerous selection process since it is very subjective and could be based on whims and moods.

- *Voting.* Each member of the team votes for several solutions. The solution with the greatest number of votes is selected.
- *Advantages and disadvantages.* In evaluating design solutions it is normal practice to consider advantages and disadvantages of a particular solution. Simple evaluations can be accomplished in an informal manner, but it is always good practice to write down the positive and negative aspects of a particular solution.
- *Prototype and test.* A solution selection is made based on test data. This choice may prove to be an expensive approach since the item needs to be manufactured before the selection decision is made.
- *Decision matrices (morphological analysis).* This is a more formal selection procedure because each solution is rated against prespecified selection criteria. These may be weighted. Care should be taken in using this method, since subjective views may creep into the decision process. This method is discussed in detail later in this chapter.

Many experienced designers intuitively evaluate the solution options, which may be satisfactory for small projects with fairly simple selection criteria. Large and complex projects require a more formal evaluation procedure, which in some cases may need to be documented. The procedure may also need to be shared by the design team, which also may necessitate a more formal evaluation style.

2.9.5.1 Case study: removal of plywood from floor panels [10]

A project was given to a group of design students to design a small machine that would separate plywood facing from galvanized steel-covered floor panels. The floor panels can be seen in Plate 2.6. The plywood facing was adhered to the galvanized sheet steel floor panel by an epoxy resin. Research showed that this could not easily be separated by chemical means or high temperature and a more mechanical means would be required. After producing a PDS, the students proposed several methods to remove the plywood. Their list of methods with advantages and disadvantages appears in Table 2.1.

PLATE 2.6

Plywood Facing Adhered to a Galvanized Sheet Steel Floor Panel

Table 2.1 Methods, Advantages, and Disadvantages for the Plywood Removal Case Study

Method	Advantages	Disadvantages
Electric bandsaw	Possibility of keeping both steel and wood as reusable components Cheap to purchase Able to run from 240v source	High maintenance costs (epoxy resin will cause high wear on saw blades) High temperature may warp the ply, therefore negating the advantage of keeping it intact Slow separation process Sawdust
Industrial plane	Quick separation Easily automatable process High initial cost	Expensive High maintenance (lots of moving parts) Dust/debris No chance of saving wood
Cutting wire	Low initial cost Intact plywood Low maintenance cost	Possibly not effective Slow Inaccurate
Hydraulic machine blade	Safety (stationary blade) Design simplicity Relatively cheap Low wear and therefore low maintenance costs Quick process	Fairly high setup costs

Even though the parameters for each solution are reasonably simple, it is not easy to intuitively evaluate the best method. A more formal process could be used in maximizing advantages and minimizing disadvantages. Several methods are available, but perhaps the most popular is the morphological evaluation chart shown in Figure 2.11.

The morphological chart allows the designer to evaluate each of the parameters the different solutions might give. If each parameter is given a value, in this case, a maximum of 10 marks, a total at the end will show the score and hence the rank. In the preceding case it was the hydraulic machine blade that was deemed to be the best design option.

There must be a note of caution here, since allocating marks is a very subjective activity and can be subconsciously weighted by the designer. Furthermore, the designer may have a particular vision in mind, which will also subconsciously throw the mark allocation.

The morphological evaluation chart is a very powerful tool when used properly and its shortcomings are understood. It is able to sift the various parameters within several solutions and put them in order.

The features or parameters can be arranged so that the most important are on the left of the chart and ranked so that the least important will be toward the right of the chart. In this way the chart can be manipulated. Instead of 11 columns, the first five most important features can be evaluated.

2.9.6 Use of the morphological chart

It is important to use the chart properly in order to overcome the subjective nature of marks application. Should the list of solutions be much longer than just the mere four we've mentioned, it would be normal to choose the top three ranked solutions for further evaluation.

Morphological Chart to Evaluate the Best Method of Separating Plywood From Galvanised Steel													
	Capital Cost	Maintenance Cost	Separation Speed	Accuracy	Power Source (240V)	Automation Possibility	Temperature Generation	Waste Debris	Re-usable Materials	Efficiency	Safety	Score	Rank
Band Saw	8	5	3	5	10	4	3	5	7	6	7	63	3
Industrial Planer	6	5	8	8	10	9	6	2	8	8	8	78	2
Cutting Wire, Powered	7	4	3	4	10	3	3	5	7	5	7	58	4
Hydraulic Machine Blade	6	7	9	8	10	9	9	8	3	9	9	92	1
Feature Importance	27	21	23	25	40	25	21	21	25	28	31		

Excellent: Max. Marks 10
Poor: Min. Marks 0

FIGURE 2.11

Morphological Evaluation Chart [1]

Should the designer choose the top-ranked solution and say, "This is it!"? There is a danger that even though ranked number one, a solution might not possess some qualities that are absolutely essential for the design challenge. It is therefore important that more stringent evaluation takes place over perhaps the top ranked three solutions. This can be done by the designer manually checking the outcome of the top-ranked solutions against the PDS.

Considering the morphological chart in Figure 2.9, the number-one ranked solution is the hydraulic machine blade; number-two ranked is the industrial planer. Both solutions will fulfill the PDS. The main difference is that the industrial planer will create a great deal of small friable debris, which can be bagged, is easily shipped, and is ideal for recycling into chipboard. On the other hand, the hydraulic machine blade produces broken, flat, large pieces of plywood that cannot easily be recycled and for which the only disposal solution may be by burning.

According to the morphological chart in Figure 2.9, the top-ranked solution is the hydraulic machine blade, yet in selecting this tool, the designer is creating a problem of the disposal of the broken pieces of plywood. Should the designer instead choose the number-two-ranked solution, the industrial planer, the problem of disposing of the debris is already solved.

In considering the design in a whole-life sense, it is more appropriate to adopt the number-two-ranked solution than the number-one-ranked solution. The planer will adequately fulfill the needs of the PDS, and it gives the added benefit of sustainable disposal of the debris.

2.9.7 Benefits of a structured selection and evaluation method

The response of the marketplace to a product critically depends on the selected solution, but the solution selection can also dramatically influence the manufacture and cost of the product as well as the environmental impact.

For example, the solution for the flywheel system mentioned earlier was to create a flywheel with a fairly low speed, which allowed the materials procurement and manufacture process to be kept to that of normal practice, thus reducing the environmental impact and improving the value of sustainability.

The alternative option would be for a high-speed flywheel that would require magnetic levitation bearings, composite flywheel, vacuum chamber, and a long lead time to market. There were many other elements of a special nature were required, making this option very expensive and much more unsustainable than the low-speed option.

A structured solution selection process helps maintain objectivity throughout the selection process and guides the production development team through a critical, difficult, and sometimes emotional process. A structured solution selection method offers the following potential benefits:

- *Products that meet the PDS.* Concepts are explicitly evaluated against customer-oriented needs.
- *Better manufacturing.* Precise evaluation with respective manufacturing criteria improves the products' ease of manufacture and helps match the product to processes.
- *Competitive design.* The design team will attempt to match or exceed key metrics compared to competition.
- *Reduced time to market.* A structured method tends to coordinate all players within the design team to all those outside the design team, such as manufacturing engineers, industrial designers, marketers, and project managers. This results in increased certainty and better communication.
- *Documented decision process.* A structured selection method results in an archive and rationale behind selection decisions. The record is useful for introducing new team members or for quickly assessing changes in customer needs. Historically it may be used as a start for a new but similar project.

2.9.8 Concept design

The concept design is an important way-marker for the whole design process. It marks the culmination of a great deal of investigative work and solution generation coupled with practical engineering knowledge as to the best method of achieving an appropriate outcome to the design challenge.

The concept design illustrates the best option to meet the design challenge with the available technology, funding, and skills. It is an opportunity to check the function and performance of the design against the requirements of the design challenge. The concept design specification should therefore be checked against the PDS, and of course they should match completely. In the real world, a 100% match may not be possible, but those elements of the specifications that do not match should not be detrimental to the functioning of the proposed product. Should the mismatch be unacceptable, a return to solution generation is advised.

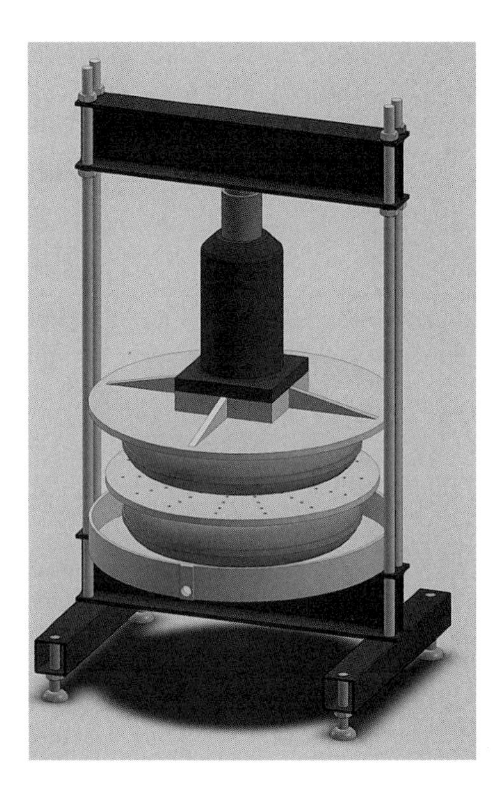

PLATE 2.7

Typical 3-D Model (Apple Press) [6]

The concept design is also important to show prospective clients the attributes of the new product. This can be shown in many ways. The presentation can be prepared as a poster, a verbal presentation with slides, a set of 3D models (see Plate 2.7), a physical model such as that which an architect might present (as shown in Plate 2.8), and many other communication methods appropriate to the project.

The concept design is often best shown as a visual representation. There is nothing that engages the human mind more than a good picture. This visual representation should, however, be more than just a picture; it should be annotated in a way that shows off the best features. It is actually a "point-of-sale" device created to entice the client.

A large amount of development money may ride on this presentation and convincing the client that the project is worthy of investment. The presentation should be geared toward the understanding of the client, who is likely to be an entrepreneur rather than an engineer. Information to which the client would be receptive is more financial and functions related, such as graphs, rather than stresses, strains, power consumption, and other technical details. The client would be more interested in overall performance and return on investment.

PLATE 2.8

Physical Architectural Model [2]

The client may be accompanied by other members of his or her team, such as the accountant, the production manager, and the chief designer. The accountant would ask questions such as, "How much will it cost?" The production manager will ask questions relating to whether the machine tools and manufacturing facilities are available. The chief designer will look carefully at the technicalities of the design and assure himself that due process has been completed in terms of stress analysis, materials selection, power consumption, and other practical details.

The client will look to his or her team to answer questions that are outside her field. The presentation and presenter should have answers to these questions so that the client can make a funding decision.

2.9.9 Concept design specification

The concept design specification (CDS) is developed at the same time as the concept design. Refer back to the classic design approach in Figure 2.7. The CDS should form a list of functions, performance data, requirements (such as fuel consumption) to make it function, envelope size, weight, and other technical details that will combine to give external people the opportunity to understand the new product.

Metrics should also be quoted, since these are the measurable elements by which the performance will be rated. Metrics here should be compared against those of the PDS, which were formulated directly from the brief. The two specifications should match. If there are any great differences, an iterative return to the solution creation process would be in order. The CDS should also cover an estimate of capital cost, maintenance, and probable running costs. Since the design specification can also be used as a selling tool, it could also incorporate chief benefits of the proposed new product.

As the concept developed during Phase 2 of the design process, the CDS should have been constantly compared to the PDS. The CDS should meet every point of the PDS. In practice this may not happen, but the differences should be so small as to be negligible. If there are large differences that are unworkable in the end product, the concept design should be reevaluated and modified to create a newly modified concept design and newly modified CDS.

A CDS for the conceptual apple press shown in Plate 2.7 is listed as follows:

Concept Design Specification for an Apple Press

Technical Details

- Cut apple capacity: 50 Ltr
- Load application: 10 tonne
- Applied pressure: 620 KN/m^2
- Yield approximate: 65% liquid yield by volume
- Approximate mass: 100 kg
- Envelope size: height, 1.6 m; width, 0.8 m; depth, 0.8 m
- Throughput (approximately): 75 kg/hour
- Apple juice outflow (approximately): 50–60 Ltr/hour

Operation

1. Load two pressing bags with chopped apple.
2. Place the pressing bags on the base platen with the perforated plate between the two.
3. Insert the pressure plate on the top.
4. Insert the hydraulic jack between the pressure plate and the top beam.
5. Operate the jack handle.
6. Apple juice will flow from the bags into the catchment tray and out through the drain hole.

Maintenance

After completion of the pressing and before the press is packed away for the season, it requires hosing down to remove debris and congealed apple juice. The pressing bags could be washed, but it is recommended that new pressing bags be purchased.

Benefits

- Adjustable upper beam height to allow for a range of jacks
- Location ring on underside of upper beam (easy location of the jack)
- Jack can be operated by one person at approximately 1.2 m from the ground
- Easy insertion of fruit through the use of pressing bags
- Low-cost pressing bags easily obtainable
- Easy removal of pulp after pressing
- Perforated plate between bags to aid in pressure distribution
- Perforated plate assists flow to the collecting dish
- Guidance on pressure plate to prevent rotation during pressing
- Simple construction
- Low cost, estimated at £150
- Easily cleaned with a hosepipe

2.9.10 Sustainability

When material is extracted from the Earth, energy is required. Quite often this energy is derived from fossil-based fuel, which will have a "carbon footprint" environmental impact. Whatever the means of extracting and processing material for a product, it will take some form of energy, whether from a fossil-fuel source or from a natural, renewable source. To measure the environmental impact "sustainability" of the product, it is necessary to consider all the energy (embodied energy) used in extraction, processing, transport, usage, and eventually disposal. This calculation can be quite complicated and is discussed in a later chapter.

As an example of measurement of sustainability, use has been made of a CES Edu-Pack, which is a large database of materials and processes equipped with a very efficient tool for measuring embodied energy. The results can be seen in Figure 2.12. The graph shows the embodied energy subdivided into the main areas of energy usage: material processing, manufacture, transport, usage, disposal, and end-of-life potential, which covers recycling, reuse, refurbishment, and the like.

Figure 2.12 clearly shows the embodied energy for the apple press. It is clear that most of the energy is used in preparing the raw material. A little more is used in manufacture, and still more is used in transport. The apple press is manually powered, so in use it does not use direct energy. Disposal has been set as totally recycled, and it can be seen that even this element requires energy, but there is an enormous end-of-life potential since in recycling the material it replaces raw material that would otherwise be extracted from the Earth.

In the past very little consideration was given to sustainable elements of a product. Designers rarely looked at sourcing materials from an environmental point of view. Sourcing has always been driven by

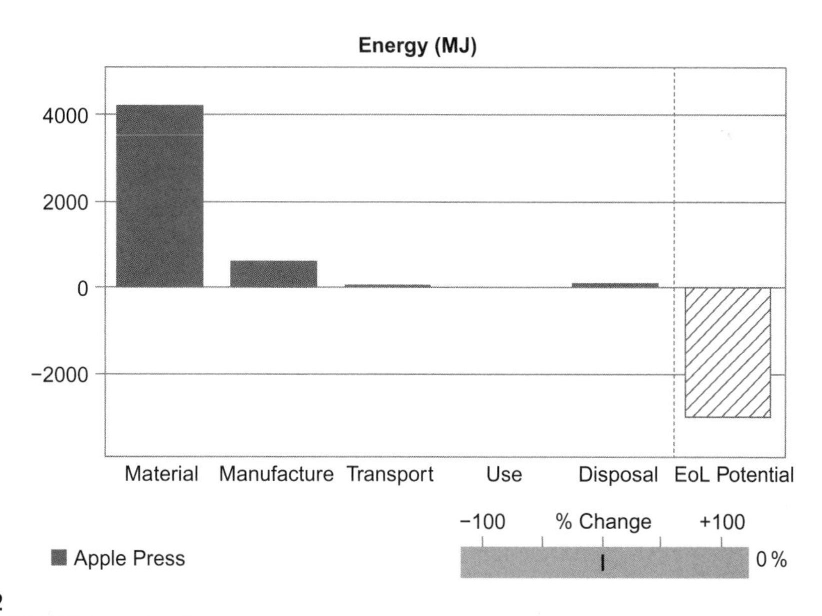

FIGURE 2.12

Embodied Energy for the Apple Press Shown in Plate 2.7 [1,9]

cost reduction. It is proposed that future concept designs and concept design specifications include an embodied energy value.

The designer has travailed through the investigation in Phase 1, and at this point, the end of Phase 2, has completed the generation and selection of solutions. This work has culminated in the creation of the concept design and the description of the concept specification. The next step, Phase 3, converts the concept design into a real "nuts and bolts" product.

The designer must realize that design is a complex business involving attempts to optimize a complicated set of relationships that will inevitably lead to a compromised solution.

Often elements and parameters are in conflict—for instance, high strength and low weight are necessary in many products, particularly so in aircraft construction. Compromise decisions have to be made here, since strength is an important feature of an aircraft structure, as is weight. There must be compromise between the two opposing elements.

There is seldom a unique solution to a design problem. Often several acceptable design schemes can be evolved to meet a given specification. The designer's experience and the use of evaluation tools will assist in judging which solution is the most suitable.

It has been said that the designer has to think of everything. It is also true that the designer must have in mind several goals throughout the design progress. The designer must constantly ask the following questions:

- Can it be made?
- Does it work?
- Does it look good?
- Is it sustainable?
- What does it cost?

2.10 PHASE 3

2.10.1 Phase 3 detail design: initial considerations

Phases 1 and 2 are really development phases that are mostly theoretical. It should be noted though that many hours from highly skilled individuals will have been injected into the project. Sometimes the design phase and easily consume 30% of the overall budget depending on the project.

During Phases 1 and 2, feasibility analysis will have been conducted so that a close approximation to the PDS could have been put forward as a final design concept. Stage 3 is where serious resources start to be used in the prototyping and manufacture phase.

To begin Phase 3, the designer must perform several duties at once. He must create the shape of each component and in many cases perform analysis that will guide him toward selecting the correct wall thicknesses, materials, finish, and so on. In some cases the designer may, for instance, create an image of a shaft within the CAD system, but in analyzing highly stressed parts of the shaft he may realize that these proportions have to be increased in size. The designer will then go back to the CAD image and change sizes. This iterative process may continue for every component within the product. The designer must ensure that before he commits the model to the manufacturing process, all aspects are correct. The consequence of not doing so would be wasted resources, time, and money.

This quest of ensuring that the detail design is correct is probably the most crucial phase of the whole design process. If there is a small mistake here, the dire consequence may result in breakage and could possibly cost someone's life.

2.10.1.1 Example: accident with a rock drill

A rock-drilling rig similar to the one shown in Figure 2.13 operates with a 250 kg gearbox sliding up and down the mast. When the drill string sinks into the ground during the drilling operation, the drill operator unfastens the drill string from the gearbox and raises the gearbox to the top of the mast to insert another length of drill pipe. The hydraulic controls for raising and lowering the gearbox are at the side of the drill head.

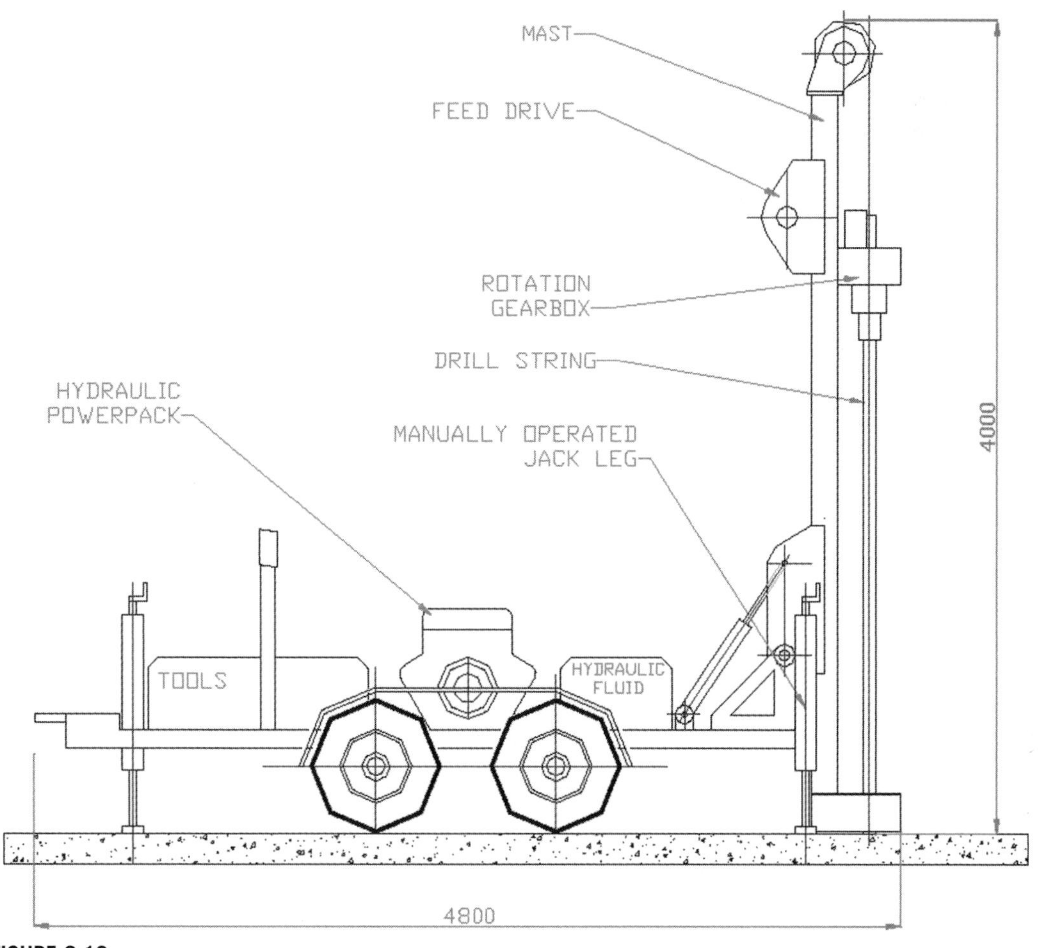

FIGURE 2.13

A Rock Drill of a Similar Type to the Example [1]

On one occasion the driller was lifting the whole drill string out of the ground. This amounted to a mass of approximately 2.5 tonnes, since it was lifting the mass of the gearbox plus the mass of the drill string. The whole 2.5 tonnes were pulled up the mast on a chain fastened via a chain anchor to a hydraulic ram. Since the chain anchor is a highly stressed component, it needs to be made of a high-strength material.

In this case the chain anchor parted and the whole 2.5 tonnes whistled down the mast, breaking the bottom of the mast and causing great damage to the rig. The fault was traced back to the design office, where the designer had merely inserted on the drawing a low-strength material. This seemingly innocuous mistake could have maimed or even killed the drill-rig operator. Fortunately he emerged uninjured but shaken.

People have different perceptions of design, but this incident highlights the fact that designers have a massive responsibility to "get it right." Nothing can be left to chance, and every consequence must be considered.

2.10.2 Modeling and detail design approach

The method via which a designer approaches the detail design is often based on personal taste but is also dependent on industry practice.

Some designers might want to draw, in 2-D, the whole product with all the parts in place. This is called a *layout drawing*. A gearbox layout may mean the inclusion of gears, shaft, seals, bearings, and the like and will include annotation, sizes, and other notes, such as surface finish and tolerance bands. It will also include specifications of components. An example of a layout drawing is shown in the gearbox layout in Figure 2.14.

Consistent with emerging practices, a more modern approach is to build up the components in 3-D within an appropriate software system. Parts would be individually created as 3-D solid parts and then fitted together, rather like a physical build process. An example of the gearbox casing as a 3-D component is shown in Figure 2.15.

Many component suppliers of items such as bearings, seals, and motors offer designers a free technical image download of their product. This could be 2-D or 3-D image, depending on the designer's requirements.

Normally included in any 3-D software package is an interference interrogation tool, which alerts the designer should any parts overlap other parts. Once notified, the designer would then change the original component part model so that components do not occupy the same space as other components.

This assembly process creates in 3-D a complete product, achieving a similar function to the 2-D layout but with all the components modeled. The term *modeling* here means that all the components are of known size and have been saved in separate files. Due diligence now needs to be performed on each component. This includes stress analysis, tolerances application, surface finish, and correct material selection. Often a 3-D software package will include basic finite element generation for a particular component and the means to stress-analyze the component so that the designer can judge its strength. Other tools, such as dynamic analysis, thermal analysis, and fluid flow analysis, are also available. More specific and high-level software of this type can also be purchased as an add-on to the main 3-D software drafting package.

Should 2-D drawings be required, they can be quickly transformed from 3-D models into the drawings.

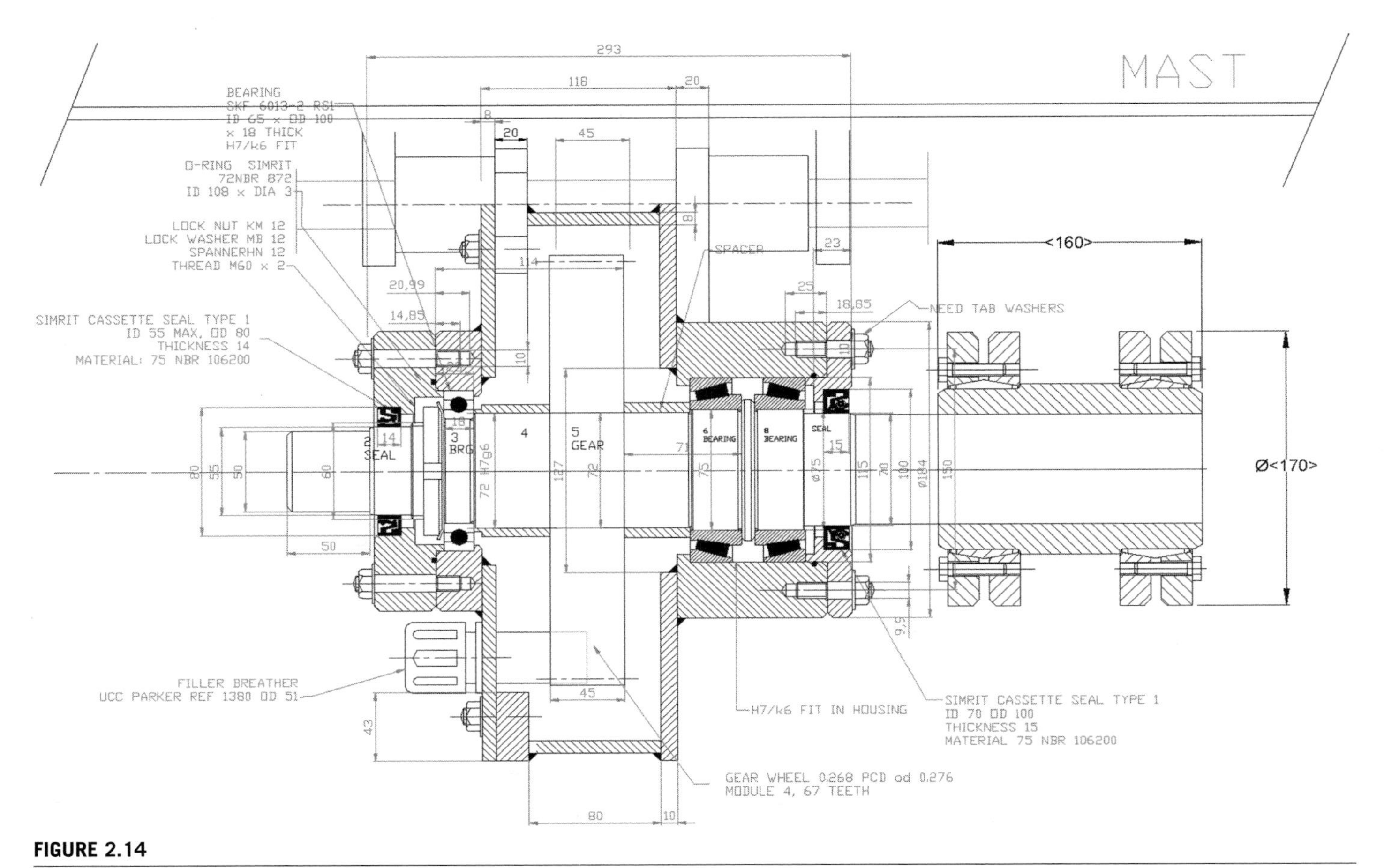

FIGURE 2.14

Layout of Gearbox [1]

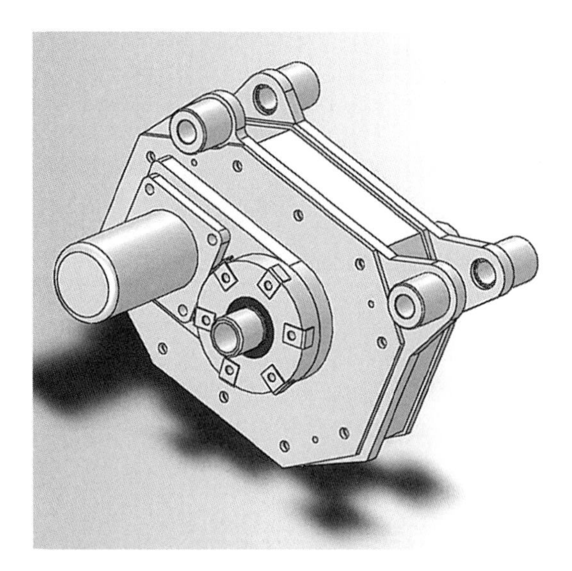

FIGURE 2.15

Gearbox Casing as a Full 3-D Component [1]

Similar information is required in the modern era, as was required when only 2-D drafting was available. The provision of the support data in such a system would have meant that the designer would manually create parts lists, materials specifications, and bought-out part specifications, to name just portion of the information required.

Within the 3-D software, once the model is formulated, the digital information can be used in many ways, with much of the information generated automatically. Parts lists are only one element of automatic information processing. Various departments in the manufacturing organization require data in different formats. The production team requires specific data, as does the marketing team. The digital 3-D software allows for manipulation and easy dissemination through e-mail and storage on remote Internet-based servers.

The analysis package in most 3-D drafting software allows for easy analysis of fairly complex systems. In the right hands, this is a very powerful tool. It can also be a tool that is too easy to use for a novice who perhaps does not completely understand the engineering. There have been some notable disasters where the fault has been traced back to the designer believing the results from the analysis without checking, a problem termed *computer-aided failure*, or CAF. The analysis algorithm will generate answers to some very complex questions, but if the information input is incorrect or the algorithm has been set up with errors, the outcome will be flawed no matter how complex the analysis. In the words of a notable academic at the University of Huddersfield in the United Kingdom, "Computers are stupid. They will only do what you tell them to do!" [4]

Any designer is advised to perform hand calculations or find some means of proving the analysis performed by the computer. Often this proof may be in the form of formal tests or merely informal experiments to prove the data from the analysis. The human intuitive, analytical mind applied to a clinical answer is usually required to make sense of the result and to spot deficiencies that the computer analysis alone isn't equipped to understand.

2.10.3 The design process: information output

Information from the design process and the results of Phase 3, the detail design process, need to be disseminated in a form that can be readily understood.

Analytical results will help form the size and shape of components. They will give the designer the confidence and ability to define the shape and specify materials for particular components. In finalizing the size of a component—say, a shaft—the designer may also have applied commercially sought components such as bearings and seals. These will now be precisely specified and would normally include the company name and the company part number.

This is covered in detail in chapter 5, which discusses the processes and logic of detail design, drawing and communication.

2.11 FINAL DESIGN SPECIFICATION

With reference to the classic design approach illustrated in Figure 2.7, we can see that the design journey is almost complete. As the final model, drawing, and parts list is published to the rest of the company, the design team should be in a position to create the final design specification.

The FDS is a description of the product, including performance data and consumption data. It should include costings for prototypes and for production versions, though some of this information may be formulated in conjunction with other departments. The final design specification should match the product design specification and more precisely match the original brief that was identified by the client.

2.12 PRODUCT SPECIFICATION

The product specification is a much cut-down list of performance data an example of which can be seen in Figure 2.16. It is data that the customer sees when he opens the product packaging. The product specification should give basic details of how the device performs and should indicate inputs, such as fuel, required to make it perform. Typically the product specification will include:

- Performance data
- Fuel input
- Energy efficiency
- Energy usage
- Connection details and specification
- Instructions on the use of the product
- Sundry data gleaned from the prototype and testing process

A typical product specification for a large commercial cooker could be as follows in Figure 2.16:

Product Data [8]

Supplied ready set for group H natural gas. Conversion kit for LPG is packed with the cooker.
INSTALLER: Please leave these instructions with the user
DATA BADGE LOCATION: Inside base drawer of cavity

Connections

Natural gas: 20 mb Electrical: 230/400 V, 50 Hz
Butane: 29 mb
Propane: 37 mb
See appliance badge for test pressures.

Dimensions

Overall height: Minimum 905 mm, maximum 925 mm
Overall width: 900 mm
Overall depth: 600 mm to facia, 650 mm over handles
Minimum space above hot plate: 650 mm

Ratings

Oven	Full	Divided
Fan element:	3.31 kW	1.65 kW
Top element:	3.49 kW	1.75 kW
Browning element:	2.11 kW	1.06 kW
Bottom element:	1.38 kW	0.69 kW
Hotplate	**Natural Gas**	**LPG**
Wok burner:	3.5 kW	3.5 kW
Large burner:	3 kW	3 kW
Medium burner:	1.7 kW	1.7 kW
Small burner:	1.0 kW	1.0 kW

Oven Efficiency

Energy efficiency class: A
Energy consumption based on standard load: 0.99 kWh
Usable volume (liters): 115
Time to cook standard load: 44 min
Surface area of grid: 2,400 cm^2

FIGURE 2.16

Typical Product Specification for a Large Commercial Cooker [8]

2.13 PROTOTYPES

In some cases it may be necessary to build prototypes for testing purposes. A prototype is the first full-scale, functional model and is often built to prove the concept and to confirm the claims of the design team. Testing highlights areas of uncertainty and gives feedback data to designers. Testing can show that the functioning of the device might not be as originally envisioned. In such a case, the iteration design process is applied and the elements that are lacking are modified. When a passenger vehicle is manufactured, for example, it is rigorously tested so that any deficiencies can be eradicated and data such as fuel consumption in urban areas or on motorways can be added to the product specification.

References

[1] A.D. Johnson (Senior Lecturer), University of Huddersfield, Huddersfield U.K., 2013.
[2] Open source.
[3] D. Warburton (Research Student), University of Huddersfield, Huddersfield U.K., 2012.
[4] R. Taylor (Senior Lecturer), University of Huddersfield, Huddersfield U.K., 2012.
[5] Scribd.com, www.scribd.com/doc/50459583/Example-Product-Design-Specification, 2012.
[6] M. Melia, K. Hinchliffe (Research Students), University of Huddersfield, Huddersfield U.K., 2012.
[7] C.L. Dym, P. Little, Engineering design: a project based introduction, Wiley, 2009.
[8] AGA Rangemaster Group, Clarence Street, Royal Leamington Spa, Warwickshire, CV31 2 AD, U.K.
[9] Granta Design: CES EduPack software, 2013.
[10] P. Lockwood, T. Armitage, J. Halpin, A. Ausin (Research Students), University of Huddersfield, Huddersfield U.K., 2012.

CHAPTER

Sustainability and Its Application Within Engineering Design

3

3.1 WHAT IS SUSTAINABILITY?

In 1983 the U.N. General Assembly became aware of a heavy deterioration of the human environment and the natural resources that the human race required to survive. It fell upon Javier Perez de Cuellar, Secretary General of the United Nations, to create a commission to investigate aspects of sustainability for the whole world. The commission's brief was to rally countries to work and pursue sustainable development together. The result of this initiative was that the *Brundtland Commission* was quickly established, led by Gro Harlem Bdtland, the former prime minister of Norway.

In 1987 Brundtland published the first part of the *Brundtland Commission Report*, entitled "Our Common Future" [5]. The report was wide ranging, researching and interviewing politicians, industrialists, academics, nongovernmental organizations (NGOs), the general public, and anyone who wanted to contribute to the findings. The report's findings extended through many diverse areas of sustainability and the impact of humans on ecosystem Earth, highlighting the result of a depleted ecosystem on humankind. The areas of interest can be summarized as the three main pillars of sustainable development: economic growth, environmental protection, and social equality.

More important, the Brundtland Commission defined sustainability as follows:

> *Sustainable development is development that meets the needs of the present without compromising the ability of future generations to meet their own needs.*

The report contains two key concepts:

- The concept of *needs*—in particular, the essential needs of the world's poor, to which overriding priority should be given
- The idea of limitations imposed by the state of technology and social organization on the environment's ability to meet present and future needs

The concept of needs is really a geopolitical requirement and is very valid. From the engineering design discipline point of view, the second concept is very relevant and should be in the back of every designer's mind when he or she is creating new products. The statement highlights the fact that Earth has a limited capacity to support an ever-expanding human race with all its technology and social systems.

After considering the definition of sustainability and then realizing that ecosystem Earth has limitations, the question should be asked: Is sustainability achievable?

To answer this question, perhaps a slightly different definition of *sustainability* should be considered: True sustainability is development and use of products and services whereby *zero* resources are taken from the Earth.

Sustainability in Engineering Design. http://dx.doi.org/10.1016/B978-0-08-099369-0.00003-0

3.2 IS SUSTAINABILITY ACHIEVABLE?

An unknown industrialist once said, "Everything costs money; everything has an environmental impact." Any product development requires resources in terms of materials and the energy to manipulate those materials. It is certainly true that sustainability can be improved by recycling materials, reusing appliances, repairing components, and reducing material usage. These are known as the 4Rs, listed as follows:

- Recycle
- Reuse
- Repair
- Reduce

It is true that many people and institutions feel that by accommodating the 4Rs approach, they have done enough. Certainly adoption of the 4Rs assists sustainability massively, but it largely addresses the end of the life of a component or product, merely enabling materials to be reused in some way. The adoption of the 4R approach does not consider the environmental impact of a product prior to its end of life. Significantly, the extraction of raw material from the Earth, its manipulation in manufacture, and especially its use in the form of a product may contribute more to environmental impact than any strategy that could be adopted at the end of that product's life.

The extraction of materials, their manipulation, their use, and their eventual disposal will always require energy. No matter what sustainable strategies are in place, the one single requirement that will always be needed is energy to extract and create the product. It is this energy that is and will prove to be the greatest need for future generations.

In humans' ever-increasing quest for new products, there will always be a requirement for materials and there will always be a requirement for energy. Sustainable strategies may assist in reducing the environmental impact, but the reality exists that it may *never* be possible to achieve true sustainability where there are no new resources extracted from the Earth. Appropriate sustainable strategies may help us come close to attaining the sustainable "Holy Grail," however.

Open and closed cycle loops

The current open-loop cycle described above and shown in Figure 3.1 is clearly not sustainable over the long-term. It fails to take account of issues associated with material resources or with the energy associated with transport of materials. Furthermore, it also fails to account for the end-of-life issues associated with a product.

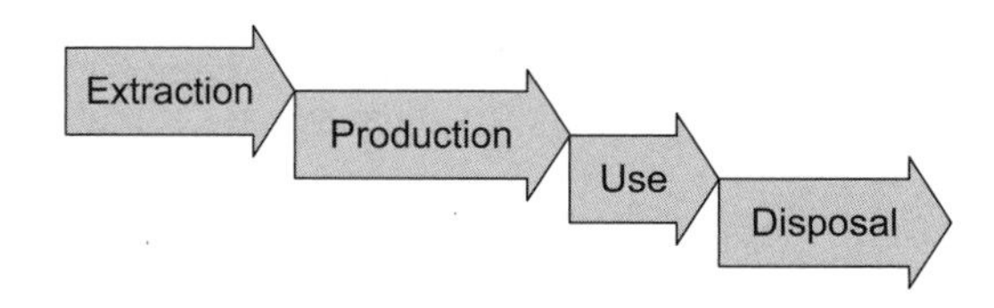

FIGURE 3.1

Open-Loop Material Cycle [6]

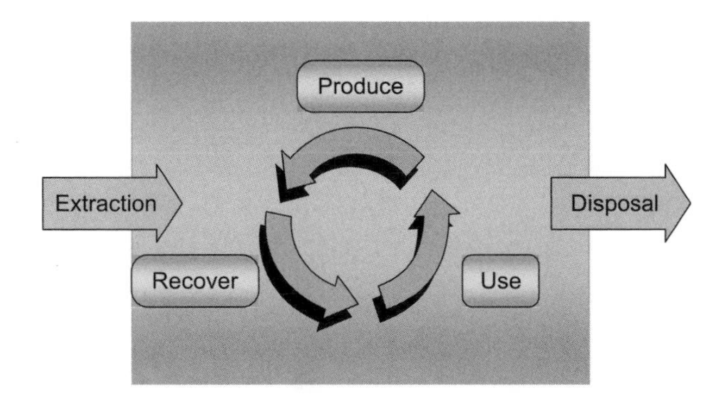

FIGURE 3.2

Closed-Loop Material Cycle [6]

The implication of the closed loop cycle indicated in Figure 3.2, is that materials (and some energy) embedded in a product can be partly or wholly used as raw materials for the next generation of product. Clearly, the closed loop system is in actuality only partly closed: like air conditioning systems, it requires input of fresh materials as an admix to the recycled or re-used ones, and produces waste in the form of materials and heat or carbon emissions. It is in the power of the designer radically to radically increase the recycled or re-used proportion of the mix.

The effects that the closed loop material cycle has on the environment will include:

- A reduction in materials use overall, leading to extension of the reserve life for key materials
- A reduction in materials going to land-fill
- A reduction in embedded energy and carbon as typically less energy is needed to refurbish or remanufacture than to manufacture from scratch
- A reduction in energy used for transport as refurbishment and repair tend to be local activities

3.3 SUSTAINABILITY: PAST AND PRESENT

Recent decades have shown an increase in awareness of environmental impact and the need for sustainability. This has been due to a larger awareness of the impact, brought about largely because the impact could be measured.

Until the Middle Ages, the impact on the planet of human infestation was fairly minimal, since no one used electricity or internal combustion engines. The advent of the Industrial Revolution in Europe and the United States began the need for more and more energy. In the beginning, mechanical power was generated from steam engines, which were used to drive farming machines, diggers, engineering machine tools, and entire mills in the steel, textile, and many other industries. These powerhouse machines spawned other industries, such as engineering, coal mining, and ship building. The prosperity of towns, and indeed countries, became dependent on steam power. All this power came at a price:

PLATE 3.1

Pollution in Manchester, U.K., Circa 1850 [1]

pollution! Natural waterways became clogged with industrial slime and debris. The air around industrial centers became laden with soot and other unpleasant chemicals. Perhaps the term "dark Satanic mills" came to represent this pollution, though it seems that originally William Blake, in his poem "Jerusalem" (1804), intended to draw attention to Britain's industrial might and its propensity for manufacturing weapons of war.

The pollution caused by steam power was tangible. Airborne smog and chemicals affected health and longevity of the population. A typical industrial landscape is shown in Plate 3.1.

Electric motors and the internal combustion engine gradually replaced the steam engine, but it was the Clean Air Act in the United Kingdom (1956) and in the United States (1970) that reduced the air pollutants dramatically, forcing steam power into near oblivion.

The generation of energy through coal- and oil-powered power stations and through the use of the internal combustion engine created its own pollutants. These pollutants, such as carbon dioxide, carbon monoxide, nitrogen oxide, and sulfur, were more dangerous because they could not be seen. At the turn of the 21st century, millions of internal combustion engines burning fossil fuels (oil derivatives) were used daily worldwide. The emissions created enormous health problems in large urban areas, but perhaps more concerning was the increase in global warming, largely attributed to the increase in carbon dioxide in the atmosphere. Furthermore, polluting chemicals derived from the burning of fossil fuels have been shown to deplete the protective ozone layer in the stratosphere.

Measurements of these pollutants and their effects on Earth's ecosystem have shown a definite deterioration. As the evidence mounted of human impact on global climate change, so did scientific, political, and popular concern. Although pollution was the original focus of environmental concern, measurement of the environmental impact has evolved and broadened the concern.

Sustainability generally deals with the causes of environmental impact rather than merely the results. The broad view of sustainability takes in many disciplines, including the built environment, geographic sciences, mechanical engineering, sociology, psychology, finance, and business.

Mechanical engineering, with allied disciplines such as electrical and chemical engineering, is at the forefront of new product development. As a society, humans have become reliant on appliances and other products. Every item we purchase is used and eventually discarded in some way. Each product has been designed and manufactured by artisans and engineers, and its production, use, and disposal naturally impact the environment. To apply longer-term sustainability, the design and manufacturing engineers will need to deliver the sustainable products; it is those individuals who have the knowledge and the opportunity to do so.

Individuals and companies often talk about sustainability, with little idea of how this can be achieved within their organizations. However, some individuals as well as organizations have embraced sustainability for many years and have become very successful. The real key to achieving sustainability is to educate and empower the majority of people and organizations to adopt practical approaches to sustainability in terms of mechanical engineering design.

Architects and builders have long been building sustainable structures. Even early man built dwellings that were self-sustaining. There are many modern examples of sustainability in the built environment. Perhaps some of the better ones can be found in the recycling of building materials. Plates 3.2 and 3.3 show the reuse of building materials applied to the Citadel Walls in Ankara, Turkey. This can perhaps be described as an overenthusiastic reuse of building materials.

The geophysical environment has also been actively enhanced in the application of sustainability projects. Beach groynes are an excellent example of sustainability of coastline. Plate 3.4 shows beach groynes on the beach near Bournemouth, United Kingdom. These wooden structures are built like fingers out into the sea and perpendicular to the shore, thus preventing long shore drift and preserving the shoreline.

Some excellent examples of sustainability can be found in mechanical engineering, but, it could be argued, in a limited sense and that not enough is being done, since most of the sustainability focus is applied to recycling.

Recycling of steel and other metals is well practiced, as is the recycling of some plastics. Steel recycling in the United States (see Figure 3.3) achieved some dizzying heights, reaching recycling rates of 103% in the early part of the 21st century [4], though averages were nearer 80%.

PLATE 3.2

Roman Lintel Reused in the Citadel Walls, Ankara, Turkey [1]

PLATE 3.3

Example of the Reused Byzantine Sculptures in the Citadel Walls, Ankara, Turkey [1]

PLATE 3.4

Beach Groynes, Bournemouth, U.K. [1]

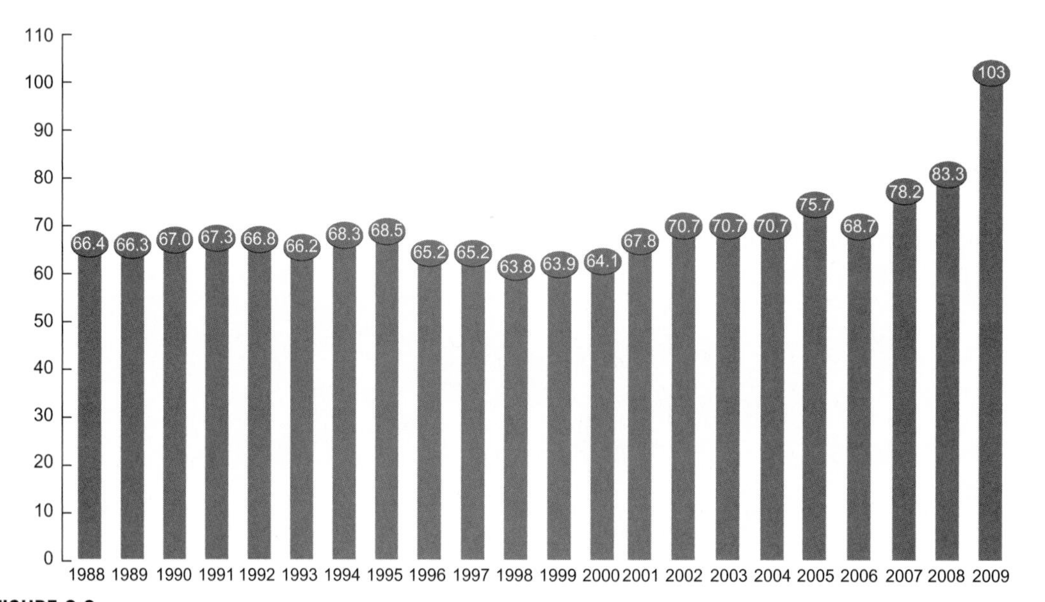

FIGURE 3.3

Overall Steel Recycling Rates in the United States, 1988–2009 [3]

Vehicle manufacturers normally aim for 90% recycling of vehicles at the end of their lives. This includes plastics and other components. In the United States, steel recycling from vehicles reached rates of 121% of output in 2009, with averages close to 100% [4].

Vehicle manufacturers have also been attempting to design and manufacture the sustainable use of vehicles by optimizing new designs to reduce mass. Design optimization is a major tool in the sustainable design armory. For instance, a reduced-mass vehicle requires a smaller engine and smaller brakes, tires, and so on. Merely by reducing the overall mass, parts become smaller and fuel use is reduced, leading to fewer emissions.

Power plants have also been the target of design engineers' creativity in the development of leaner internal combustion engines and, more specifically, hydrogen engines and electric vehicle drives.

Though there has been much foresight on the part of individuals and institutions to develop sustainably engineered products, the question has to be asked: Is this enough?

It is a fact that resources are still being stripped from the Earth at an alarming rate. Worldwide steel production was around 127.5 million tonnes in 2010 [5], with only around 70 million tonnes being recycled. Steel, however, is the world's most recycled commodity. If design and manufacturing engineers are to have a conscience, the response to our question must be, We can do more!

Recycling is only one aspect of sustainability and generally pertains to the end of life of a product. Figure 3.3 shows the design and manufacturing sustainability model, which considers the whole life of a product rather than just the end of it.

Many global companies have sustainable policies in place, but each has its own approach and its own agenda. This is easy to understand when there is such a diversity of products and industries. There is a great need for a cohesive and coherent approach so that all designers can work toward similar goals and create a significant effect.

3.4 THE CLASSIC DESIGN AND MANUFACTURE MODEL

The classic design and manufacture model shown in Figure 3.4 has been used for centuries. The designer receives the brief, creates the concept design, and converts the concept design into manufacturing drawings, which are then used by the manufacturing function to create the product.

This particular model is used by many companies without thought to the source of the material, the energy used in manufacture, and during the products' use. Furthermore, no consideration is given to the disposal of the product. Often the product is merely thrown into a bin or into land fill.

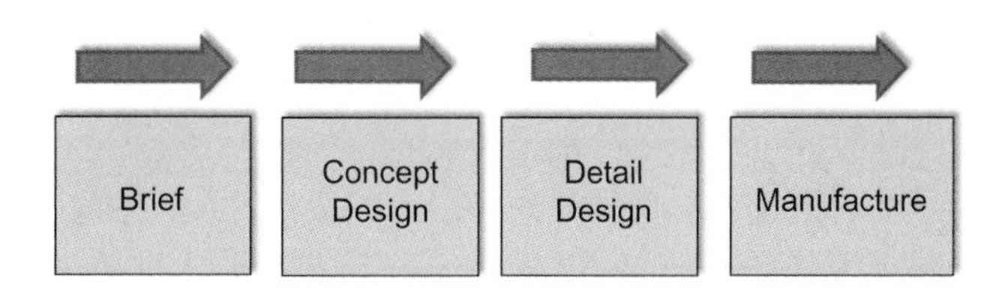

FIGURE 3.4

The Classic Design and Manufacture Model [1]

To order to apply sustainable techniques, the whole life of the product needs to be considered. This is within the purview of the designer. It is the designer who can:

- Specify the source of the raw material
- Specify the method of manufacture
- Create a product system that is environmentally friendly in use
- Design into the product a sustainably disposable set of components

It is the designer, therefore, that is the key to future sustainable products. It is therefore essential to change the mental attitude of the designer, as did Genichi Taguchi in the 1960s with his approach to quality manufacturing.

3.5 THE TAGUCHI APPROACH TO QUALITY MANUFACTURING

Genichi Taguchi was a manufacturing and statistical engineer who, in the 1960s, promoted theories relating to the quality of manufactured goods. He noticed that the quality of manufactured goods was usually left to the manufacturing craftsmen, which meant that every craftsperson was essentially building every product as a single item. This approach was time consuming and expensive. Taguchi suggested that quality could be specified at the design stage, before components were manufactured. He also recognized that to achieve designed-in quality, the mindset of the designers had to be changed.

As designers were eventually made more aware of the "new" design method, manufactured goods were produced in huge quantities and became more available at a lower cost and higher quality. Engineering designers created better components by introducing standards, tolerances, and precision machine tools, which led to large-quantity production methods. The Taguchi approach revolutionized batch production because it produced consistent quality during manufacture.

3.6 THE TAGUCHI ANALOGY APPLIED TO SUSTAINABLE ENGINEERING DESIGN

Taguchi suggested that quality should start with the designer. He should specify exactly how the product should be made and to set quality specifications. The manufacturers would adhere to these specifications so that a quality product should emerge.

The Taguchi analogy can be applied to *sustainable engineering design* (SED). Sustainability cannot be confined to individual elements of the product life process. It has to be considered at the very beginning of the design stage, and it is therefore within the realm of the designer to apply sustainability principles to the whole life of the product. It is now the designer's responsibility to design sustainability into his or her new creations.

Important Concept: It is the responsibility of the designer to apply sustainability principles to the whole life of the new product.

It is the designer who is key and who must envisage and design components using sustainable techniques, equipment, and methods.

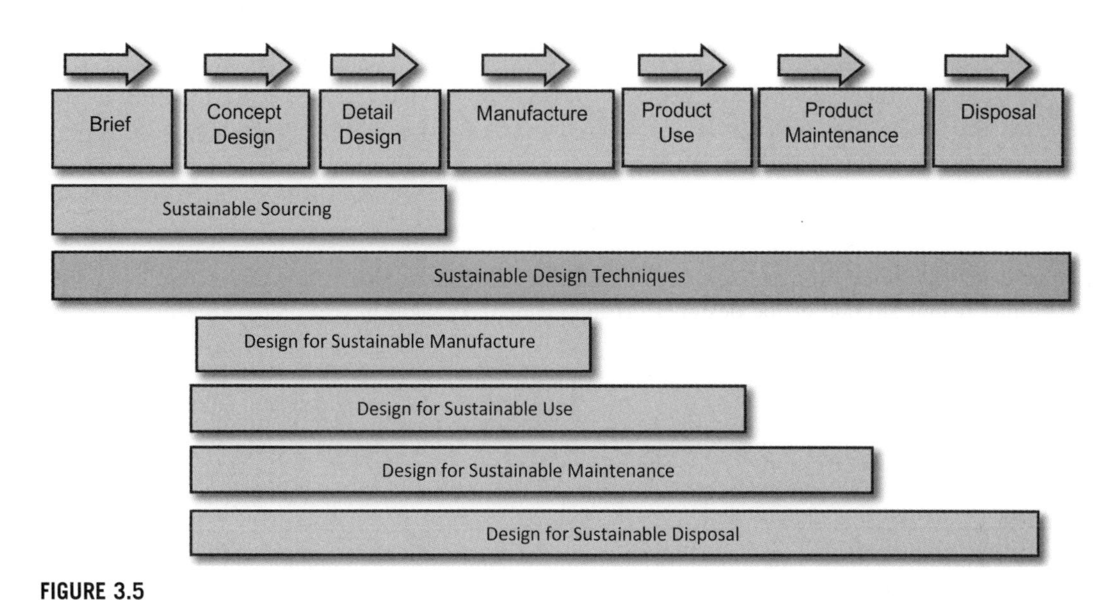

FIGURE 3.5

The Sustainable Engineering Design Whole-Life Model [1]

It must be acknowledged that although some engineering designers have sustainability in the forefront of their design practice, the majority of engineering designers may only pay lip service to SED. To achieve true SED, the designer's mindset has to be modified.

The engineering design process can no longer be related purely to cost and social benefits. The engineering design and manufacture process must now accommodate cost, social, and sustainability benefits.

A new design and manufacture model can now be formulated that combines the original design and manufacture model with sustainable elements, as shown in Figure 3.5.

Only a few years ago, reduction in pollution was the watchword for environmentalists. This was a crude yardstick for what has now developed into *sustainability*. The term *sustainability* encompasses many disciplines, from farming to the creation of buildings, from protection of the natural environment to power generation. Within the engineering design discipline the approach to sustainability has evolved into a multifaceted one. The new model shown in Figure 3.3 now encompasses the whole engineering process, from sourcing of materials to disposal of the product at the end of its life. It is the whole-life-cycle approach.

3.7 SUSTAINABLE SOURCING (ECOSOURCING)

The practice of sourcing material spans a vast area. It not only examines the extraction of raw material but also its transport. Sustainable sourcing could be from a certified source, as are some timbers, but may include sourcing recycled materials.

3.7.1 Transportation

It is inevitable that raw materials will almost always be hewn from the ground and then transported to the processing point using fossil fuels to provide the energy to propel the transporters. This practice often occurs over very long distances.

The current common practice is for Western organizations to source products in the Western Pacific Rim: China, Japan, and Korea. This is largely done on the basis of reduced cost, since the cost of labor is relatively lower in the Pacific Rim than in the West. When large quantities can be produced and freight containers can be filled, it is relatively cheaper to transport materials, even over long distances.

It should be remembered that the environmental impact of producing these goods is roughly similar in the Pacific Rim as it would be in the West. The real impact on the planet's resources comes in the burning of fossil fuels used in the energy generated for transportation. Responsible sourcing would mean that designers would specify local suppliers, thus reducing the environmental impact of transporting goods over long distances. An added benefit is that local industries would thrive.

Some commodities have to be transported since they are available only in another part of the world. In such cases the question would be, How can these goods be transported using sustainable methods? Before the advent of steam propulsion for ships, for example, only wind power was available. Wind power is 100% sustainable. In the modern era, perhaps sailing ships could be resurrected and employed, or perhaps modern sailing versions that used the natural elements for propulsion could be used. This is no pipe dream.

Examples such as the *MV Beluga* (see Plate 3.5) have shown that the wind can be harnessed to provide part of the necessary propulsive power for large modern freighters. For example, Gerd Wessels of Wessels Reederei says, "There is enormous free wind-energy potential on the high seas. With *Skysails* [7] we can reduce energy by 30% on a good day, giving at least 15% annual fuel savings." Flettner rotors [8], rotating sails, have also been fitted to freighters with some success.

Similarly, Eco Marine Power [9] has designed solar-powered craft, from small ferries to freighters. Plate 3.6 shows a drawing of Eco Marine Power's solar ferry *Medaka*. Critics may scoff at using sail or solar power for freight, but the question should be asked, What energy source would be used if there were no more fossil fuels? Smaller vessels such as the *Medaka* may be powered purely from solar-

PLATE 3.5

Skysails' *MV Beluga* [6]

PLATE 3.6

Solar Power Ferry *Medaka* [9]

Eco Marine Power: 2012: http://www.ecomarinepower.com/en/solar-ferry-medaka

generated electricity. Large freighters currently use hybrid solar power and diesel, the solar energy contributing approximately 20% to 30% of the total energy required to drive the ship.

3.7.2 Techniques

Techniques could be changed to accommodate processes that gave a sustainable benefit over current techniques. A recent emergent technology is that of rapid prototyping, which has grown alongside the development of 3-D computer models and has usually been associated with the 3-D printing of actual-sized plastic models.

New techniques in this area are those that create 3-D components using laser-fused metal powders. This technique effectively reduces time to manufacture and reduces transport costs and environmental impact to almost zero, since the component can be formed with no waste at the assembly plant.

3.7.3 Managed Sources

All raw materials should be labeled with a *sustainable source value* (SSV). The main feature of an SSV label would be to inform the designer of the environmental cost of the raw material. This may at first appear to be expensive and complicated, but the system already exists in the form of managed exotic timber, which carries a certificate of authenticity of sustainable sourcing. In the design of street furniture (external seating), local U.K. governing authorities generally have a policy of specifying that the timber for a seating element (usually Iroko) is sourced sustainably. This guarantee is certificated and shows where the timber was grown and how it will be sustainably replaced. With such a system in place for other materials such as steel, designers could select a material according to its sustainable impact.

3.7.4 Material Flow Systems: Open and Closed Loop

This concept, introduced in the 1990s, is now being embraced by, among others, the European Union (EU). Joke Schauvlike [10], President of the EU Environmental Council, is of the opinion that "We must deal with our materials, and with our energy, more efficiently. At the end of their life we must be able to reuse materials as new raw materials. This is called completing the cycle." This approach was discussed in economic and energy terms by Clift and Allwood in an article titled "Rethinking the Economy" [11].

It can be seen that the present linear design-manufacture systems model (see Figure 3.1) is not sustainable over the longer term because manufacturers take no account of the issues of raw material extraction and transport discussed previously, nor of the end-of-life issues once the product is no longer usable or is obsolete.

In the closed-loop material flow system model (see Figure 3.2), materials and components are recovered and reused, reducing material inputs and outputs as close to zero as possible. This produces a *ladder* of sustainable end-of-life disposal techniques, which, as we've seen, are commonly termed the 4Rs:

- Recycle
- Reuse
- Repair
- Reduce

The 4Rs will be dealt with in more detail in later pages; however, the use of recycled material in a new product can be considered as prudent sustainable sourcing of raw product.

3.7.5 Recycle

Recycling of end-of-life components and materials has, in some quarters, been a way of life for thousands of years. Throughout the world peasants have built their houses with building materials appropriated from ruined castle walls. In the modern era, many products are now recycled, and whole industries are being built on the recycling nature of many products. Older vehicles at the end of their life are stripped of their useful components before being crushed and resmelted. Various components taken off those vehicles are then resold. There is a thriving "breakers" business for breakers yards and, more recently, Internet sales of appropriate parts.

A great success story is that of recycled vehicle tires. The tires are retrieved from the vehicles and processed by stripping and granulating the remaining rubber. This material can then be reused in a diverse range of items, including speed humps (see Plate 3.7), children's playground flooring, artificial tarmac for driveways, bullet-absorbing walls for firing ranges—the list is seemingly endless.

Other commodities that can be recycled include glass, cardboard, newspapers, plastic bottles, wood, rainwater, and horse manure. This list is not exhaustive, but it does show the diversity and possibility of recycled products.

3.7.6 Designers' Duty

It is the duty and responsibility of the designer to source materials from sustainable sources, or at least from sources that have a reduced impact on the planet's resources. This emphasis would reduce the SSV.

PLATE 3.7

A Typical Speed Hump Manufactured from Recycled Rubber Tires [12]

3.8 DESIGN FOR SUSTAINABLE MANUFACTURE (SUSTAINABLE MANUFACTURE VALUE, OR SMV)

The designer or design team is the entity that selects the manufacturing process. The designer can select processes, techniques, and materials. Industries differ in their manufacturing techniques and, indeed, in the materials they use. For example, the fabrication industry normally uses material that is black finished (hot rolled) mild steel and standard sections. Attachment methods are normally through electric arc welding and other welding variations. The machine tool industry, on the other hand, often combines cast-iron low-carbon and medium-carbon steel with high-precision machining methods. Designers in each of these industries are specialists who know intimately what can be achieved and how to achieve a particular design shape through the available manufacturing methods.

The formation of components has for centuries mostly relied on the removal of material to create a shape. This process results in a great deal of waste, both in material and in energy required to remove the material. Casting components defines the shape with much less waste, but even this process requires a great deal of energy to produce molten material and then machine to final size. Sometimes the energy-expensive process cannot be avoided, but the designer should fix his gaze on reducing energy and material waste. Focus should also be directed toward the selection of materials that can be processed easily and selection of manufacturing methods that are not energy hungry.

3.8.1 The Smart Factory

Sustainable manufacturing may be expanded beyond component manufacturing. Factories can improve their energy usage. Larger organizations are beginning to realize that the old methods of manufacturing in large factories can be a drain on the environment. In such factories, a great deal of energy is used, which is not only costly to the company, increasing product cost, but also is environmentally draining.

It is difficult to change the practices of an older factory; often this is not done because the task is so daunting. However, with the correct program in place, small changes can make a large difference over time. New build factories offer the opportunity of creating *smart factories*. For example, the department store chain Marks and Spencer commissioned a clothing manufacturer in Sri Lanka to build and renovate a £7 million factory along environmentally friendly guidelines [14].

Brandex, the largest Sri Lankan apparel exporter in 2007, converted a 30-year-old factory into an eco-friendly plant. The plant has reduced the company's carbon footprint by 77%, from 2,076 metric tons to 494 metric tons. The plant has achieved its green credentials by applying the following techniques and systems:

- Skylights in the roof, which reduce the need for overhead lighting
- Light-emitting diode (LED) light systems (low-energy lights at each workstation), which further reduce the need for overhead lighting
- Water recycling, which makes water available for toilet flushing and for external irrigation; sewage treatment on-site using anaerobic digestion helps recycle water
- Rainwater harvesting makes up 15% of the recycled water
- Harvesting of biogas from anaerobic digestion of sewage creates enough gas to power the kitchen
- Waste recycling
- New-build used bricks made from stabilized earth (better insulation)
- Low-energy evaporative cooling systems replacing air conditioning, reducing energy requirements

It is estimated that the plant will be 40% more energy efficient than other factories and will use 50% less water.

3.8.1.1 Renault Tangiers Smart Factory [15] See Plate 3.8

Renault has built a factory at Tangiers in Morocco that will reduce its carbon dioxide emissions by 98%. Operational from 2012, the plant has achieved these reductions by two main methods:

- Water recycling. There will be no water released into the natural environment.
- Electrical power is totally generated by renewable energy sources such as wind, solar power, and hydroelectricity.

The main consideration thus far has been conservation of energy where energy may escape from buildings, but smart factories combine much more than just energy escape or use. The term *smart factory* is a global term referring to many of the smart practices that can contribute to sustainability in manufacturing. A smart factory actually encompasses three basic areas of sustainable manufacturing:

- Virtual factory
- Digital factory
- Smart factory

These three areas combine many aspects of sustainable energy use during the design and manufacturing process. The advent of computing technology has brought about more efficient and almost instantaneous communications. It is inevitable, therefore, that computer technology in its many forms has become the basis of a digital creation and management, ensuring that the process from inception of a design to the final distribution of a product is controlled using advanced computing technology.

PLATE 3.8

Renault Smart Factory, Tangiers, Morocco [15]

Computing technology provides the means to achieve some high-level goals, the main theme being high production, with more efficient overall management leading to energy efficiencies, which reduce the environmental impact.

Designs are concluded with near-complete prototypes on screen as 3-D images. This precludes the building of prototypes, with all the effort, cost, and material use that this process would normally require. Control of distribution reduces fuel and therefore reduces carbon emissions. Finally, control of the factory environment reduces environmental impact, with enhanced factory efficiency and improved manufacturing efficiency. The outline of the smart factory is shown in Figure 3.6.

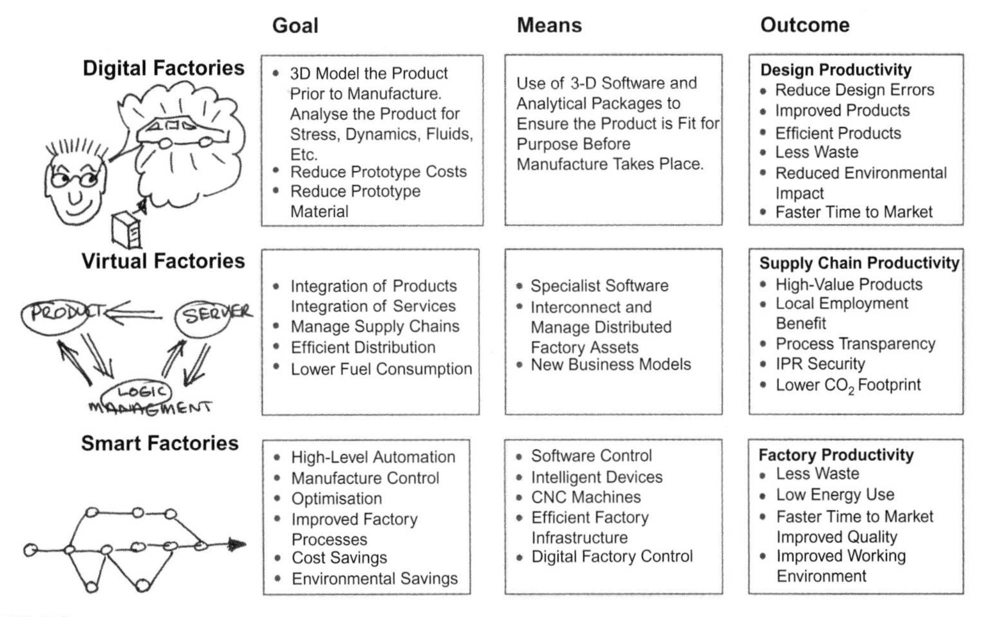

FIGURE 3.6

Smart Factory Based on Information Process Technology

3.8.1.2 Smart Factories

Figure 3.6 shows that smart factories of the future (and some of the present) aim for reduced waste, reduced energy, faster time to market, and better quality. These goals can be achieved through more automation as well as control and optimization of factory processes.

3.8.1.3 Virtual Factories

Factory efficiency can also be improved by good planning and good management. This is the *virtual factory*, the main essence of which is to manage supply chains and create value by integrating products and services. Should this be achieved, much of the trial and error of logistics and planning will be removed. Efficiencies will be gained by precise application of energy in terms of transport, combining products, and so on.

3.8.1.4 Digital Factories

The advent of 2-D CAD, 3-D CAD, and digital control for manufacturing has transformed manufacturing. It is now possible for a designer to create a 2-D image on specialist software, transform it into a 3-D model, perform stress and other analysis, and send it via wireless to a machine tool that will manufacture the component.

This is a far cry from the days before computers, when draftsman drew the image on paper, which was then sent through all the factory systems such as planning, purchasing, and production before it even reached the machine tool, where the operator then used his skills to manufacture the component.

In real terms the advent of the *digital factory* has improved quality, improved use of components, improved life of components, reduced cost, and made the time from conception to product very short.

Before computers, the concept-to-product time for a normal family car was perhaps five or six years. In today's digital factory, using these specialist engineering software tools, it is now possible for a car to be complete only 12 months after its conception. The reason is that digital information can be shared across a company or even between companies so that the design of, say, an engine can be instantaneously shared with the designers of the vehicle body, thus keeping everyone informed and design work on the complicated engineering projects can be completed in parallel.

We can see that energy is the key to the healthy state of the human race on planet Earth. In view of the burgeoning human population, we are walking a tightrope in terms of providing enough energy while demand for it keeps increasing.

Some resources are diminishing, but other resources could be better used. In view of the difficulty of creating energy, it also makes sense to conserve as much energy as possible. This brings us back to the sustainability model and the sustainable life value (SLV), which encompasses the total energy used to create and use a product from sourcing the material to its eventual disposal.

It is therefore necessary for engineers to focus not only on creating energy from renewable sources but also conserving and storing energy.

The use of natural light, intelligent building management systems, recycled waste disposal, LED lighting, rainwater harvesting, and biowaste for the generation of biogas are all ways in which manufacturing plants can better their *sustainable manufacturing value* (SMV), which is a rating of manufactured products that use processes that minimize negative environmental impacts, conserve energy and natural resources, are safe for employees, communities, and consumers, and are economically sound. These achievements were implemented by the Brandex Group in Sri Lanka [13], which recently converted its 30-year-old factory to meet "green" factory standards. With this conversion, Brandex achieved 77% carbon emissions reduction and 46% energy use reduction.

Packaging reduction and the use of recycled materials are also major methods of improving the SMV. Mattel, the toy company, has for some time been instrumental in reducing the source fiber for its packaging and has reduced the amount of product packaging, thus not only saving on cost but improving the SMV.

3.9 DESIGN FOR SUSTAINABLE USE (SUSTAINABLE USE VALUE, OR SUV)

For certain classes of machinery and equipment, life as a working unit is arguably the element in the product's life that has the most impact on sustainability. The energy used during this period therefore can be termed "Sustainable Use Value" or SUV.

In the field of construction equipment and road transport, the energy consumed by a machine in use during its lifetime far outweighs the energy consumed in its production. Large plant manufacturers, for example, have optimized the design of their machines over the years to use lean-burn diesels, minimizing engine size and emissions by using flywheels to reduce peak demand. This is the approach adopted by Caterpillar Industrial Power Systems under Chief Technology Officer and Vice President Gwenne Henricks. At the CEA Conference in 2011, Henricks clearly indicated that design for sustainable use and extended product life cycles is the key challenge facing the industry as it develops new products.

Similarly, emerging technologies have allowed radical improvements in electric car charging, reducing the use of relatively low-efficiency internal combustion engines, at least for short journeys. Future developments show great promise for electrically powered vehicles, with some analysts suggesting that this technology will replace the internal combustion engine in a few short years.

Designers have to take responsibility for the environmental impact of their equipment. The complete nonuse of fossil fuel power may not be practically achieved in the short term, but it may be significantly reduced. This can be done in several ways, as discussed in the following sections.

3.9.1 Design Optimization

One optimized aspect of vehicle design is that of reduced weight. Mechanical structures can be optimized for strength, buildings optimized to accommodate earthquake oscillations, ships optimized for speed through water, and so on.

Since the first fuel crisis in the 1970s, vehicle manufacturers have been optimizing the mass in their passenger vehicles. The reduction of mass in a passenger vehicle means that the engine can be small, the brakes can be smaller than usual, tire size can be reduced, and many other features required to move the mass can be diminished. The drive for this mass-losing optimization is to reduce fuel consumption, but the consequences mean that other elements are reduced, therefore reducing the environmental impact in manufacture and also during the vehicles useful life.

There are many modes of optimization. Again, taking a passenger vehicle as an example, town cars are optimized to carry passengers and have small-power engines but are maneuverable so that they can easily negotiate an urban environment. Because they generally travel at low speeds and carry multiple passengers, urban vehicles are optimized for short journeys and for carrying passengers. Conversely, a sports car is designed for speed, with a powerful engine, low-profile wheels, and a streamlined body shape so that it slips through the air. Both vehicles carry passengers but are optimized in different ways to achieve a different result in use.

Optimization can also be applied to achieve sustainable use in selecting appropriate power systems and methodology in use that improves the *sustainable use value* (SUV). In this case a large item of plant would perhaps use a lean-burn diesel engine that has been conditioned to reduce emissions. The engine may even burn biofuels, which generally tends to further reduce emissions.

3.9.2 Incorporate Equipment That Gives Back

Emerging and young technologies such as solar and wind power can easily be incorporated into many products. New-build houses, for instance, could incorporate solar panels (photovoltaic, or PV, panels) on their roofs. Vehicles could also be fitted with PV panels and make use of the air they disturb in traveling by incorporating micro wind generators.

3.9.3 Reduce Energy Use

There are many options for reducing energy use, some of which are discussed here. Design equipment that is lighter in weight allows the application of smaller power units. Specify lean-burn internal combustion power units. Wherever possible, use electric drives. As renewable power availability grows, electric drives will become the power source of the future. So far our examples have considered applying power plants, which use less energy, but we must not forget that conserving energy is just as important. Insulation against heat loss or even extraction of waste heat are excellent methods for improving efficiency. Excess heat can be converted to usable energy via a heat exchanger.

3.9.4 Use of Natural Energy

Energy is the driver for any usage process. It makes sense to use naturally generated power and low-energy solutions where possible. It is the designer's responsibility to select the lowest-energy option and to design that option into new products, thereby improving the SUV. One of the many ways this could be done is by applying electrical drive units such as those in electrically powered vehicles and other transport vessels. Hydrogen engines, still in their infancy, have a zero impact on the environment and could become an alternative power source of the future.

3.9.5 Energy Storage

No matter how clever the application, it is inevitable that there will always be "take" from the environment. Therefore, devices have to be built with the capacity to generate energy for those processes that demand it. Some of these devices that actually generate energy are dealt with elsewhere; however, energy storage must be considered. Large chemical batteries are useful and efficient in use, though their manufacture and eventual disposal can take a heavy toll on natural resources.

An alternative to chemical storage is the use of *kinetic energy storage* devices. These devices are essentially flywheels that rotate at high speed, storing kinetic energy that can then be converted back through generators into usable electricity. Flywheel batteries can be very high-tech systems requiring a significant amount of manufacturing resource. Other systems can be low tech, made to normal engineering principles, which demand fewer resources from the environment for manufacture.

Whichever method is adopted, the resources used in its manufacture are given back during the use of the device. These storage devices are able to be charged when there is low demand and therefore available energy at, say, 4:00 in the morning. When there is high demand in the national electricity grid, say, at 7:00 in the evening, the role of the flywheel can be reversed so that it becomes a generator and energy can be introduced back into the electricity grid. This strategy smoothes out the electricity demand peak and reduces the necessity to generate electricity on demand, thus reducing resources required for generation.

In a slightly different application, a large bank of flywheel batteries could store the output from several power stations when demand is low and return it to the grid during high-demand periods. This scheme would mean that a quantity of flywheel batteries could actually eliminate a power station.

Batteries will also come into their own when high energy demand is required by a single user. A typical example is that of an aluminum-smelting foundry that uses enormous quantities of energy during the smelting process. A bank of flywheel batteries could absorb energy at times of low demand and inject it into the smelting system when operationally required.

The emerging technology of electrically powered vehicles requires not only infrastructure but also a quick means of recharging. A bank of flywheels in strategic locations would provide that means, perhaps domestically or while a customer is parked at the supermarket. Energy storage devices possess a very low SUV simply because the resources needed to create them are greater than compensated during the life of the device.

3.10 DESIGN FOR SUSTAINABLE MAINTENANCE (SUSTAINABLE MAINTENANCE VALUE, OR SMaV)

3.10.1 The Need for Maintenance

The aim of any designer striving to build sustainability into his/her products must be to extend the life of the product. This idea is quite controversial, since our throw-away society that developed during the 1960s entreated designers to build a finite life (design life) into their products. For instance, a washing machine will perhaps have a finite life of, say, three years. Manufacturers argue that if products last for extended periods, eventually the market will be saturated and there will be no more sales. Though this may happen in part, it is inconceivable that humanity will no longer require new appliances. Instead the emphasis will move toward maintaining existing equipment.

It is normal practice for high-value products to be designed with maintenance in mind. It would be inconceivable that a passenger vehicle or an aircraft would be discarded simply because it had developed a fault. Anyone who has ever owned a passenger vehicle will understand that scheduled maintenance is necessary to keep the vehicle running efficiently.

Internal combustion engines are able to be maintained by changing oil filters, spark plugs, and so on. Brakes are designed so that brake pads and shoes can be replaced. Even the drums and the discs can be removed and replaced. It is quite common to see vehicles still running after 20 or even 30 years. Often these vehicles are maintained by enthusiasts who support a thriving industry that revolves around the maintenance of elderly vehicles.

As the designer creates new products, it is his responsibility to design sustainability into them so that not only just high-value products can be maintained, but also the smaller, low-value products such as toasters, food mixers, and the plethora of small items the consumer would normally discard after they become faulty.

Some excellent examples of small sustainable products are shown below. In Plate 3.9; the Aldis projector is shown and the Dualit toaster is shown in Plate 3.10. Both these examples were originally subject to "accidental" sustainable engineering design. However, they correctly embrace sustainable design principles. It is significant, however, that Dualit is a thriving company with an approach to design that allows easy maintenance, which is very sustainable.

The Aldis projector shown in Plate 3.9 was built during the 1950s, prior to the Plastics Revolution. The components are die-cast and fastened by screws. Though the projector may not have been designed and manufactured with sustainability in mind, it can nevertheless be taken apart completely so that components can be replaced, cleaned, and returned to service. This kind of construction can be extended to many other products so that maintenance is the first thing to consider when the device breaks down. The consumer's mindset also needs to be changed since the current approach is to throw the product away when a part breaks.

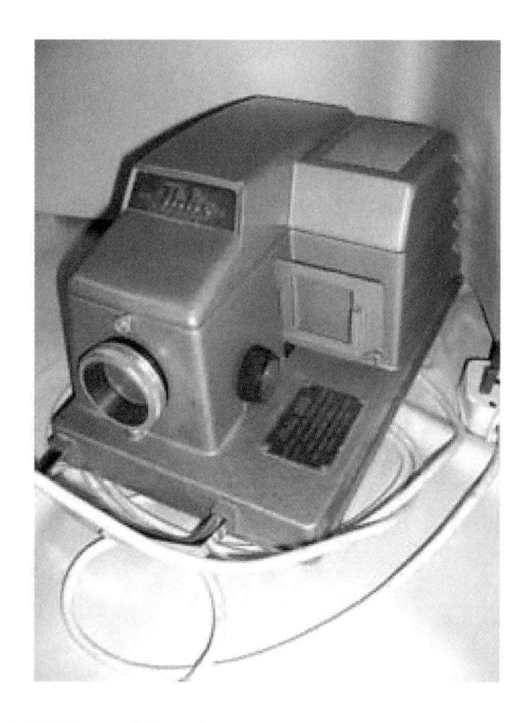

PLATE 3.9

Aldis Projector: Die-Cast Parts Allow Easy Maintenance [1]

PLATE 3.10

Dualit Toaster: Designed for Easy Maintenance [16]

The Dualit toaster shown in Plate 3.10 is part of a classic range of kitchen appliances that have been designed and built with ease of maintenance in mind. In 1946 Max Gort-Barton launched the first flip-sided toaster. From the beginning the original toaster and the range that followed were designed to be easily maintained and to impart a long life. Perhaps sustainability was not Max Gort-Barton's original aim, but the brand has come to represent a long-lived, maintainable household product.

It is certainly true that with the Dualit range, when components wear out they can easily be replaced by the customer. This is true sustainability through maintenance. The toaster and the other products in the range possess a very large *sustainable maintenance value* (SMaV). SMaV is the energy used in maintaining the product but should be offset against the energy saved by extending the products' life through maintenance.

Sustainable longevity derived from regular maintenance can be applied to many to engineered components. The water well rock-drilling machine shown in Figure 3.7 uses a rotation gearbox that slides up and down the mast as the drill string buries itself in the ground. The gearbox also pulls the drill string out of the ground after the hole has been drilled. At the top of the hole, very sharp abrasive debris is forced upward, hitting the underside of the rotation gearbox. The lower gearbox seal within the gearbox, as indicated in Figure 3.8, suffers accelerated wear due to the arduous conditions.

As the debris wears the seal, gearbox oil finds its way past the seal and, on mixing with the debris, forms a very abrasive paste that eats into the shaft, causing even more wear and further oil leakage. At this point the gearbox needs to be maintained by changing the seals and refurbishing the shaft. The damage caused to the shaft can be severe, and material normally needs to be deposited on the shaft at the contact point of the seal. The deposition, usually chrome, is then machined back to the correct size and the shaft replaced in the gearbox, along with new seals and fresh oil.

Often drilling operations occur in very remote areas, such as the African bush or perhaps high in the Himalayas. Maintenance is not easy in the field; therefore it has been known for drill crews to discard a leaky gearbox and replace it with a spare. Sometimes a gearbox is returned to a workshop, but this could be thousands of miles away. The solution would be to design a gearbox so that it could be maintained in the field. To this end the gearbox designers use a specialist cassette seal comprising a lip seal and a sacrificial sleeve arrangement, as shown in Figures 3.8 and 3.9.

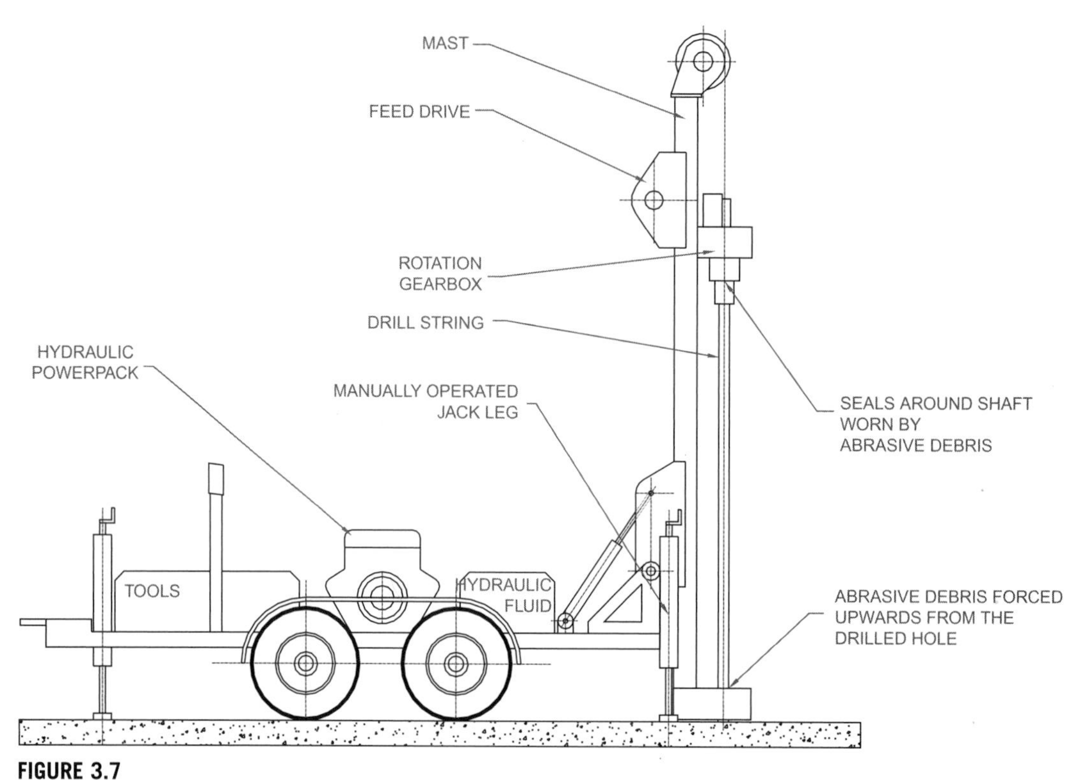

FIGURE 3.7

Trailer-Mounted Water Well Rock Drill Used in Remote Areas

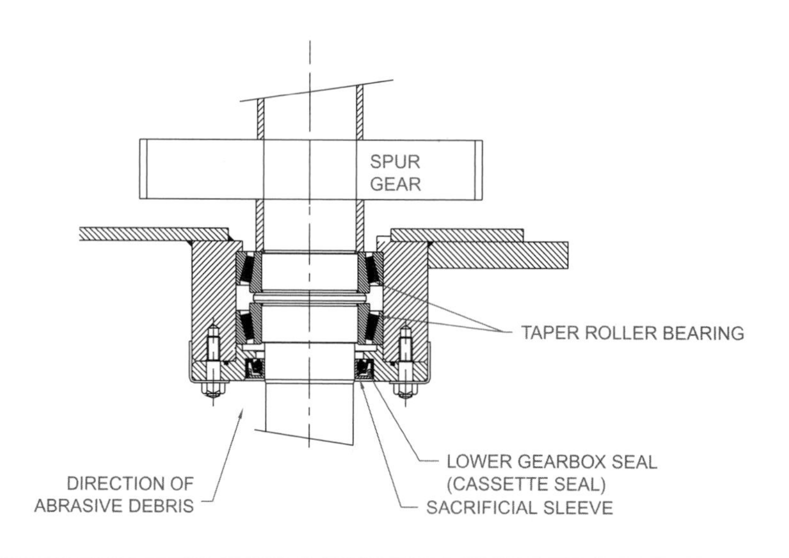

FIGURE 3.8

Portion of Gearbox Showing Lower Gearbox Seal [1]

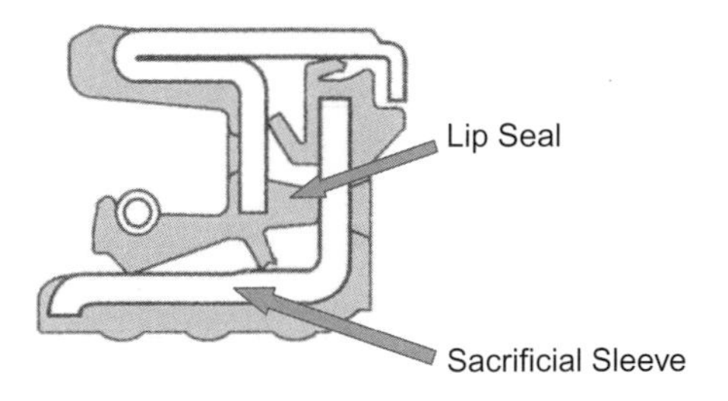

FIGURE 3.9

Example of Cassette Seal Showing Sacrificial Sleeve [17]

The application of this specialist seal ensures that the gearbox can be stripped to its basic parts in the field. The cassette seal is merely replaced, since there is no damage to the shaft. Carrying a small seal as a spare saves a great deal of time, effort, and money. The most important aspect of this is that the gearbox can be maintained, prolonging its life and reducing the energy required for its transport and later refurbishment.

In this case a little thought on the part of the designer reduced environmental impact dramatically; prolonging the life of the gearbox reduced downtime and improved the serviceability of the whole drilling rig.

Gaskets are flat seals that follow the contours of a gearbox, allowing the fastening studs to protrude. These are expensive to produce, since the required tooling needs to be manufactured. Gaskets also do not travel well because they become creased and damaged. Gaskets can be used to seal rocker box covers on engines and gearbox lids plus many other low-temperature sealing applications.

Should the gearbox in our example need to be serviced in the field, it will require a new gasket between the gearbox lid and the gearbox body. Rather than use a fiber gasket, it is possible to use Polytetrafluoroethylene (PTFE) chord, which winds around the fastening studs and along the seal profile of the gearbox. When the lid is fitted and tightened, the PTFE chord squashes and spreads, forming a perfect seal. Just as the cassette seal can be easily carried by a service engineer, so can a roll of PTFE chord. As before, the use of PTFE chord is a small measure with huge consequences and savings, ensuring easier maintenance, especially in remote areas.

3.10.2 Lubrication

During the design of mechanical parts, a good deal of emphasis is placed on tolerances, surface finish, and seal and bearing selection, but often lubrication of the moving parts is left until last, almost as an afterthought. Lubrication is really "liquid engineering," since without lubrication moving parts would quickly become hot and seize. The introduction of not only the correct lubricant but also the lubricant delivery system could prove to be one of the most important elements of a working machine.

Wind generators are often situated on hilltops or out at sea in very arduous conditions. Designers aim for a 20-year life span, but one of the acknowledged weak points in any wind generator is the bearing system of the turbine shaft.

A bearing failure may be catastrophic for the wind turbine, since the generated heat could create a fire, endangering the whole machine, sometimes with spectacular results, as shown in Plate 3.11.

PLATE 3.11

Wind Turbine Fire [19]

Failure within the turbine nacelle may be due to several reasons, from electrical failure to mechanical failure of moving parts.

Bearing companies have devoted a great deal of effort to producing long-life bearing systems, but equally important is the lubricant. Typically this lubricant is a form of grease application or a recirculating oil system. Specialist lubricants have been developed after much research so that they will perform no matter what the environmental or loading conditions. The consequences of a bearing or lubricant failure are that the whole nacelle has to be dismantled. This is no easy task and can become very costly, especially when the nacelle is 30 m off the ground and out at sea. The obvious answer is to provide longevity in terms of good design, correct selection of components, a easy maintenance and an efficient lubrication regime.

3.11 DESIGN FOR SUSTAINABLE DISPOSAL (SUSTAINABLE DISPOSAL VALUE, OR SDV)

The designer is the creator of the product and has the influence to create a sustainably friendly disposal technique. In the past the designer's mindset has always been to reduce cost, but he/she should now refocus slightly on *sustainable disposal*. There are several ways that a product at the end of its life may be used or disposed of in a sustainable way. As we mentioned before, sustainable end-of life disposal techniques are commonly termed the 4Rs:

- Recycle
- Repair
- Reuse
- Reduce

Many of the 4R components overlap. For instance, when a product is to be reused, it would normally have to be refurbished to bring it back up to near its original condition. Repair may also involve some refurbishment, as do some forms of recycling.

3.11.1 Recycling

Thus far the material sourcing that we have considered has been from an original source. However, this need not be the case, since materials can be garnered from several other sources, perhaps the most obvious being from recycled materials.

Much of the procedure for recycling involves taking in end-of-life products, extracting similar materials, and reforming them into raw materials that can be used in place of the original materials in a remanufacturing process. Vehicle tires are an excellent example in that the rubber is granulated and used in forming various rubber products, including soft flooring for playgrounds and road speed bumps.

As we have seen, some materials, such as building materials, have been a recycled source for thousands of years. In more recent years steel has been successfully recycled and is now the world's most recycled material. There has also been a surge in the variety and diversity of recycled materials. These include shoes and clothes, electrical appliances, glass, nonferrous metals, and vehicle tires. It is estimated that up to 90% of discarded items and products [18] can be recycled or reused.

Many processes in manufacturing would require fewer raw materials. Unfortunately, recycled materials are often a mixture of similar materials, which presents difficulties in remanufacturing since impurities or various material types may threaten product quality. The recycling of glass is quite common, and it is relatively easy to recycle; however, recycled glass arrives at the recycling plant as broken shards but in many colors such as brown, blue, green, and clear. This colored glass cannot be mixed when the end product is required to be a clear glass bottle.

This mixture of materials highlights the main problem with recycling in that materials are mixtures of similar products rather than pure. Separating the parts of the recycled mixture is often achieved by hand and is obviously labor-intensive. Some inventive companies have created products that can be produced from a mixture of similar materials. Excellent examples of such products are:

- Plastic roof tiles from a variety of recycled plastic types, e.g., polypropylene, polyethylene, acrylonitrile butadiene styrene (ABS), etc.
- Simulated wooden planks made from a variety of recycled plastics and used as seating planks in street furniture
- Paving stones manufactured from multiple colors of granulated glass

Materials gleaned from recycling processes are less costly and use less energy than the original source materials. The use of recycled materials also means there is reduced extraction of original materials, thereby cultivating sustainability. The practice of recycling also creates a local economy, since recycling of materials can take place and be processed locally.

3.11.2 Repair and Refurbish

Die-cast components and products were the norm in the 1950s. Items were held together with screws and could be dismantled and repaired. During the early 1960s, the use of plastics became popular for toys, kitchen implements, garden tools, household devices, and many other products. These were normally "snap-together" items and were almost impossible to dismantle without breaking the product and hence rendering them difficult to repair. This era was the beginning of the "throw-away society." The mindset of "throw away and buy another" has to some extent started to turn toward "refurbish and reuse," but a large shift in the mindsets of both designers and consumers is still required to make this attitude pervasive.

Refurbishment means that a product is not thrown away but restored so that the product's life can be extended. Economic recessions are great events for focusing both consumers' and manufacturers' minds on reducing costs. Rather than companies buying new equipment after discarding old products, economic recessions tend to focus companies on refurbishing components.

During a recent recession in the United Kingdom, the civil engineering industry suffered greatly. New housing construction and large civil engineering projects were cancelled. This had a rippling effect among equipment suppliers to the industry. A West Yorkshire (U.K.) manufacturer of brick and block crane attachments (HE & A Limited) found a lucrative market in refurbishing equipment and supplying spares as the new-equipment market evaporated. Plate 3.12 shows a typical brick/block crane attachment.

This is a product that has been manufactured for several years with no real consideration for sustainability or refurbishment. The company created a new design of brick/block clamp that was designed specifically with refurbishment in mind. The clamp may be welded if it breaks and components replaced when worn. It can be restored to a working product with much less energy input and with a much smaller impact on environmental resources than the manufacture of a new product. This is an excellent example of refurbishment, giving an extended life and providing a very low *sustainable disposal value* (SDV). Sustainable Disposal Value is the energy required to dispose of the product and should be offset against the energy saved by avoiding the procurement of new raw material.

An excellent example of repair is that of motorcycles used in India and Pakistan. In these countries the favored individual transport is the 70 cc or 100 cc motorcycle, shown in Plate 3.13.

Here the designers have taken the initiative and designed a vehicle with a low-resource-impact value. These motorcycles are designed with simple parts, low cost, ease of repair, and relatively low impact on resources when manufactured and also a low impact in use. They can be refurbished

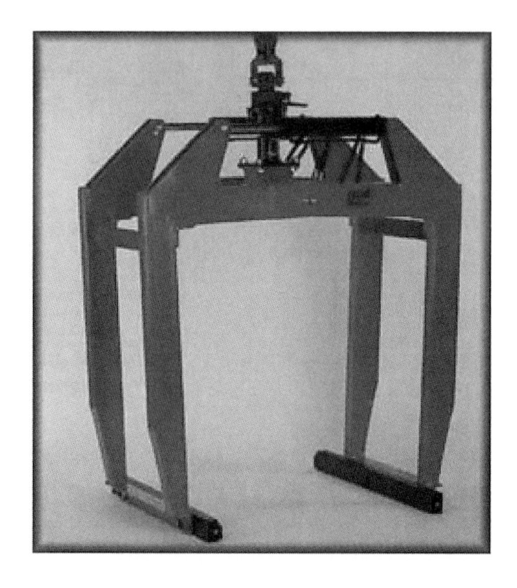

PLATE 3.12

Brick/Block Clamp [20]

PLATE 3.13

70 cc Motorcycle [1]

and repaired as long as parts are available. It is a simple task to remove the faulty parts since the product has been designed for easy maintenance. This is an excellent example of *cost-led sustainability*, or conversely, an example of sustainability driving costs down. Since these motorcycles can be repaired, refurbished, and maintained, the life of the vehicle is almost infinite and provides a very high SUV and a very high SDV.

3.11.3 Reuse and Refurbish

Reused and refurbished products and materials use less energy in restoration than if the same material had been extracted fresh from the earth. Designing machinery and equipment such that it can be repaired and refurbished gradually became less common as companies sought to save on costs by reducing the labor content and by moving production to cheaper labor sources. This mindset developed into the throw-away society. Indeed, much of design thinking is set to give the product a finite life so that when the user deems that the product is obsolete, he just throws it away. In this way manufacturers ensure that newer versions of their products are always in demand.

As Dr. Walter R. Stahel of the Product Life Institute in Geneva, Switzerland, has said, "It is our contention that legislation, lobby pressure and tax-based initiatives will drive a resurgence in equipment designed for ease of refurbishment and reuse, and that forward-thinking producers will use a positive marketing message similar to campaigns such as Fair Trade to begin to place a premium on sustainably designed products." Stahel explores how moving from disposable products to service, refurbishment, and repair delivery could lead to restructuring of a post-industrial economy.

Energy use would partly be substituted by mainly skilled labor as repaired products and recycled materials would create substitutes for primary material. Activities that are labor- rather than capital-intensive are less subject to the economies of scale that characterize the chemical and material industries. Thus Stahel's concept of the performance economy also embraces enhancement of a local economy and the maxim, "Do not repair what is not broken, do not remanufacture something that can be repaired, do not recycle a product that can be remanufactured."

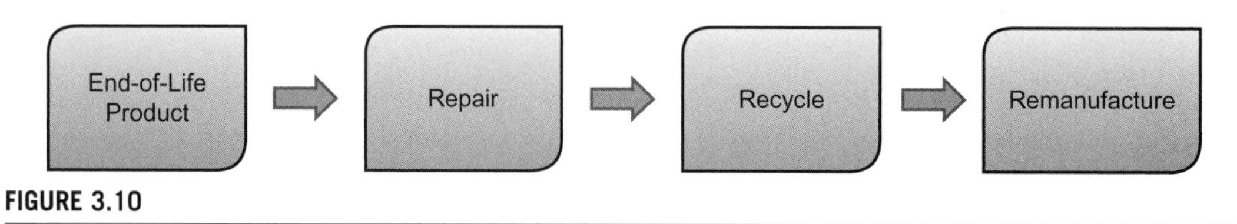

FIGURE 3.10

Hierarchy of End-of-Life Actions [1]

Stahel's concept introduces a hierarchy of end-of-life actions that designers should now consider when designing a new product. The flow diagram shown in Figure 3.10 indicates each level of the hierarchy.

Stahel's concept, though excellent, is not detailed enough to be practical. Consideration needs to be made as to the type of product, the material, and its value. Though sustainability has to be in the forefront of designers' minds, they cannot ignore cost, and to refurbish low-value items may be commercially unviable.

Some components such as bearings and seals are sacrificial. They are in place to protect the high-value components such as shafts, and upon maintenance or refurbishment these particular components should be removed and replaced with new versions. The fate of the old components is that they will be recycled. Furthermore, the material and indeed the component itself needs to be considered when making decisions as to disposal. And enhancement of the hierarchy of end-of-life actions presented in Figure 3.9 can now be revised as shown in Figure 3.11.

The responsible designer with sustainability in mind should therefore consider each level of the hierarchy as the design progresses and as each component is included within the design.

As each component is considered, it may seem to be against sustainability principles to sacrifice and discard certain components. Components such as seals and bearings are generally of low value, and it would be expensive to try to repair or refurbish these. The sacrificial components perform several tasks in their operation. They support major components such as shafts and prevent damage by providing a wear surface. The frugal refurbishment engineer may want to reuse and reinsert these used seals and bearings, but consequences of these actions should be carefully considered.

It is true that components such as seals and bearings are of very low cost relative to the value of the machine in which they are components. This machine may be a fundamental part of, say, an internal combustion engine production line, producing £1 million worth of engines each day. Should the reused bearing or seal fail, the production line will stop, losing a great deal of money. It is therefore false economy to reuse all components, since incorrectly appropriated reuse can make larger systems unreliable, incurring not only cost penalties but also sustainability penalties.

To plan the correct route to disposal, the designer has therefore to make some decisions relating to the components used in the design. This is complicated since each component is individual and may function perfectly well, even if worn. It is therefore necessary to consider the component, the material, the function of the component, and its value. The generic end-of-life decision chart shown in Appendix 3.1 assists in this complex process. A decision chart was devised for major components of the brick and block clamp shown in Figure 3.12; this chart shows a schematic of the brick and block clamp illustrating the location of the major components.

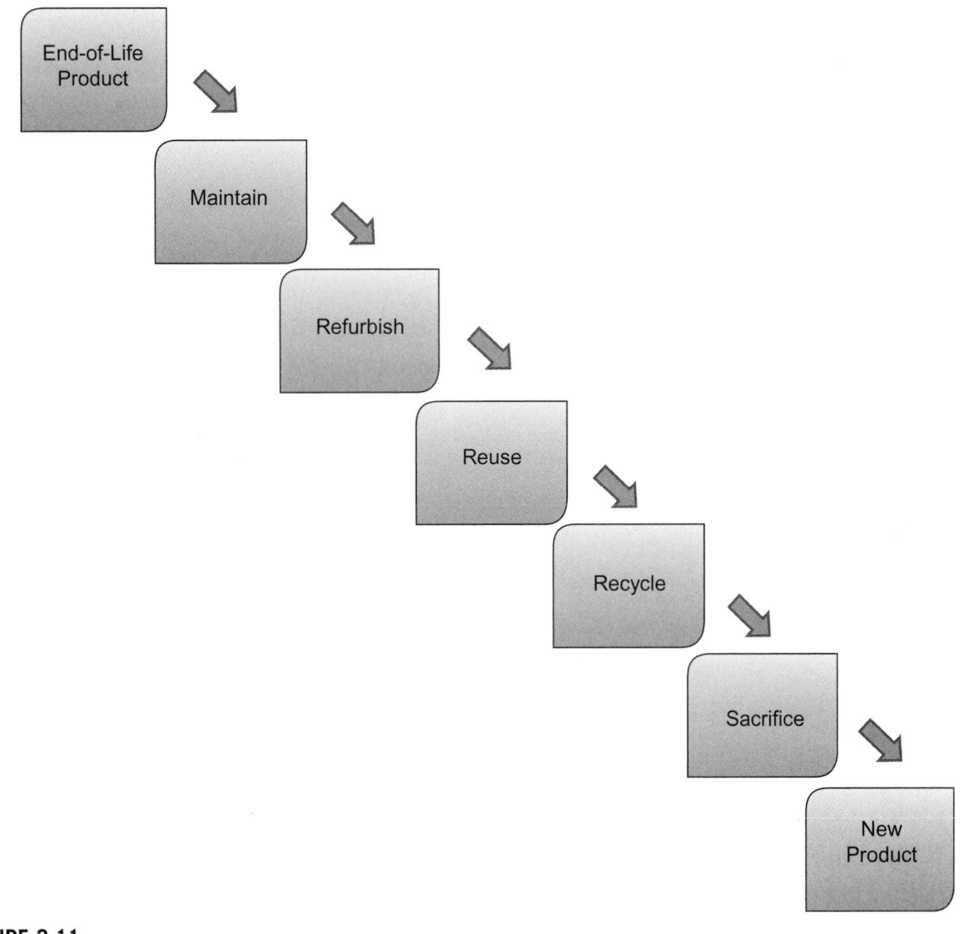

FIGURE 3.11

Model for End-of-Life Decision Actions [1]

The chart shown in Figure 3.13 has several functions. It shows the design decisions and indicates to the designer how various components may be disposed of at the end of their life. When complete, a design should have a technical specification file in which all relevant documentation, analysis, parts, and so on are kept. This is necessary throughout Europe in particular in order to verify a *Conformité Européenne* (CE) mark. The decision chart should also be kept in this file to provide information to the refurbishment team to help them proceed with each component.

3.11.3.1 Pivot Pin

Pivot pins provide hinge points, but they are also a high-load item. Upon refurbishment the designer has elected to sacrifice the pin and replace it with a new component. The pin should be recycled.

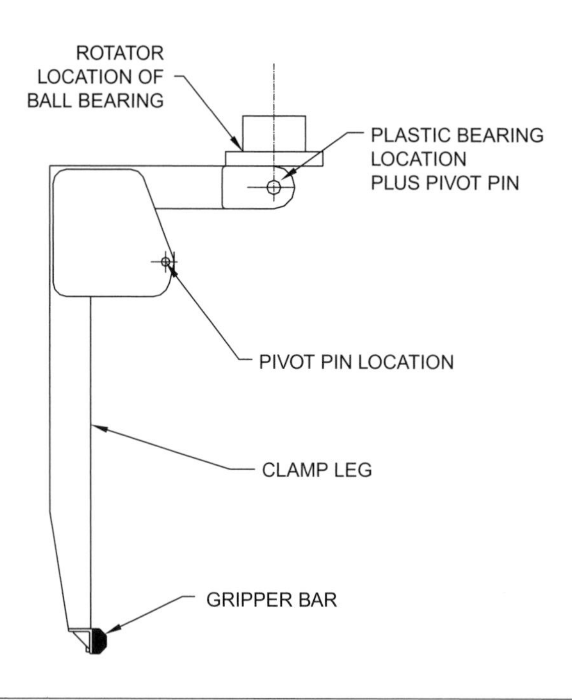

FIGURE 3.12

Schematic of the Brick-and-Block Clamp [1]

3.11.3.2 Self-Lube Plastic Bearing

Engineering plastics are a major sacrificial element in a great deal of machinery, especially machinery that works in the outside environment. These bearings are designed to be easily removed; therefore, upon refurbishment they are replaced with new bearings, whereas the old bearings are recycled.

3.11.3.3 Ball Bearing

This major component, more expensive than the plastic bearing, is housed in the rotator. The consequence of the ball bearing's failure is that the clamp cannot be operated. In this case the designer has decided that with 50% of its life still remaining, it could be maintained and reused. If it is deemed that the bearing has served out its life, designer suggests that it be removed and replaced and the original bearing recycled.

3.11.3.4 Clamp Leg

The clamp leg is a high-value item, so the designer has no intention of discarding it. At 50% of its design life, the designer suggests that it should be maintained; at 100% of its supposed life, he has allocated refurbishment and reuse. At some point the clamp may be returned, having been bent out of shape through misuse. In this case refurbishment may not be possible and new leg parts may need to be manufactured. The old component will then be recycled.

General Title:______	Brick and Block Clamp Refurbishment Profile							
End-of-Life Decision Chart				Date	xx/xx/xxxx			
Component Name	Material	Relative Value Low = 1 High = 10	State of Wear, Life Usage					
			25%				Maintain	
Pivot Pin	Steel	1	50%				Refurbish	
	EN8		75%				Reuse	
	(080M46)		100%				Recycle	
			Sacrifice				New Component	
			25%				Maintain	
Self-lube	Plastic	1	50%				Refurbish	
Bearing	Nylon		75%				Reuse	
			100%				Recycle	
			Sacrifice				New Component	
			25%				Maintain	
Ball	Steel	2	50%				Refurbish	
Bearing			75%				Reuse	
			100%				Recycle	
			Sacrifice				New Component	
	Steel		25%				Maintain	
Clamp	Hollow Section	9	50%				Refurbish	
Leg	Black Plate		75%				Reuse	
	Fabrication		100%				Recycle	
			Sacrifice				New Component	
			25%				Maintain	
Gripper	Nitrile	3	50%				Refurbish	
Bar	Rubber		75%				Reuse	
			100%				Recycle	
			Sacrifice				New Component	

FIGURE 3.13

End-of-Life Decision Chart

3.11.3.5 Gripper Bar

The gripper bar is made from rubber and is the contact between the clamp and the brick product. It is really intended as a sacrificial part of the clamp, but on its refurbishment the designer has suggested that if it is 50% worn, it may be reused. At 100% worn the sacrificial rubber gripper is destined to be removed and replaced with a new component. The old component will then be recycled.

3.11.4 Reduce

This is perhaps the oldest of the sustainable design techniques. By optimizing the design, the reduced use or perhaps the most efficient use of materials can be applied to a new product. This approach will reduce the amount of energy associated with manipulation, manufacture, and transportation. This idea has been seen particularly in packaging improvements, where a revised approach can improve the material usage efficiency without compromising the product life, safety, or security.

The main thrust here is to use only materials and services that are necessary to complete the task. In a recent business trip from the United Kingdom to Orlando, Florida, the author's attention was drawn to the difference between the size of engines in the airport shuttle bus. Both in the United Kingdom and in the United States, the shuttle bus held 12 passengers. The engine capacity of the U.K. shuttle bus was 2 L, whereas the engine capacity of the Florida shuttle bus was 5.8 L. The question has to be asked: If the 2 L capacity of the U.K. shuttle bus is adequate, why cannot this engine size be applied to the Florida shuttle bus?

Sometimes this "overdesign" can be attributed to mindset, where the designer does not even consider that he is overdesigning the product. To him this design work is normal practice. This overdesign could also be attributed to the design approach of "if it looks right, it is right." This is a very subjective approach under which a design may be created and is inevitable that overdesign will be the result, since humans tend to make things bigger and stronger than needed.

Careful optimization is therefore required, using means such as analytical tools to refine the materials applied to a product. The great benefit of this approach is that it will also reduce the environmental impact through reduced material usage and reduced energy needed for manufacture.

A reduction in engine size leads to a lighter vehicle with smaller brakes, reduced body structure, and reduced fuel consumption and emissions. A great example of optimized vehicle design is that of racing cars of all formulas. Here the emphasis is on acceleration and speed, but to achieve those goals a reduction in weight is required. Some of the best engineers strive to shave grams from every component in their racing vehicles.

3.12 GIVEBACK

No matter how products are reused, refurbished, or recycled or how cleverly the environmental usage impact is reduced, the plain fact is that resource use is merely being slowed. There will always have to be some form of "take" from the Earth's resources.

Giveback is a technique whereby designers actually build devices that "give back" to the environment or they design giveback components into current products. Solar power panels on car roofs or micro wind generators built into vehicles are just two ideas that could be explored. Most vehicles are left outside for much of their lives; PV panels set into the roofs of those vehicles and built-in micro wind generators could produce energy that could then be stored. Trading this stored energy into a central repository could give discounts for the liquid fuel or the recharging of an electrically powered vehicle.

In another application, solar panels could be incorporated on the roofs of new buildings. Imagine the energy-generation possibilities that solar panels would offer if introduced on the millions of homes in the United Kingdom. The power generation would be enormous and would reduce our reliance on power stations.

Consider a hypothetical application whereby all the 26 million houses and all the 31 million cars in the United Kingdom were fitted with modest solar panels. The energy generated would amount to 0.0028% of the total U.K. annual power consumption. This seems at first to be an insignificant figure but it has a value of £8.8 Billion. Unfortunately, humans are currently conditioned to react to cost rather than energy usage. See Appendix 3.2 for more precise analysis.

Energy can now be generated from multiple renewable resources, from wave energy to wind energy, but the great challenge is integrating this energy into the electricity grid. Renewable energy is basically intermittent, since the wind only blows on windy days and the sun only shines during the day (although in the United Kingdom even this is sometimes debatable).

Once generated, the energy needs to be stored. As part of collaboration between the University of Huddersfield, U.K., and ESP Ltd., a battery flywheel system has been developed (see Figure 3.14). These kinetic energy storage devices have the capacity to store large amounts of energy in a spinning flywheel. At peak times these devices can be tapped to provide the electricity grid with power. Several thousand of these in a single facility could provide enough storage capacity to eliminate a power station. This is an excellent example of low *sustainable giveback value* (SGBV) (See page 99) since a flywheel does not take from the environment during its use and its presence creates efficiencies in the electricity supply system. It presence also avoids other large systems, such as power stations, from being built, with their inevitable enormous drain on the environment.

An excellent example of SGBV is the World Trade Center Building in Bahrain (see Plate 3.14). This building incorporates wind generators. The building rises 240 m. The shape of the towers is designed to funnel the wind onto the wind turbines, generating 675 KW in total, which is up to 15% of the total power consumption of the building.

FIGURE 3.14

Idealized Flywheel Battery 20KWh Storage [20]

PLATE 3.14

Bahrain World Trade Center Showing Wind Turbines [21]

3.13 THE MEASUREMENT OF SUSTAINABILITY

3.13.1 Sustainable Measurement Using Carbon Dioxide

Carbon dioxide is always painted as the ogre of greenhouse gases, but it is not the largest constituent of these gases. The proportions of the main greenhouse gases are as follows [22]:

Water vapor: 36–70%
Carbon dioxide: 9–26%
Methane: 4–9%
Ozone: 3–7%

We can see that moisture is the biggest greenhouse gas, yet the common assumption is that it is harmless. As a liquid, water is life giving, but as moisture in the atmosphere it prevents sunlight from reaching the Earth and also acts as a blanket, keeping heat within the envelope of clouds.

Though carbon is certainly a large proportion of the whole of greenhouse gases, perhaps the question should be asked, "Why is carbon dioxide used as an environmental yardstick?" The focus on carbon dioxide cannot be considered wrong, since it is a useful tool, but it does not really cover the whole picture. The use of carbon dioxide as a means of measurement assumes that all the expended energy is derived from fossil fuels. In reality, renewable energy sources are incrementally replacing the use of fossil fuels. This change is further enhanced by the introduction of mobile power plants such as those used in electric or hydrogen-powered vehicles. It is envisaged that the use of fossil fuels will diminish so that eventually the favored energy source will be electric, whereby the power is derived from natural sources (see Figure 7.2).

3.13.2 Energy as a Measurement Parameter

In reviewing the sustainable design options, it is clear that the conversion of materials into products uses energy. One of the byproducts of energy use is carbon dioxide; however, energy use itself is a much clearer indicator of resource use and is a much more accurate indicator of environmental impact.

Energy is used in one form or another to extract raw materials, convert them into products, drive them during their useful lives, and dispose of them at the end of their lives.

It is therefore proposed that each element of the sourcing-conversion-use-disposal process is given a sustainable value. We have mentioned these elements and can now assemble them into a tool with which designers and environmentalists can judge the sustainable impact of a product. The definitions are listed here:

Sustainable life value (SLV)
Sustainable source value (SSV)
Sustainable design value (SDeV)
Sustainable manufacturing value (SMV)
Sustainable use value (SUV)
Sustainable disposal value (SDV)
Sustainable giveback value (SGBV)
Sustainable maintenance value (SMaV)

These indicator values should be kept as low as possible, since the lower the value, the lower the impact on Earth's resources. SGBV is a new introduction and should be as high as possible, since this is a give-back value and will help to negate the "takeout" values of the rest of the model. See also Figure 7.3.

These values combined give the overall sustainable impact tool, or *sustainable life value* (SLV). The SLV model is shown in Figure 3.15.

SLV is derived from the addition of SSV, SMV, SMaV, SUV, and SDV and is a measure of the environmental resource impact during the life of a product.

- SSV represents the embodied energy required in creating the raw material to manufacture.
- SDeV represents the energy used during the design process.
- SMV represents the energy required to manufacture the product.
- SMaV represents the energy required to maintain the product. This value is likely to be very low relative to other values such as the SUV.
- SUV represents the energy required during the useful life of the product. In the case of a vehicle, this could be the largest energy use during the vehicle's life.
- SDV represents the energy used in the eventual disposal of the product.
- SGBV is a measure of how much resource is returned and can therefore be deducted from the resource impact (SLV), as the model in Figure 3.15 shows.

3.14 ACTUAL MEASUREMENT OF SUSTAINABILITY

The model in Figure 3.15 breaks down the life of a product into manageable elements but does not allow quantifiable energy measurement in joules. Measuring embodied energy within a product is an enormous task, since it involves the energy required to extract and manipulate the raw material plus

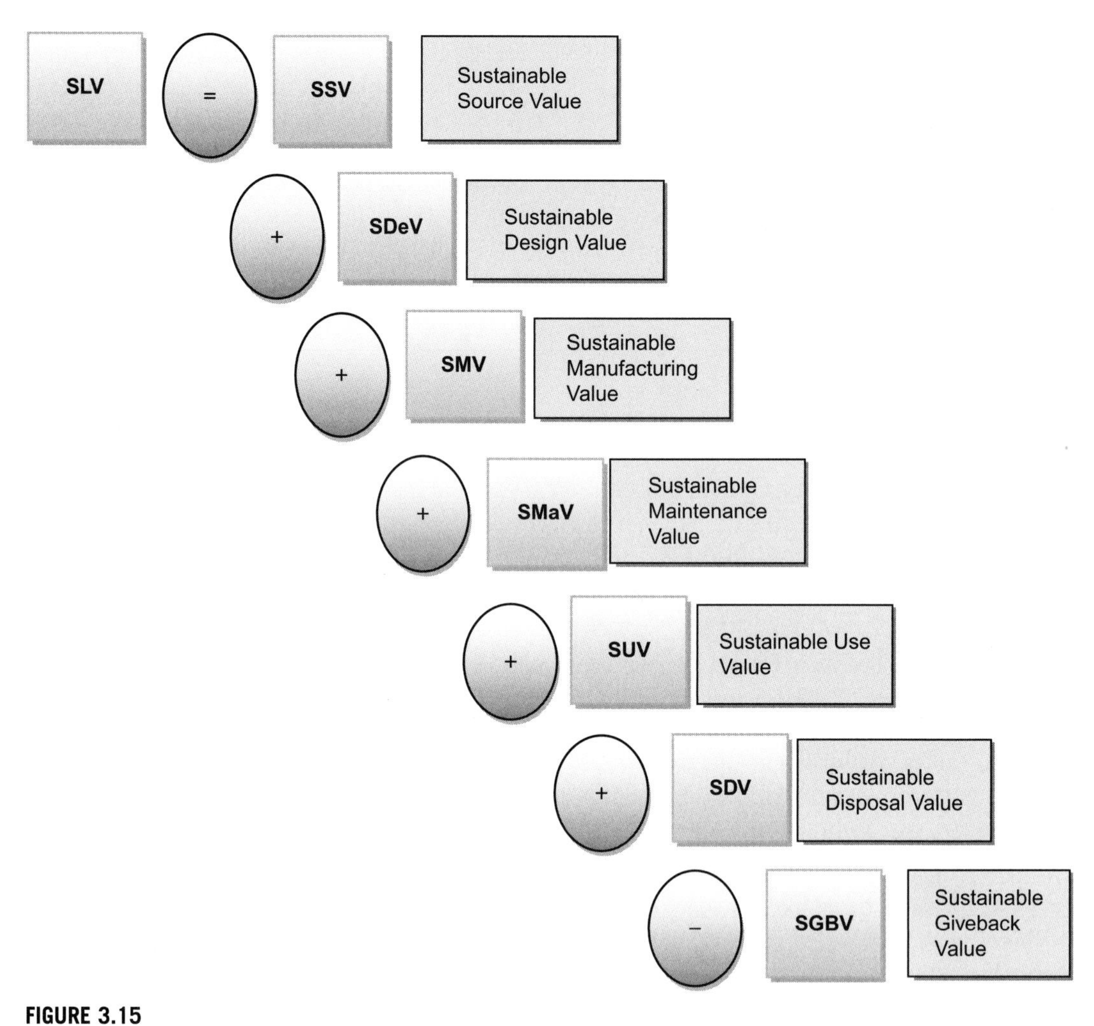

FIGURE 3.15

Sustainable Life Value Model

the energy required to manufacture, transport, use, and finally dispose of the product. Within this matrix there are no answers, such as manufacturing process and manipulation difficulty of the material.

There are several tools available that attempt to achieve this difficult task—Eco-rucksack and Envirowise, to name two of the more popular versions. Each uses a slightly different approach, and not all give a complete picture since they tend to specialize in a particular environmental problem. Perhaps the most useful sustainability tool for mechanical engineers is that offered by Granta Design Ltd. of Cambridge, U.K. [23].

Granta Design offers the *CES EduPack*—basically an exhaustive database of materials and manufacturing processes is aimed at the educational as well as the commercial market. Its main

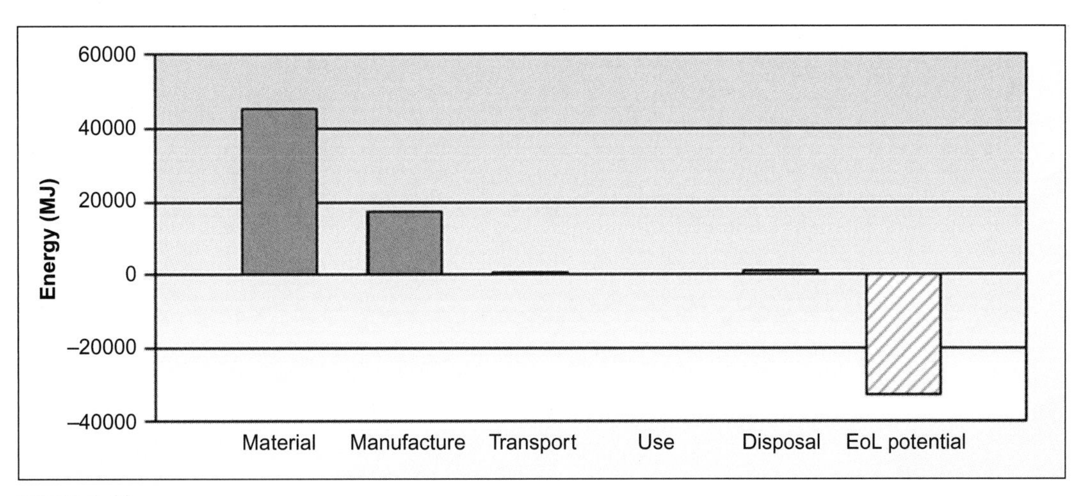

FIGURE 3.16

Embodied Energy for the Flywheel Rotor [23]

advantage is in its interrogation technique, whereby almost any parameter—strength, cost, mass—can be compared against any other parameter.

In terms of sustainability, the software offers an "eco-audit" section, which, for a given component, will measure the embodied energy. This covers the whole life cycle of the product from extraction through disposal, including transport. Furthermore, this software will instantly convert the embodied energy in joules into carbon footprint in tonnes, assuming, of course, that all the energy is fossil fuel derived.

Figure 3.16 shows a typical CES EduPack appraisal of the flywheel system mentioned earlier, where a sustainable analysis has been conducted on the rotor. This shows a block graph where the embodied energy in MJ is displayed against the elements of the flywheel life. Essentially this tool gives the SLV for the component. CES EduPack gives a report that's complete with all the parameters input for a component. A typical report is included in Appendix 3.3.

The rotor has a mass of 1,600 kg and, as expected, will have a large environmental impact in extracting and preparing the material before manufacture. The manufacture of the component being cast and then machined requires more embodied energy compared to the transport phase. Here a distance of 500 km was assumed. Since the flywheel is a store of energy and does not require energy to be directly generated, it is assumed that energy in use will be zero.

At the end of life, which may be 20 years, energy will also be required to dispose of the rotor, but this can be completely recycled, and therefore disposal energy is minimal. The end-of-life (EOL) potential energy is assumed, since the rotor will be recycled, and therefore it will not take such a great deal of energy to reconvert the material into a usable commodity. The EOL potential is therefore an estimate of energy that is now not used in extracting raw material from the Earth.

Assuming that the entire energy requirement is derived from fossil fuels, the same report converts the embodied energy in MJ into a carbon footprint as a single low in Figure 3.17.

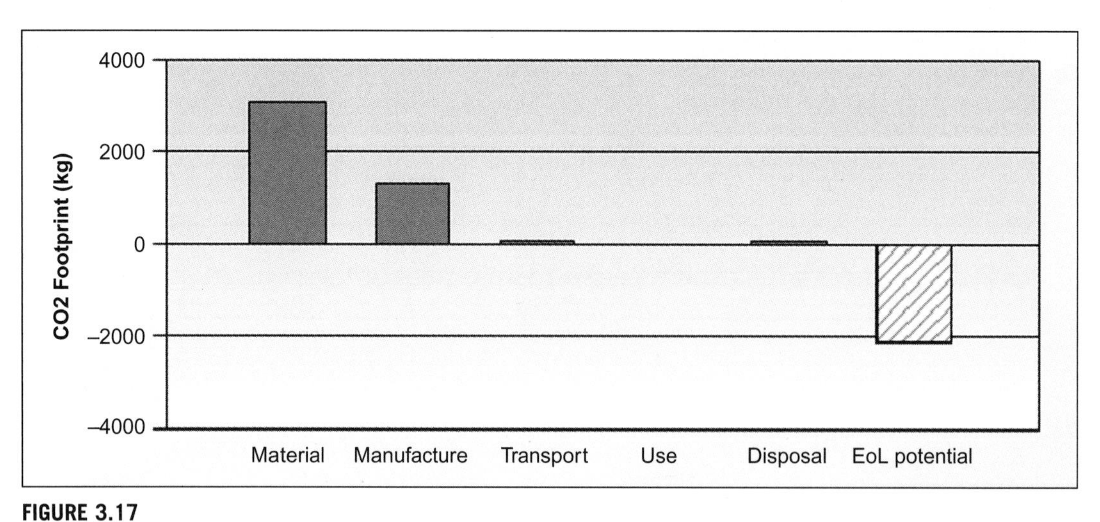

FIGURE 3.17

Carbon Footprint for the Flywheel Rotor [23]

Phase	Energy (MJ)	Energy (%)	CO2 (kg)	CO2 (%)
Material	4.52e+04	70.5	3.04e+03	68.3
Manufacture	1.72e+04	26.8	1.29e+03	29.0
Transport	690	1.1	49	1.1
Use	0	0.0	0	0.0
Disposal	1.05e+03	1.6	73.5	1.6
Total (for first life)	**6.42e+04**	**100**	**4.46e+03**	**100**
End of life potential	–3.31e+04		–2.09e+03	

FIGURE 3.18

Numerical Summary of the Eco-Audit of the Flywheel Rotor [23]

As expected, the carbon footprint values closely follow the embodied energy values, but as more energy is used from renewable resources, this carbon footprint value will fall, since the reliance on diesel, petrol, gas, and coal energy generation will eventually diminish.

Figure 3.18 presents a summary of the values applied to the flywheel rotor. This is also provided in the eco-audit report shown in Appendix 3.3.

3.15 SUSTAINABILITY COMPROMISE

The process of design is always a compromise and is highlighted by the conflict between lightweight but strong materials required for aircraft structures. Here the designer needs to make a compromise between strength and weight. This typifies the dilemma for many designers, when they need to make decisions to obtain the best outcome from their particular design.

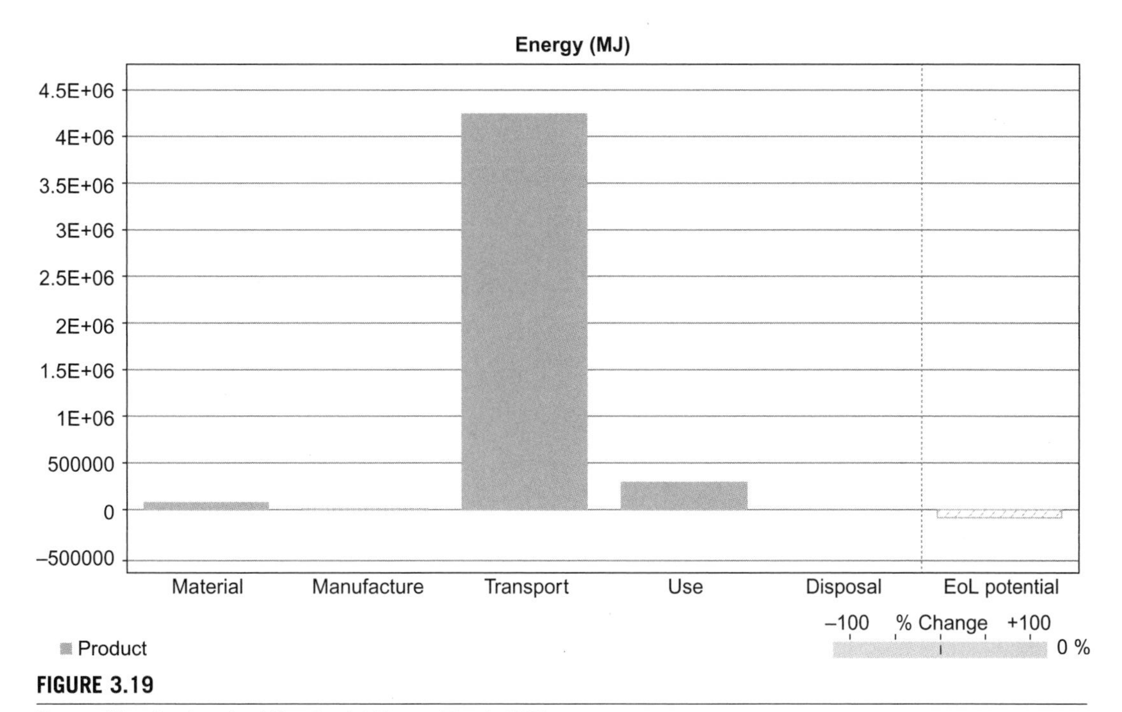

FIGURE 3.19

Energy Use During the Life of a 14-Tonne Truck [22]

Sustainability presents similar dilemmas in that designers cannot design a product that is completely sustainable. The designer needs to make judgments as to where he can reduce environmental impact as much as possible. In Figure 3.16 the greatest energy use and therefore the greatest impact occurs during the material procurement phase. To reduce this impact, further versions may use recycled materials thus reducing the environmental impact.

Figure 3.19 shows the general life cycle for a 14-tonne truck. It has been assumed that the truck is diesel powered and will cover 1,000,000 km during its lifetime.

The energy used during the life of the truck, shown in Figure 3.18, is a very different profile than that of the flywheel shown in Figure 3.15. Since the truck is built for transporting goods, it can be expected that the transport phase is large compared to the material procurement or manufacture phases.

This particular truck is powered by a diesel engine and will create a large carbon footprint. Figure 3.20 shows the carbon footprint created during the same life cycle for the same truck.

As expected, the profile for the carbon footprint follows that of the embedded energy. The designer can glean a great deal from this profile, since it is quite clear that efforts should be focused on the transport phase in order to improve the SLV. Effort could also be focused on the *material procurement phase* or the *use phase*, but by far the greatest impact could be made by focusing on the *transport phase*. At first it could seem that making inroads into the transport phase might be quite daunting, but the designer can concentrate on the following possible actions:

- Reduce the mass transported. Material Optimized design would be useful.
- Rethink the transport mode: rail, barge, aircraft.

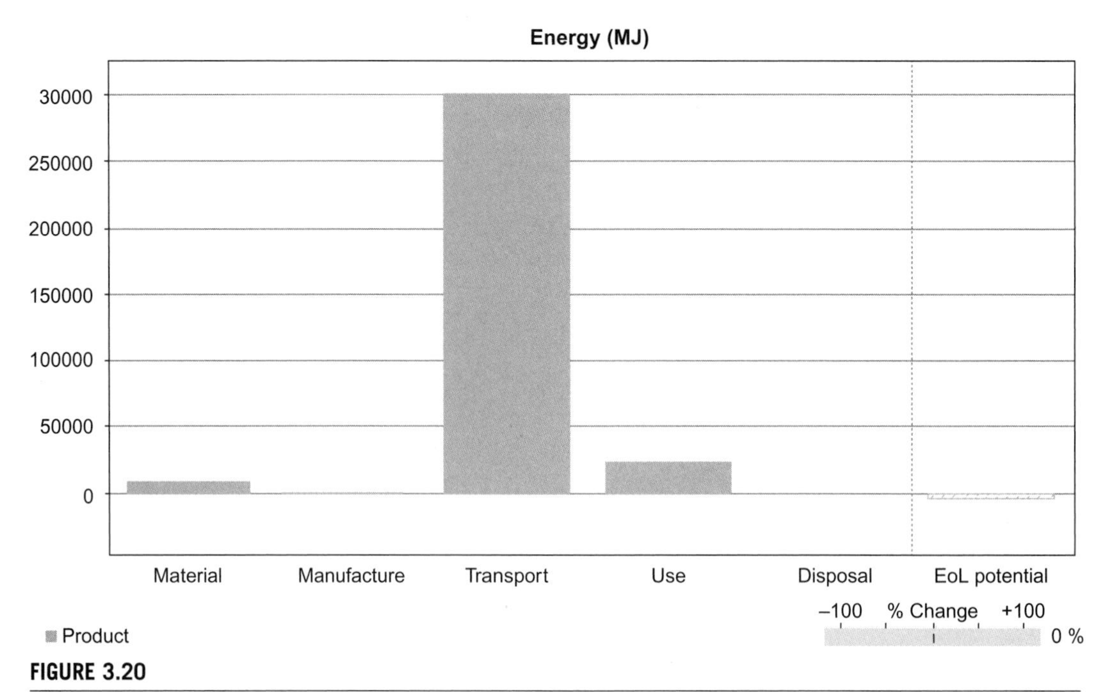

FIGURE 3.20

Carbon Footprint of the Life of a 14-Tonne Truck [23]

- Reduce the distance of transport. Employ trucks in a localized economy.
- Reduce the speed.
- Improve the CD factor, thus lowering drag and improving fuel efficiency.
- Change the power source to a more environmentally friendly type, e.g., hydrogen, electric, etc.

These actions are all viable options, but viewing the carbon footprint diagram in Figure 3.20, it seems reasonable that diesel engines should be discarded in favor of a power plant with a smaller carbon footprint. In the short term, a lean-burn diesel engine could be incorporated, but longer term an electric drive may be the answer. The electric drive could take advantage of renewable energy fed directly into the grid and used to recharge the vehicles batteries. This would massively reduce the carbon footprint, even though energy usage levels would remain similar.

Sustainable compromise, therefore, runs alongside design compromise. The designer cannot create a perfect product; there will always be several areas that designers must trade off against each other to achieve the optimum product efficiency.

3.16 CONCLUSIONS

In this worldwide consumer society, it is inevitable that products will be produced, used, and disposed of, but every product reduces Earth's resources. Perhaps this process cannot be prevented, but it can be slowed substantially.

The task falls to our engineering designers and engineering design teams to instigate a change in attitude and approach toward sustainable engineering design. As in the Taguchi model for quality engineering, it can only be the designer who can instigate the shift in attitude and the change in design practice. Design is the key to sustainable engineering.

The model outlined here reduces transport distances, reduces manufacturing resources, and reuses, recycles, and repairs goods. Adopting these elements will lead to a localized economy model that reduces financial costs and provides local, specialist employment.

We can see that an element of giveback will help reduce the impact of a product by reducing the SLV and that the implementation of giveback technologies such as flywheel batteries can greatly offset products' impact on Earth's resources.

The use of carbon footprinting is useful but is viewed more as a commercial tool and does not embrace the whole life cycle of the product, from sourcing to disposal. It is proposed that a much more useful and accurate measurement is energy use, since this is involved at every stage of the product's life.

The introduction of a measurement tool, SLV, is a great step forward in the measurement of resource impact. The introduction of CES EduPack gives a quantitative approach to sustainability, allowing the designer and other engineers to make specific judgments and become empowered to select sustainable avenues based on quantifiable assumptions.

Summarizing:

- Every manufactured product reduces the Earth's resources.
- This process cannot be prevented, but it can be slowed.
- *Design* is the key to sustainable engineering.
- It is the *designer* who can shift attitudes and change design practice.
- The SLV model outlined in this chapter:
  - Reduces transport distances
  - Reduces manufacturing resources
  - Reuses materials
  - Recycles materials
  - Repairs goods
- Adopting the SLV model will lead to:
  - A localized economy model
  - Reduced financial costs
  - Local employment

3.17 SUSTAINABLE ENGINEERING DESIGN: NECESSITY OR LUXURY?

Earth's resources are dwindling at a heavy rate; over the last 30 years in particular there have been much debate and suggestion as to how the use of resources could be reduced. Years ago a few lone practitioners began to act, but still the extravagance continued. Recent years have seen increased climate change and a greater push toward sustainability.

The Institution of Mechanical Engineers (IMechE) continues to urge its members and partner institutions to adopt a sustainable approach to design. The initiative has focused the minds of mechanical engineers and initiated action. Many global companies are developing their own strategies, but there is a great need for a cohesive and coherent approach so that all designers can work toward similar goals and have a significant effect. This text proposes a complete model, from sustainable sourcing to sustainable disposal, and further suggests a much needed measurement method, that of sustainable life value (SLV) coupled with an advance measurement tool, CES EduPack.

As global resources dwindle, it is no longer possible to sit on the sidelines and watch them disappear. Earth's resources are no longer a luxury. It has become necessary to protect those resources while still attempting to provide products and services for the planet's inhabitants.

References

[1] A.D. Johnson, Huddersfield University UK, 2013.
[2] www.Liv.ac.uk, Liverpool University, UK, 2013.
[3] American Iron and Steel Association, www.steel.org, 2012.
[4] Steel Recycling Institute, www.recycle-steel.org, 2012.
[5] Our Common Future: Report of the World Commission on Environment and Development, World Commission on Environment and Development, 1987. Published as Annex to U.N. General Assembly document A/42/427, Development and International Cooperation: Environment, August 2, 1987.
[6] Sustainable Engineering Design Necessity or Luxury A.D. Johnson, et al., 2011.
[7] Skysails GmbH, www.skysails.info, 2012.
[8] German Aerospace Research Council (DLR), www.dlr.de, 2012.
[9] Eco Marine Power, www.ecomarinepower.com/en/solar-ferry-medaka, 2012.
[10] Ellen MacArthur Foundation, www.ellenmacarthurfoundation.org, 2012.
[11] Clift, Allwood, Rethinking the Economy, Chem. Eng. (2012).
[12] Road Safety GB, www.roadsafetygb.org.uk/news/892.html, 2012.
[13] Towards a Sustainable Industrial System, University of Cambridge and Cranfield University, U.K, 2010, ISBN: 978-1-902546-80-3.
[14] www.triplepundit.com/2008/08/eco-friendly-sri-lankan-factories (accessed 2012).
[15] Factories of the Future and Next ICT Calls, Dr Erastos Filos, FoF ICT Coordinator: NCP Info Day, 13 May 2011, Brussels, Belgium.
[16] Dualit Limited, County Oak Way, Crawley, West Sussex, RH11 7ST, telephone: +44 (0) 1293 652 500; fax: +44 (0) 1293 652 555; e-mail: info@dualit.com, 2012.
[17] Freudenberg Simrit GmbH & Co., KG, D-69465 Weinheim, Germany, 2012.
[18] City of London MBC, www.cityoflondon.gov.uk, 2012.
[19] DailyMail.co.uk, Article 2071633-0F1B4D700000057, 2012.
[20] Unver, Ertu, University of Huddersfield, U.K., 2012.
[21] Sami T's Flickr photostream: 2012 2139274930_5c7be4866d_z.
[22] U.S. Environmental Protection Agency, Greenhouse gas emissions, www.epa.gov/climatechange/emissions/index.html, 2012.
[23] Granta Design Ltd of Cambridge, Rustat House, 62 Clifton Road, Cambridge, CB1 7EG, U.K., e-mail: info@grantadesign.com,:2012.

APPENDIX 3.1

General
Title:____

End of Life Decision Chart **Date**

Component Name	Material	Relative Value Low = 1 High = 10	State of Wear, Life Usage	Start	Maintain	Refurbish	Re-Use	Recycle	New Component
			25%				Maintain		
			50%				Refurbish		
			75%				Re-Use		
			100%				Recycle		
			Sacrifice				New Component		
			25%				Maintain		
			50%				Refurbish		
			75%				Re-Use		
			100%				Recycle		
			Sacrifice				New Component		
			25%				Maintain		
			50%				Refurbish		
			75%				Re-Use		
			100%				Recycle		
			Sacrifice				New Component		
			25%				Maintain		
			50%				Refurbish		
			75%				Re-Use		
			100%				Recycle		
			Sacrifice				New Component		
			25%				Maintain		
			50%				Refurbish		
			75%				Re-Use		
			100%				Recycle		
			Sacrifice				New Component		

APPENDIX 3.2

- Design into a product some element that gives energy back to the "system"
- Every car in the UK was fitted with a modest solar panel = 1.0×1014 J/year
- Every house in the UK was fitted with a modest solar panel = 1.5×1014 J/year
- Power consumption of UK per year : 8.374×1018 J/year

- Proportion of Collected Energy $= \dfrac{(1.0 + 1.5)1014 \times 100}{8.374 \times 1018} = 0.0028\%$

That is 0.0028% of total UK Power Use
That is £8.8 Billion worth of Energy

CES EduPack Eco Audit Flywheel Rotor

APPENDIX 3.3

Appendix 3.3:
Eco Audit Report

Product Name	Product
Product Life (years)	1

Energy and CO2 Footprint Summary:

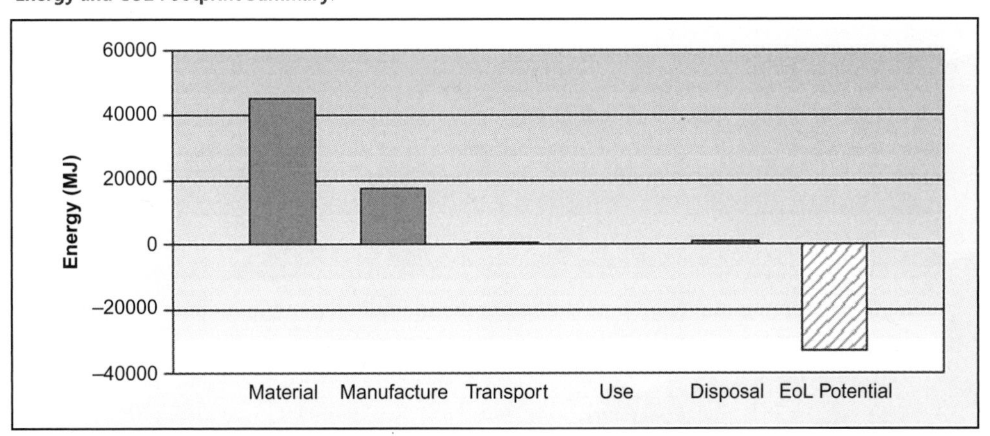

Energy Details...

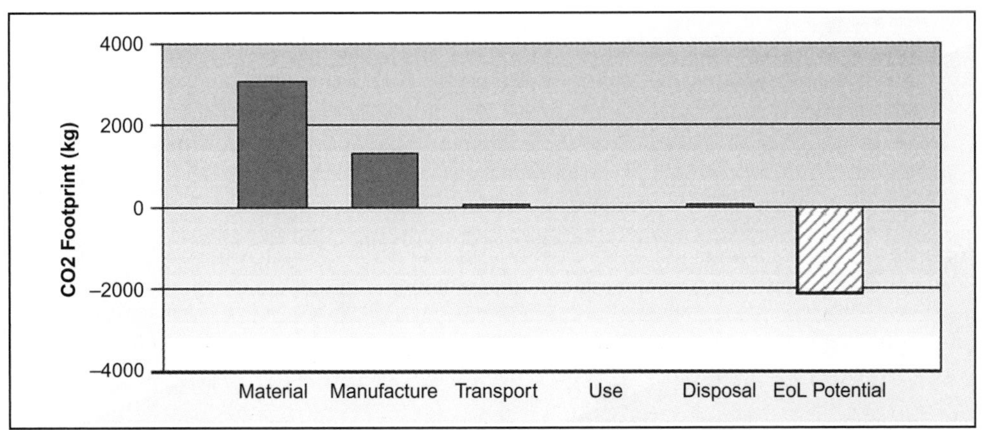

CO2 Details...

Phase	Energy (MJ)	Energy (%)	CO2 (kg)	CO2 (%)
Material	4.52e+04	70.5	3.04e+03	68.3
Manufacture	1.72e+04	26.8	1.29e+03	29.0
Transport	690	1.1	49	1.1
Use	0	0.0	0	0.0
Disposal	1.05e+03	1.6	73.5	1.6
Total (for first life)	**6.42e+04**	**100**	**4.46e+03**	**100**
End of Life Potential	–3.31e+04		–2.09e+03	

Eco Audit Report

Energy Analysis

Energy and CO2 Summary

	Energy (MJ)/Year
Equivalent Annual Environmental Burden (averaged over 1 year product life):	6.32e+04

Detailed Breakdown of Individual Life Phases

Material:

Energy and CO2 Summary

Component	Material	Recycled Content* (%)	Part Mass (kg)	Qty.	Total Mass	Energy (MJ)	%
Flywheel	Low Alloy Steel	Virgin (0%)	1.5e+03	1	1.5e+03	4.5e+04	100.0
Total				1	1.5e+03	4.5e+04	100

*Typical: Includes 'Recycle Fraction in Current Supply'

Manufacture:

Energy and CO2 Summary

Component	Process	Amount Processed	Energy (MJ)	%
Flywheel	Casting	1.5e+03 kg	1.7e+04	100.0
Total			1.7e+04	100

Transport:

Energy and CO2 Summary

Breakdown by Transport Stage Total Product Mass = 1.5e+03 kg

Stage Name	Transport Type	Distance (km)	Energy (MJ)	%
Flywheel	32 Tonne Truck	1e+03	6.9e+02	100.0
Total		1e+03	6.9e+02	100

Breakdown by Components

Component	Component Mass (kg)	Energy (MJ)	%
Flywheel	1.5e+03	6.9e+02	100.0
Total	1.5e+03	6.9e+02	100

Use:

Energy and CO2 Summary

Relative Contribution of Static and Mobile Modes

Mode	Energy (MJ)	%
Static	0	
Mobile	0	
Total	0	100

Disposal:

Energy and CO2 Summary

Component	End of Life Option	Energy (MJ)	%
Flywheel	Recycle	1.1e+03	100.0
Total		**1.1e+03**	**100**

EoL Potential:

Component	End of Life Option	Energy (MJ)	%
Flywheel	Recycle	-3.3e+04	100.0
Total		**-3.3e+04**	**100**

Notes:

Energy and CO2 Summary

CHAPTER

The Tools of the Design Process and Management of Design

4

4.1 INTRODUCTION

4.1.1 Product development: an introduction

The economic success of manufacturing companies depends on their ability to identify the needs of customers and to create products that meet these needs. The products also have to be produced at a competitive cost.

Achieving these goals is a *product development* problem involving

- Marketing
- Design
- Manufacturing
- Sales

A product is something sold by an enterprise to its customers. Product development is the set of activities beginning with the perception of a market opportunity and ending in the production, sale, and delivery of the product. The money earned by selling the product to the customer allows the company to develop future products.

Product development, therefore, is the process of developing a new product from the initial perception of need to the final product.

4.1.2 Marketing

Marketing often facilitates the identification of product opportunities within the market. Marketing defines market segments and identified customer needs. Marketing also communicates between the manufacturing company and the customer regarding timing of product release, launching, prices, and promoting new products.

4.1.3 Design

The design function plays the lead role in defining the physical form of the product to best meet customer needs. The design function also generates information relating to many other aspects of the product prior to manufacture, postmanufacture, and after the product's life has expired. The design function therefore could include the following:

Sustainability in Engineering Design. http://dx.doi.org/10.1016/B978-0-08-099369-0.00004-2

- Mechanical Engineering Design elements such as
  - mechanical components
  - electrical and control components
  - software
- Industrial Design elements such as
  - aesthetics
  - ergonomics
  - user interfaces
  - color
- Sustainability elements such as
  - material sourcing
  - low-environmental-impact components
  - usage impact
  - longevity
  - sustainable disposal
- Marketing and Sales components such as
  - envelope size
  - customer satisfaction
  - warranty
  - packaging
  - installation
  - after-sales service

4.1.4 Manufacturing

The manufacturing function is primarily responsible for implementing and operating the production system in order to produce the product. Manufacturing may also include purchase, distribution, and installation as well as the physical manufacture of the component. The manufacturing function should adhere to the specifications laid down by the design department, which should have researched components and considered the end function, something that manufacturing is unable to do because of lack of information. If sustainability is one of the major elements, then the design department should also specify the low-environmental-impact manufacturing environment.

4.1.5 Duration and cost of product development

It is astounding how much time and money are required to develop a new product. In reality, very few products can be developed in less than a year and many require 3-5 years. Figure 4.1 shows four engineered products and Figure 4.2 shows the scale of the associated product development effort.

The cost of product development is roughly proportional to the number of people on the project team and to the duration of the project. There will certainly be expenses for the project team as there almost always will be expenses related to the acquisition of tooling and equipment for production. These expenses are incurred by a core team of specialists such as design engineers, purchasing specialists, marketing specialists, industrial designers, electronics designers, and other specialists deemed

FIGURE 4.1

Four Engineered Products. (For color version of this figure, the reader is referred to the online version of this chapter.)

	Hammer	Skateboard	Passenger Vehicle	Passenger Aircraft
Annual volume	80,000	40,000	100,000	50
Lifetime (years)	50	3	6	30
Sales price £ ($)	10(15)	30 (45)	10k (15k)	100m (150m)
Development cost £ ($)	80k	250k (375k)	266m (400m)	200m (300m)
Development time (years)	1	1	4	5
Number of parts	3	22	10,000	130,000
Manufacturing investment £ ($)	100k (150k)	500k	333m (500m)	200m (300m)

FIGURE 4.2

The Product Development Effort for the Four Products.

necessary to take the project forward. Furthermore, there are secondary personnel who can be counted within the extended team. They are people such as accountants, sales specialists, legal advisers, assistance personnel, and administrators.

Figure 4.3 shows a simplistic model combining the departments that might be involved in the product development process.

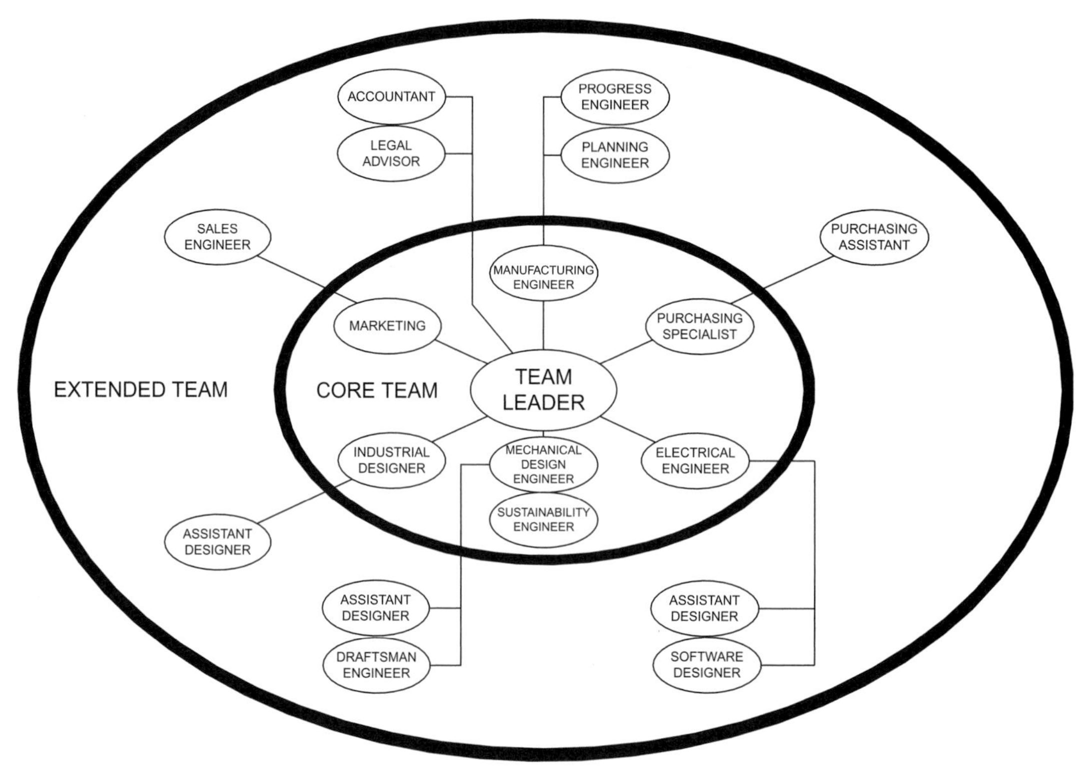

FIGURE 4.3

The Composition of the Product Development Team.

From the estimated values in the comparison chart shown in Figure 4.2, it can be seen that the more complex the product, the higher the development cost. This might seem obvious, but costs should be linked to the number of people involved in the development process. In fact, complex products need more specialist personnel input, which in itself will generate secondary duties such as legal advice and administration. Development costs are, therefore, extremely high: There is no earned value during the development process, yet there is potential for huge salary costs.

The composition of the product development team shown in Figure 4.3 highlights the need for personnel within the core design team as well as in the extended team. Typically, there needs to be a product champion who, in this case, is the team leader. It should also be pointed out that even in the simplest of projects, there needs to be a "Product Champion" who will drive the project and take responsibility for keeping it on schedule.

4.1.6 The challenges of product development

Developing a good product is difficult. Few companies are highly successful more often than not and these odds present a significant challenge for the product development team. Some of the characteristics that make product development challenging are described in the following sections.

4.1.7 Trade-offs

An aeroplane can be made lighter, but this action will probably increase the manufacturing cost. It is difficult to recognize, understand, and eventually manage such a trade-off. The Airbus A380 was such an enormous aircraft that weight and strength were of great concern. The traditional fuselage skin material was found to be unacceptably heavy when designed for appropriate strength. A composite form, aluminum alloy material, was therefore developed that met the strength and weight criteria. This exercise was very costly in terms of time and money.

4.1.8 Market dynamics

Technologies improve, customer preferences change, competitors introduce new products, and the economic environment shifts. Decision making in such a fluid environment can be extremely difficult and even disastrous. During the 1960s, aircraft manufacturers had a dilemma to face. They had to manage two functions:

- Build smaller and supersonic catering for a smaller but high-value number of passengers
- Build larger catering for a larger number of lower value passengers.

The supersonic airliner was, of course, the "Concorde." The large aircraft became the Boeing 747. History records that the Concorde has been an exciting aircraft, but the trend of the market, which was correctly read by Boeing, was to provide large aircraft to transport hundreds of people.

4.1.9 Details

The choice between using screws or snap fasteners on the enclosure of a computer can have economic implications that could cost millions of dollars when applied to large-volume products. For example, a domestic bed, manufactured from a rectangular hollow section needed to be clad in plasticized MDF so that it was conducive to bedroom decor. The cladding designer applied over 50 screws to the unit to fix the fiber board to the bed frame. The fixing operation consisted in drilling the holes, inserting threaded nuts, assembling the cladding components, and tightening the screws. It was calculated that the operation would cost £150 per bed unit, which was 10% of the manufacturing cost. Clearly this was unacceptable. The solution was to use industrial double-sided foam tape, which could be applied quickly without drilling the frame or tightening screws. The cost of this was estimated to be £20 per bed.

4.1.10 Time pressure

Any one of the above development difficulties could be easily manageable by itself given plenty of time, but product development decisions must usually be made quickly and without complete information.

4.1.11 Economics

Developing, producing, and marketing a new product requires a large investment. To earn a reasonable return on this investment, the resulting product must be both appealing and relatively inexpensive to produce.

4.1.12 Organizational realities

In reality, some organizations exhibit characteristics that lead to dysfunctional product development teams. These characteristics may include the following.

4.1.12.1 Lack of team responsibility

Team leaders may continually intervene in the details of a project without fully understanding the basis of the team's decisions. In this case, the team leader does not feel that he is able to trust the judgment of his team. This is really a failing of the manager: he should have selected a better team, since he cannot keep from tinkering with the project. This is dangerous as the team leader may have only partial information, which could debase decisions. The technical designers, who are close to the problem, should be the actual decision makers.

4.1.12.2 Function allegiances

Wherever there is a group, the human condition generates politics. Representatives from marketing, design, or manufacturing may attempt to influence decisions in order to increase their own political standing, with complete disregard for the success of the product.

4.1.12.3 Inadequate resources

A team may be unable to complete development tasks effectively because of lack of staff; a mismatch of skills; or a lack of money, equipment, or tools.

4.1.12.4 Minimal cross functional expertise in the project team

Should key development decisions be made without the involvement of primary elements involved in the development process, problems may be caused in the later stages of development. For instance, an independent design decision could be made to increase the physical envelope of the product. Perhaps an electric motor protrudes from the product further than was originally envisaged. The extra volume does not manifest itself until the packaging stage of the project when it transpires that extra space is required for shipping. The difference means that only 300 units can fit inside a container while the original plan was to ship 500 units to be inside a single container. This obviously would increase costs that were hitherto unforeseen.

Most organizations will, at some stage, exhibit some of these characteristics. They may be stifling for the project so that the project development is rendered ineffective. The key to a successful product is to appoint a team leader, a "Product Champion," who is skilled in dealing with people and in selecting the correct personnel.

4.2 DEVELOPMENT PROCESSES

4.2.1 Processes involved in product development

Most people are familiar with the idea of a physical process such as that used to bake a cake or to assemble an automobile.

In a similar way, the product development process becomes a sequence of steps or activities used to conceive a concept design and eventually leads to the commercialization of a product. Activities in the early stages of development are intellectual and organizational. Physical development is applied as late in the process as possible. While intellectual design activities can prove to be expensive, the build process is much more expensive, so delaying the build reduces expenditure.

Some organizations define and follow a precise and detailed development process, while others may not even be able to describe their process.

Furthermore, two organizations designing similar products will employ processes that are slightly different from each other. This may be due to personalities, the product team, skills mix, company resources, and so on.

It is, however, certain that whatever process is involved in product development, there will be a general approach to the process of design and manufacture. It is useful to define the elements of this development process as it will aid in analysis of the final product specification.

4.2.2 Quality assurance

A development process specifies the phases of a development project. The process also specifies points at which checks can be made on progress and quality. A careful selection of these checkpoints ensures quality in the resulting product and also furnishes the development team with checkpoints that prompt feedback to the client.

4.2.3 Coordination

Large projects, in particular, require a clearly articulated development process. Such a master plan defines the roles of each member of the development team and informs the members what their tasks are, when their contributions will be needed, and with whom they will need to exchange data and materials. In any group situation, team members usually require specific details regarding their tasks. It is always dangerous for a team leader to assume a task will be performed. Unless precisely allocated, it is likely the task will be left undone, each team member thinking that someone else will do it.

4.2.4 Planning

A development process contains natural milestones corresponding to the completion of each phase. The timing of these milestones anchors the schedule of the overall development project. These milestones are often used for reporting back to the client and are usually written into the contract. Successful completion of a milestone may also prompt the client to inject investment in order to move the project to the next phase.

4.2.5 Management

A planned development process can be set out with dates and activities and used as a specific guide for assessing the performance of the continuing development effort. By comparing the actual events to the established process, a manager can identify possible bottleneck areas. The manager must continually monitor project costs, personnel, techniques, and the technology employed. Any hold-ups in the development should be identified and accommodated.

4.2.6 Improvement

The careful documentation of an organization's development process often helps to identify opportunities for improvements and efficiency. At the completion of the project, it is beneficial for the team leaders and management team to "debrief" to analyze process flaws, problems, bottlenecks, shortages, overspends, etc. In this way, procedures may be improved so that future development projects become more efficient and, therefore, less costly.

4.2.7 The generic product development process

The generic product development process consists of six phases (see Figure 4.4).

4.2.7.1 Generic development process: overview

The process begins with the planning phase. This is the link to advanced research and technology development activities. The output of the planning phase is necessary so that marketing can define the product and link the product specification to market needs. It is only at this point that the brief can be given to the design team to begin the concept development phase.

The initial development process considers the creation of a wide set of alternative product concepts, whose benefits are then explored so that the alternatives are subsequently narrowed down. As the design investigation focuses on a narrowing number of concepts, the specification of the product expands until reliability and repeatability can be predicted in a single new concept.

The design development process is an information-processing system. The process begins with inputs such as the objectives, requirements, constraints, and awareness of the capabilities of available technologies. The various activities during the development process use the information to formulate specifications and concepts and eventually design details for manufacture and physical development. The six phases of the generic development process are discussed next.

4.2.7.2 Planning

This phase begins with a general strategy (usually corporate strategy) and includes an assessment of market needs and the technology available to meet those needs. Often the company is already manufacturing goods for a particular industry and would fairly easily be able to apply the technology in which it is expert. In the development of new products, however, the progress of a company is always dependent on its willingness to apply new technologies. In any new development, current and new technologies should be considered when the company is attempting to meet market needs. The output of this planning phase is the project mission statement, which, among other things, includes a project definition.

The planning phase really outlines a general design specification and includes the following:

- Mission statement
- Requirements (market needs)
- Constraints
- Identification of technologies
- Target market
- Business goals
- Key assumptions

	Marketing	Research	Design	Manufacture	Finance / legal / other	Sales	Management
Planning	Define market needs Define market opportunity Define market segments	Define available technologies	Understand the brief Define product architecture Research the technology	Identify production constraints Set supply chain strategy	Provide planning goals	Plan infiltration into market segments	Allocate project resources
Concept development	Collated customer needs Identify lead users Identify competitive products	Determine technologies Investigate other options	Investigate design options Develop basic concepts Build and test experimental prototypes	Estimate manufacturing cost Assess production feasibility	Economic analysis Investigfate patent issues	Investigate packaging requirements	Overview progress Client update
Design development	Develop plan for product options Develop plans for extended product family	Define and test performance of components	Generate alternate options Define subsystems Refine designs	Identify suppliers Perform make-by analysis Define final assembly method	Identify service issues Facilitate make-by analysis	Plan for product options	Overview progress Client update
Detail design	Develop marketing plan	Continue investigation into parts and methods	Define component geometry Assign tolerances Refine design	Define production processes Design tooling Quality assurance processes			Overview progress Client update
Prototype testing & refinement		Continue investigation into parts and methods	Reliability testing Life testing Performance testing Regulation approval Implement design changes	Facilitate supplier buildup Refine assembly processes Train work force Apply quality assurance		Develop promotion and launch materials Facilitate field testing	Overview progress Client update
Build-up to production			Evaluate early production output	Begin operation of entire production system		Place early production with key customers	Overview progress Client update

FIGURE 4.4

Generic Product Development Process Tasks.

Perhaps the most important element in the planning phase is the creation of the *Product Design Specification* (*PDS*). This is really the route map given to the designers so that they can match the new product to the market needs.

4.2.7.3 Concept development

In this phase, the needs of the target market are identified from the Product Design Specification. The design team generates several design concepts, which are then evaluated for "best fit" with the market needs and PDS. Some proposed solutions will be closer to the market needs than others. The evaluation process selects one or more concepts for further development and possible prototype build and testing.

When a concept is generated, it should be accompanied by a *Concept Specification* that contains a refined description of the form, function, and features of a product. Within the concept specification, there should also be a feasibility study, an analysis of the competition, and an economic justification for the project. The *Concept Specification* should closely match the PDS and, hence, the market requirements.

Once the concept has been defined, it may be used to market the product to entrepreneurs or the company's board of directors in order to raise investment money for further development.

4.2.7.4 Design development

After creation of the concept or several optional concepts, the design development phase considers the practicalities of design and manufacture. The development plan identifies probable departments such as manufacturing, design, and purchase. The product is also usually broken down into its composite parts. For instance, a car would be broken down into body, suspension, engine, wheels, gearbox, etc. This would enable each subsystem to be analyzed and designed separately.

The result of design development phase would closely specify each product's subsystems and generate a final assembly process. Information gleaned during this particular exercise can be used to convert the concept into real-life materials.

4.2.7.5 Detail design

The detail design phase defines the complete specification of the geometry, materials, and tolerances of all the parts through the provision of detail drawings, assembly drawings, and general assembly drawings. It also includes the specification of all the bought-out parts complete with preferred suppliers and component designations. The result of this phase is the complete and precise physical description of all the parts in the product. These drawings can then be issued to the manufacturing function for precise manufacture and assembly.

The detail drawings require an index in order to be understood and coordinated. Along with the detail drawings, therefore, there should be a general *parts list*, which will act as a "map" and precisely instruct the manufacturing function to group components into assemblies and, further, assemblies into the full product. Figure 4.5 shows a portion of a typical parts list, which lists all the parts along with raw material sizes and specifications. It should be noted that this particular parts list shows elements of three subassemblies.

The detail design phase takes the concept design and applies it to the material components and, in effect, plans the manufacture and assembly of the product. During this process, the designers consider the selection of materials, manufacturing techniques, machine tools, and processes such as forging,

2.0	**2**	**Cantilever Weld Assembly**	BL 1010		#
2.1	2	Cantilever	BL 1001		#
		BMS Plate 684 × 265 × 8 thick			
		CANTILEVER DXF Profile	BL 1020		#
2.2	2	Saddle	BL 1002		#
		BMS Plate 342 × 75 × 5 thick			
2.3	8	M8 Dome HD M/c Screw × 50 long Chromed/SS	BO		
2.4	2	Rigidity bracked	BL 1067		#
		BMS Bar 213 long × 40 × 5 thick			
3.0		**Back Frame Assembly**	BL 1046		#
3.1		Primary Wall Bracket	BL 1082		
		BMS Plate 100 ×180 × 5 thick			
3.2	4	Wall Bracket	BL 1044		#
		BMS Plate 190 × 5 thick			
		Size A 408 long true			
		Size A 508 long true			
		Size A 608 long true			
		Wall Brackets Profiles	BL 1057		#
3.3	3	Rear Frame Spacer	BL 1045		
		BD MS Strip 2000 × 50 × 3			
3.4	1	WEB	BL 1065		#
		BMS Plate 100 ×180 × 5 thick			
6.4.5	12	M8 Lockwasher	BO		
6.4.6	12	M8 Nyloc Nut	BO		
4.0		**Bed Frame Assembly**			
4.1	2	Side Stringers Weld Assembly	BL 1011		
4.2	2	Side Stringers (Opposite Hands Required)	BL 1012		
		ERW Tube 50 × 50 × 14 SWG wall × 1800 long			

FIGURE 4.5

A Typical Parts List Showing Elements of Three Subassemblies.

injection molding, and casting. Further processes such as chromium plating and painting are added to the specification.

It must be emphasized here that every single component and process required to complete the product and which, therefore, adds cost to the product requires a detailed specification at this stage.

The detail design often relates to the product, but once the product has been defined precisely in terms of technical drawings, other manufacturing elements may be pursued such as specialist machine tools, jigs, and other specialist manufacturing equipment.

4.2.7.6 Prototype testing and refinement

Almost all products require testing and further refinement. Any new product should be carefully tested to match the specifications and requirements originally laid down. In some cases, manufacturers allow a protracted period for this so that durability and longevity can be ensured through the tests.

The phase involves the construction and evaluation of many reproduced versions of the product.

Early prototypes are usually produced as close to the production version as possible but, in many cases, the first prototype is several steps away from the production version. The first prototypes are required to prove that the device functions as predicted by the design team. Many development refinements may have to be made in order to creep closer to the production version.

Evaluation of a prototype involves assessing the following elements:

- Functional requirements (does it work as it should?)
- Manufacturing requirements (can it be manufactured easily?)
- Assembly requirements (can it be assembled efficiently?)
- Cost requirements (can it be manufactured to a target cost?)
- Serviceability (can it be serviced to protract its life?)

The first prototype is important as each of the mentioned elements can be analyzed and improvements instigated. The development of a new product is a learning process and the first prototype is a crucial step. Improved prototypes can then be built from the initial lessons learned.

Later prototypes are usually much nearer the production version and are often called "preproduction prototypes." These secondary prototypes are usually used to evaluate the product's performance and reliability prior to the manufacture of the production version tooling. Later prototypes are subjected to the same scrutiny as the first prototype, each evaluation adding to the wealth of knowledge leading to the final product.

4.2.7.7 Production build-up

During this phase, the product is made using the intended production system. The purpose of the ramp-up is to train the work force and to work out any remaining problems in the production processes. Furthermore, logistics system is evaluated, perhaps to alert satellite storage depots and book appropriate transport.

Products produced during this phase are sometimes supplied to preferred customers so that they can be carefully evaluated to identify any remaining flaws.

In 1976, the Ford Fiesta was named "The Bobcat." It was supplied to various agencies throughout the UK for testing and discovering teething problems. In fact, even after the Fiesta was launched in 1977, there were problems with leakage through the air intake grill resulting in a major recall to effect repairs.

At some point, the production build-up is converted to the full-production version. Usually, the launch date is the trigger point for converting the preproduction version to the full-production version.

4.2.8 The approach to concept development

Concept development involves many interrelated activities from the investigation and understanding of customer needs to the generation of possible solutions. Eventually, the specification of the best solution for manufacture is formulated. There is usually an element of research throughout the process, from the start of the design to the manufacture. The research is obviously more intensive at the beginning of the design project and may involve the research for new technical solutions and new components, as well as information research.

Rarely does the entire process proceed in a purely sequential fashion. It is unusual to complete one task before continuing with the next.

In practice, activities may overlap in time, running parallel to one another. For instance, at the beginning of a project, the search for a suitable internal combustion engine may be initiated at the same time as that for a screw compressor. Design is an iterative process requiring new information before the design can progress. When a new piece of information is gleaned, the design may have to be changed. Such is the case when the designer creates a design with, say, a hydraulic ram. Several days after the initial enquiry to locate a particular hydraulic ram, the designer may receive technical information showing that the original ram enclosure is not large enough. The designer has then to reformulate the design in order to accommodate the ram design that can be obtained.

New information may enter the design arena even while the design is ongoing, resulting in the backtracking of the design when new information becomes available or results are learned through experiment. Unfortunately, the nature of design and the reception of information often cause the design team to step back to repeat an earlier activity before proceeding.

This repetition and backtracking of activities is known as "development iteration."

A rock drilling company was tasked to design a rock drill (to drill water well boreholes) at normal elevations from sea level up to an elevation of 3000 m. The drill rig was designed with a particular naturally aspirated engine to give enough power for this elevation. The customer then reapplied the design specification so that the rig would be capable of working in the Himalayas at elevations of 5000 m. At this altitude, there is a great deal of power loss because of the oxygen-depleted atmosphere.

The design team then had to backtrack to select a new engine of greater power, which delivered the performance required by the original specification. This process involved backtracking to redesign other elements of the design, such as larger fuel tanks, strengthened chassis, larger volume engine bay, new subchassis, new engine mounts, and stronger drive couplings.

The process of concept development includes several key activities, which are discussed in the following sections.

4.2.8.1 Identifying customer needs

It is essential that the customer's needs are met. Though it may be difficult to meet all the customer's needs, the product must appeal to the market in order to sell. There would be no point in going through the design process if there was no market for the product.

The goal of this activity, therefore, is to understand the customer's needs and to effectively communicate them to the development team. Once the customer's needs have been identified, they need to be collated in a list of requirements organized in a hierarchical order. These needs may be weighted relative to the perceived importance of each requirement.

The Product Design Specification (PDS) provides a precise description of what the product has to achieve, and converts the customer's needs into technical terms and creates targets from the very outset of the project.

These specifications may be fairly broad at the beginning of the project but after proceeding through several iterations and stages of the development process, these specifications become more and more refined.

At the beginnings of the project, the brief is perhaps no more than a target wish list and can be considered a list of requirements. This initial specification builds into a more precise target, which, of course, is the Product Design Specification (PDS). As a design progresses, other specifications mark strategic points within the design process. The progressive list of specifications is as follows:

- Brief
- Product Design Specification (PDS)
- Concept Specification
- Final Design Specification
- Product Specification

The Final Design Specification (FDS), however, is a precise description of what the product can achieve. The FDS may be quite lengthy, but it is actually for internal audit so that the technical design team along with the other functions, such as marketing and sales, can precisely match the product outcomes to the original market requirements.

The Final Design Specification is often far too complex to accompany the product into the marketplace. It could confuse the customer and, at worst, could give away vital information to the competition. From the Final Design Specification, a Product Specification is formulated. This is a shortened version giving only the information that is relevant to the functioning of the product.

An example of a Product Specification, which is the specification for an electric arc welder, is shown in Figure 4.6. It gives the power supply voltage, the welding current range, etc. Figure 4.7 is an excerpt from the handbook of a CD player. It gives the kind of disc format, the voltage, the wow and flutter values, volume levels, etc.

Both these specifications possess the bare minimum of information to allow the customer to make an informed choice, and restrict the information competitors may use.

4.2.8.2 Subtasks within the process of concept development

The process of concept development can be subdivided into smaller tasks. These are discussed in brief in the following sections.

4.2.8.2.1 Concept generation

The goal of concept generation is to thoroughly explore the range of solutions and match them with the particular customer needs.

Concept generation involves a mixture of external research, creative problem solving, and systematic exploration. There may be several design elements which may be assembled into a complete product.

Product Specification:				
ITEM	**ARC100T**	**ARC130T**	**ARC160T**	**ARC250T**
POWER SUPPLY VOLTAGE	230V/50HZ	230V/50HZ	230V/50HZ	230V/400V 50HZ
WELDING CURRENT RANGE	35-80A	55-130A	55-160A	65-250A
TRANSFORMER COOLING	FAN COOLED	FAN COOLED	FAN COOLED	FAN COOLED
WEIGHT (net)	11.3kgs	14.2kgs	16.6kgs	24.5kgs
ELECTRODE DIAMETERS	1.6, 2.0, 2.5mm	2.0, 2.5, 3.2mm	2.0, 2.5, 3.2, 4.0mm	2.0, 2.5, 3.2, 4.0, 5.0 mm

Symbols and Technical Data

EN50060	European standard relative to welders for limited use
	Single phase transformer
U_0	No-load voltage
50 Hz	Nominal mains frequency
I_2	Conventional welding current
mm	Diameter of electrode
nc	The number of electrodes of reference that can be welded starting with the welding machine at ambient temperature until the first triggering of the thermal protection (only stated on ARC100T, ARC130T and ARC160T)
nc1	The number of electrodes of reference that can be welded in one hour starting with the welding machine at ambient temperature. (only stated on ARC100T, ARC130T and ARC160T)
nh	Number of weldable electrodes after the thermal protection intervenes in first hour of use. (only stated on ARC100T, ARC130T and ARC160T)
nh1	The number of electrodes of reference that can be welded in one hour starting with the welding machine a regular heat level. (only stated on ARC100T, ARC130T and ARC160T)
U_1	Mains voltage
I_1 max	Maximum absorbed current
I_1 eff	Maximum effective current consumed. (only stated on ARC250T)
	Value in amps of delayed action fuses or automatic switch to be provided in order to protect the power line.
X%	(only ARC250T) Duty Cycle with reference to the welding amperage I_2
IP21	Protection grade covering
H	Class of insulation
	Thermostat (Thermal Protection)
	Fan cooled welder
	Electrode holder
	Earth Clamp
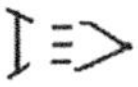	Standardised plug
	Switch

3

FIGURE 4.6

Final Specification for an Electric Arc Welding Unit.

Specifications

REMOTE CONTROL ISINSENSITIVE OR DOESNOT WORK

Check the batteriesof remote control and make sure that they are to be powerful and good conductivity.
Direct the remotecontrol to the IR sensor of the player.
Check whether thereare some obstacles between the remotecontrol and IR sensor.

ANNORMAL FUNCTION OPERATION

Turn off the power, and then turnon again.

DISC IS LOCKEDIN THE TRAY

The disc can not be taken out of the tray when theplayer is reading, soyou have to turn off the power andturn on again, and press OPEN/CLOSE key immediately toopen the tray.

Type of Disc	DVD/VCD/HDCD/CD/DVD±R/RW WMA,MP3 Kodak Picture CD	Frequency Response	CD: 4Hz 20KHz (EIAT) DVD: 4Hz 22KHz (48K) 4Hz 44KHz (96K)
		S/N radio	> 92 dB
		THD	< 0.04%
Video Format	MPEG 2		
Audio Format	MPEG 1, LAYER 1, LAYER 2 , LAYER 3	Output Terminals	Video (composite) output X 1 S-Video output X 1 2ch output X 1 Digital coaxial output X1 SCART output X 1 YPbPr output X 1
Signal Output	Color System: PAL /NTSC Audio System: DTS digital output Audio DAC 16bit/48KHz Video Output: 1 Vpp (at 75 ohm) Audio Output: 2 Vpp	Power input	AC100-240V 50Hz-60Hz 25W
		Dimensions	Body size : W260 x D230 x H38 mm Net weight : 2.3 Kg

FIGURE 4.7

Final Specification for a DVD Player.

For a passenger vehicle, there might be such elements as internal combustion engine design, suspension design, body design, and brakes design. Each element needs to be designed in parallel, resulting in a large volume of possible solutions. Overall, there may be up to 20 different preliminary concepts (minimum 3). The design team should break each one down into its subelements but should have an overview of how the elements fit together.

4.2.8.2.2 Concept selection

This is the activity that selects the very best of the concepts generated thus far. Proposed solutions may often be combined and this usually requires several iterations before an acceptable concept is created and finally considered suitable for physical development. There are several solution generation processes and tools that can aid in this process; they are discussed later in this chapter.

4.2.8.2.3 Concept testing

It is usual to emerge from the concept generation and selection stages with at least one seemingly viable concept and sometimes with several of them. Concept Specifications should be generated for each concept and compared with the Product Design Specification. The PDS represents the technical values relating to the market needs. The Concept Specification should closely match the PDS.

These comparative tests may also be related to a market research plan or could be further enhanced by feedback from a controlled selected audience.

4.2.8.2.4 Setting final specifications

In the final stages of physical development, the PDS is revisited after the concept has been selected and tested. The product requires to be rigorously matched to the needs of the customer, and in the process, trade-offs, usually between cost, sustainability, and performance, and limitations are identified.

4.2.8.2.5 Project planning

This is one of the primary activities after the brief has been received. The team create a detailed development schedule for what they believe to be the product, devise a strategy to minimize the development time, and identify the resources required to complete the project. This element is required before an economic analysis can be applied to the product and should have flexibility built in, as the development of the project could be in several directions, requiring a flexible plan.

4.2.8.2.6 Economic analysis

An early economic analysis is often performed before the project begins. This is necessary to secure appropriate funds but may prove difficult if the product is new and does not yet exist. However, when there is a rational concept to analyze, it is necessary to predict development and manufacturing costs. As the project becomes more developed and the outcome becomes more predictable, it becomes easier to predict the economic cost.

4.2.8.2.7 Benchmarking of competitive products

Research related to the main competition is absolutely essential as this provides an indication of the level of complexity of the new product and may give the design team new ideas. Benchmarking is, therefore, essential if the new product is to beat the level of technology of the competitors.

4.2.8.2.8 Modeling and prototyping

Every stage of the concept development process involves various forms of models and prototypes. They may include the early "proof of concept" models or prototypes, which assist the development team to demonstrate feasibility. These models may also be scale models, such as those used by architects to demonstrate street plans.

Some models may be shown to customers to evaluate ergonomics and style. There are many other forms of models such as financial models and design analysis models that can also be introduced.

4.3 SYSTEMATIC APPROACH TO DESIGN

The principal role of the designer is that of an originator in the process of creating new products. The initial task is to identify the "primary need" for the product, which is really a statement of the design problem. In large organizations, the primary need may be established by the marketing function, who will provide the basic need specification. In other cases, it may be the designer himself who assesses the primary need in order to develop the design.

4.3.1 The primary need

The primary need arises from various sources:

- An update in the process of manufacture
- The exploitation of new technology
- The need to improve on the product of a competitor
- The need for a new product following changes in legislation
- The need for accessories following a new development (mobile phone accessories)
- The requirement of a new product

It is important for the designer to understand that the requirements and needs set out by the customer may not be the actual requirement. Customers often have their own solutions, which they are enthusiastic to impart to the designer.

For example, one of the customer stated true needs for a particular project may show that a new machine is required to test a full vehicle under fully loaded conditions. This customer requirement could possibly be the result of his ignorance of test techniques. Investigation by test specialists may reveal that the tests need to be conducted only on a quarter car model, which would considerably reduce test equipment costs.

4.3.2 The solution process

Design solutions are not merely plucked out of thin air. There is a certain progressive process that, when applied, leads the designer through various iterations to create an optimum concept. The complete Systematic Design Technique can be seen in Chapter 2, Figure 2.7. A simplified version leading to the concept stage of the design can be seen in Figure 4.8.

Often the primary need is conveyed as a set of verbal statements, which then need to be converted into technical parameters through the Product Design Specification. The PDS provides the parameters from which solutions can be generated. Normally, the designer would be expected to provide between three and five solutions, depending on the project.

Each proposed solution requires a detailed investigation weighing the advantages against the disadvantages and ensuring that the optimum solution is gained. Often, a mathematical analysis is required to ensure that the proposed solutions possess the technical parameters required by the PDS. After the rigorous investigative and analytical process, the outcome should be a single solution, which becomes

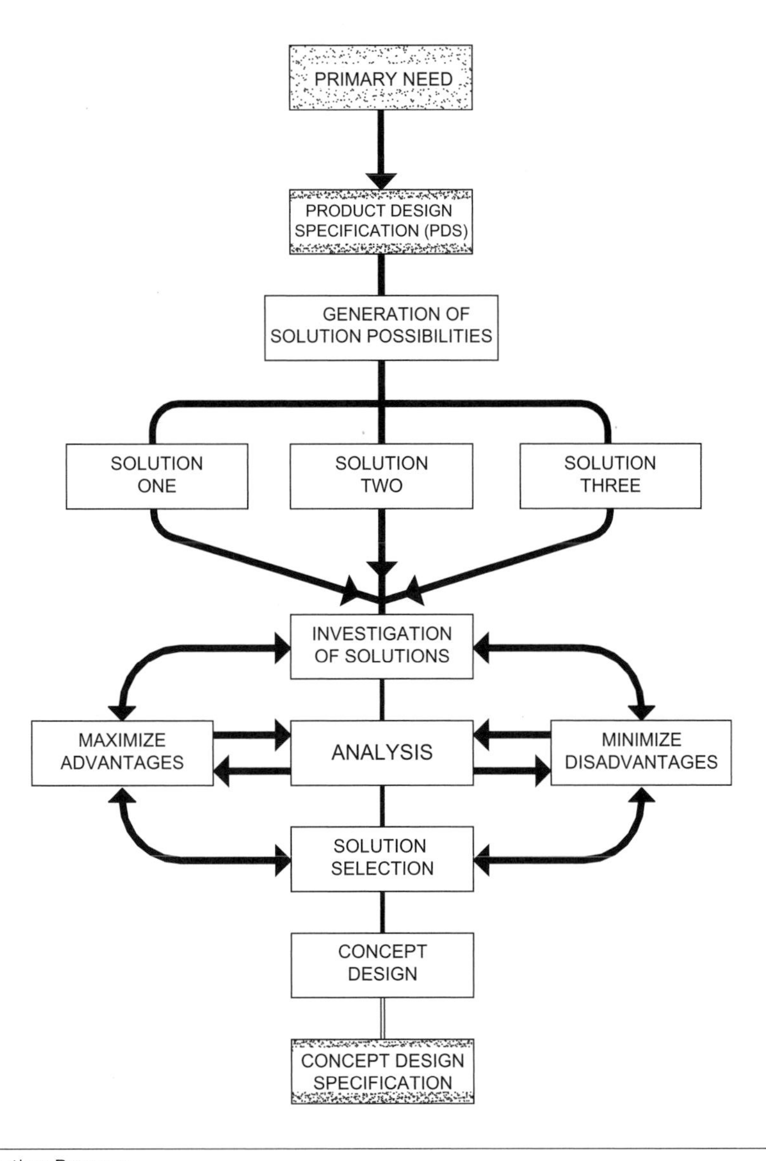

FIGURE 4.8

The Design Solution Process.

the *Concept Design*. The technical parameters are used to create the *Concept Design Specification*. The successful development of the design to the concept stage would show a close comparison between the *Concept Design Specification* and the PDS.

This systematic approach can be considered a form of stocktaking and will provide answers to such elements as

- Method of device operation
- Production procedures
- Power flow techniques and methods
- Method of control

4.3.3 Investigation, synthesis, and analysis

4.3.3.1 Investigation

When taking up the design challenge, the designer needs to apply a clear train of thought and to develop an investigative technique. Investigation is extremely important for the designer to familiarize himself with the requirements of the brief and any technological developments that may be applied to the new design proposals.

Though designers may be specialists in particular areas of industry, investigation at the start of each new design project will eradicate complacency. Design evolution cannot take place without investigation, which will naturally lead to the discovery of new materials, processes, and components.

The design of the aircraft has been the subject of intensive investigation. When the Wright brothers made their first flight in 1903, intensive investigation resulted in what was effectively a powered kite. Further investigation by scientists and engineers from all over the world resulted in huge leaps forward in the development of heavier-than-air flying machines. One of the landmark achievements resulting from this investigation and design development was witnessed in 1947 when Chuck Yeager became the first man to fly faster than the speed of sound. This was a mere 44 years after the Wright brothers first flew their powered aircraft.

4.3.3.2 Synthesis

In some circles, the act of design is considered to be artistic. In fact, the act of design is really *artistic science* and is composed of two elements:

- Synthesis
- Analysis

Analysis is dealt with later; however, it is an important part of the design process as it tests the physics and feasibility of the design through analytical means.

Synthesis is the artistic or creative process that the designer uses, first to imagine and then to model through sketches or drawings, or using physical model.

Synthesis involves collecting design information, ideas, appropriate technologies, possible components, etc. The sketching and modeling of alternative proposals formulates the design in the mind of the designer, allowing it to be shared with colleagues or clients. This synthesis is essential at the beginning of a project to integrate an enormous breadth of data. In effect, synthesizing the information is similar to filtering the most important data for use within the proposed solutions.

It is common to synthesize a minimum of three proposed solutions so that detailed comparisons can be made. A synthesized solution, when communicated through a sketch or a model, can then be analyzed for feasibility and function.

During the synthesization process, the designer's judgment plays an enormous role. Judgment is often based on previous experience and a "feel" for the requirements of the product. For instance, in the design of a new bicycle, the personal experience of the designer would tell him that a lightweight frame was essential and could not be manufactured from an "I" section girder. The more experienced designer will possess a more refined judgment.

A note of caution needs to be issued here: If a synthesized design is based purely on judgment, the result will be overdesign or "if it looks right it is right." When employing judgment, even the most experienced designers will tend to overdesign. Synthesis, therefore, has to be combined with analysis, which will refine the synthesized design into a workable and feasible one.

4.3.3.3 Example: hand mounted cycle mirror

A hand mounted cycle mirror that was light weight and unobtrusive was to be designed. The mirror was to be fitted to the cyclists right or left hand depending on the side of the road he/she was cycling.

The investigation led the designer to first consider mirror shapes and the ergonomics of attachment to the cyclists hand, which quickly led to an investigation of methods of manufacture. Injection molding was identified as the most appropriate manufacturing method. The final injection molded concept, can be seen in Figure 4.11.

The investigation of the product requirements led the designer to consider such elements as ergonomics of the hand, comfort, strap types, strength, ease of manufacture, and cost of manufacture. The design synthesis evolution can be seen in Figures 4.9 and 4.10. These figures show the thought processes of the designer as he blended the design requirements into one unit.

The design of the cable tidy clearly indicates the evolution of the design using the synthesization process. It also indicates that synthesization has to take place in parallel with investigation and analysis.

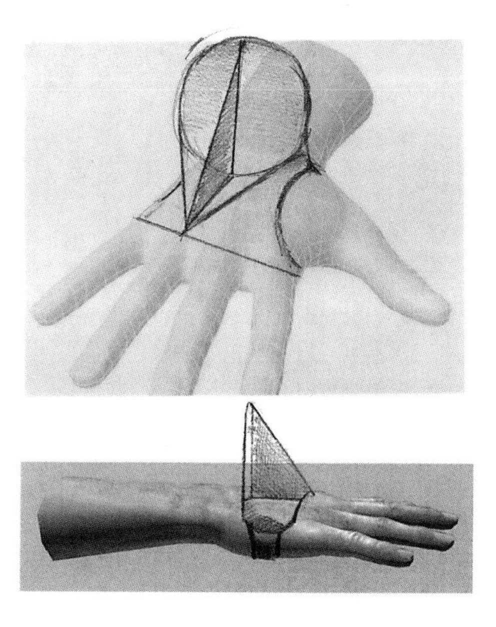

FIGURE 4.9

Injection Molded Cyclist Hand Mounted Mirror [4].

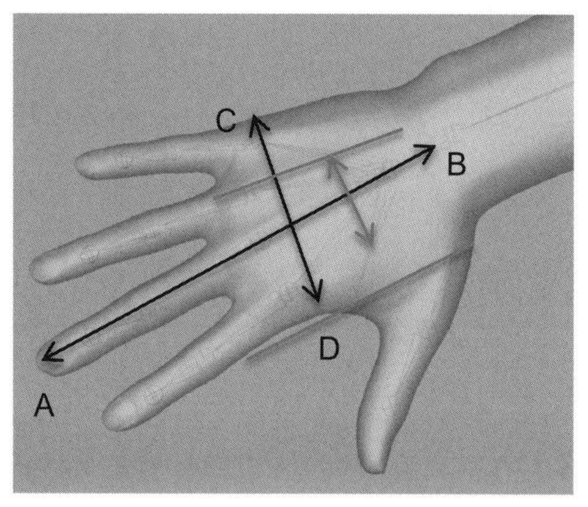

Dimension	Gender	5th Percentile (mm)
Hand length	Male	173-175
	Female	159-160
Palm length	Male	98
	Female	89
Hand breadth	Male	78
	Female	

FIGURE 4.10

Ergonomic Investigation for the Hand Mount Development [4].

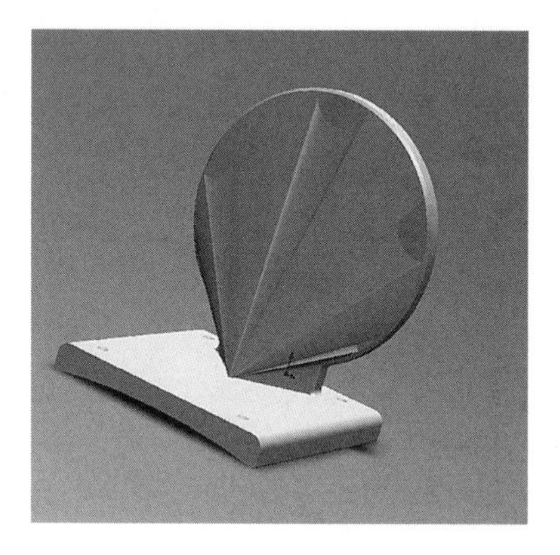

FIGURE 4.11

Injection Molded Mirror Mount for Cyclists [4].

4.3.4 Design evaluation

Evaluation of the design is always done as the design evolves but is often formalized using a systematic process. Analytical tools such as 3D CAD applied finite element analysis can be used to great effect in evaluating the synthesized shape as shown in figure 4.12. The systematic process of comparison and evaluation may be undertaken by using a points system or some sort of ranking technique. After Synthesis, analysis and evaluation the 3D CAD system allows a "photo-realistic" 3D image to be generated as shown in figure 4.13.

Evaluation points of such techniques could include the following:

- Fulfillment of function
- Reliability
- Cost
- Sustainability

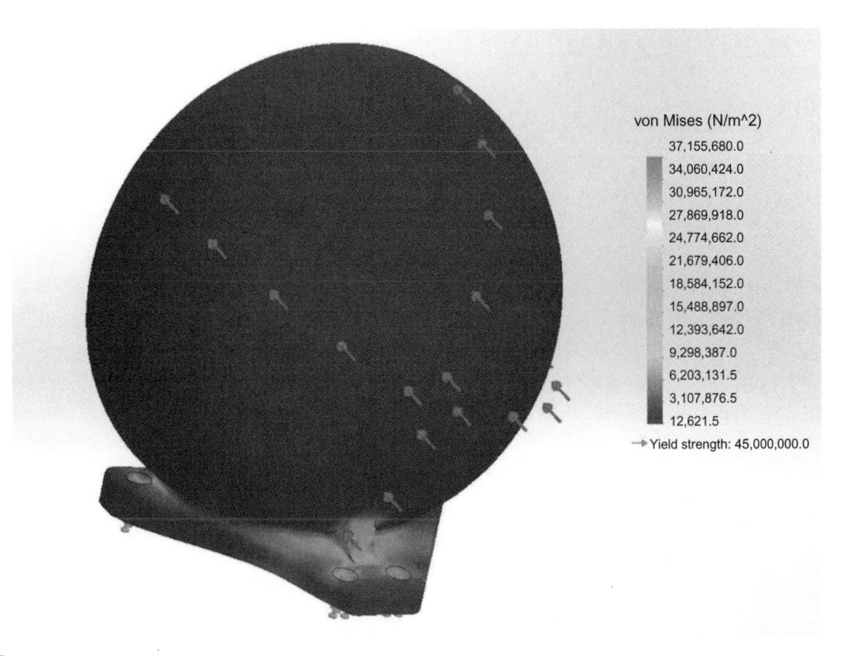

FIGURE 4.12

Strength Analysis Using Finite Element Analysis Methods [4].

FIGURE 4.13

The Final Version of the Hand Mounted Cycle Mirror [4].

- Serviceability
- Life expectancy
- Ease of maintenance
- Ease of manufacture
- Efficiency of operation
- Simplicity of layout

Ideally, the method of evaluation should be kept simple. The method of morphological analysis is an excellent tool and is very effective when used properly, but care should be taken because it can easily be subconsciously skewed. More importantly, it could be laid open to abuse were results to be manipulated to give the advantage to a favored design.

The specification of the concepts in sketch form focuses the designer on clearer, more concise thinking patterns. These designs can now be systematically evaluated using the tools shown in Figure 4.14.

FUNCTION	DESIGN A	DESIGN B	DESIGN C	DESIGN D	ATTRIBUTE TOTALS
Function	HIGH	HIGH	LOW	HIGH	32
Reliability	HIGH	MEDIUM	MEDIUM	LOW	24
Serviceability	LOW	HIGH	LOW	HIGH	24
Longevity	MEDIUM	HIGH	MEDIUM	HIGH	24
Ease of Maintenance	MEDIUM	HIGH	MEDIUM	MEDIUM	28
Ease of Manufacture	MEDIUM	HIGH	LOW	MEDIUM	24
Operation Efficiency	MEDIUM	MEDIUM	LOW	HIGH	24
Simplicity	MEDIUM	HIGH	HIGH	LOW	20
Sustainability	MEDIUM	LOW	HIGH	LOW	24
Manufacturing Cost	MEDIUM	MEDIUM	HIGH	LOW	24
Design Totals	64	80	48	70	

KEY	HIGH	MEDIUM	LOW
GENERAL ATTRIBUTES	10	6	2
COST ATTRIBUTES	2	6	10

FIGURE 4.14

Morphological Table of Design Feature Ratings [2].

The chart in Figure 4.14 shows the attribute totals, which indicate the general achievement of each of the designs. It should be noted that cost attributes have been set as inversely proportional to the general attributes.

It can be seen that most designs offer a similar value of achievement; however, they should be treated with great care as some designs offer a reduced level of the particular requirement. For instance, design C has a very low level of function, ease of manufacture, and operational efficiency even though some elements, such as simplicity, are high. Any or all of these elements could eliminate this particular design from the selection.

Perhaps the more important parameters are those indicated by the design totals. They are the scores that indicate the best options. Scoring in such a fashion can be subjective and may lead to subconscious favoritism. It is suggested that of the better options, two or three designs are analyzed closely to ensure that important features are included in the final selection.

The chart indicates that design B, scoring 80 points, is the best, closely followed by design D at 70 points. On analyzing the individual scores, it can be seen that there is a low sustainability score for design B. If sustainability has been given a high priority, then clearly this design does not meet the requirements.

The next high-scoring design is design D, scoring 70 points. This design has a low sustainability score and may suffer from the same problem as design B, but this design also has the advantage of a low manufacturing cost, which may offset the low sustainability score. Thus, in this situation, the second-highest scoring design appears to be the optimum selection.

This exercise highlights the fact that design is always a compromise and the designer has to select the best value parameters to create an optimum design. The morphological chart is an excellent tool but needs to be operated carefully.

Larger and more complex designs may require more detailed attention and perhaps may have to be broken down into individual assemblies, each one studied in a similar manner.

4.3.5 Analysis

Analysis can take several forms and requires the application of both practical and theoretical knowledge. It will include elements such as the following:

- Consideration of manufacturing tolerances (accuracy)
- The shape of the component
- Ease of maintenance
- Ergonomics
- Strength considerations (e.g., stress analysis)
- Envelope size

Analysis often involves some form of numerical calculation. The function of analysis is to define the parameters and operation of the design. For instance, an analysis of ergonomics when applied to a machine tool will consider the position of the controls as well as the forces and directions relating to how the human form may operate such elements. These parameters are often defined numerically.

In the design of an automatic mechanical vehicle parking system, the envelope of a typical vehicle had to be defined before the overall 200-space car park could be designed.

Perhaps the most obvious analytical element of design is consideration of the strengths of the components, where stress analysis defines their strength and eventual size and shape. Analysis such as this is not limited to just strengths. Some components demand dynamic, thermal, and fluid flow analytical techniques to predict performance and ensure correct function.

Synthesis and analytical techniques are often performed at the same time. A designer may consider the shape of the components alongside analytical calculations to define its feasibility. Analysis, therefore, is essential to the completion of design as it complements the *artistic* aspects of design creation.

4.3.6 Iterative design procedure

In mathematics, an iterative procedure is one in which an approximate solution is initially guessed and then fed into an iterative formula revealing a much more accurate solution. The improved solution is treated using the same procedure to reveal an even better solution. The process is continued until a solution with the required accuracy is achieved.

A systematic iterative design follows a similar path and assumes that even the best design concepts may need to be modified, both during the development process and after the launch. The trigger for an iterative process is often new information, perhaps from the marketplace, new technology, new components, or the results of analysis.

Even the modified version of components, especially complex components, may need further improvement and development until the ideal solution is achieved.

An effective iteration process, shown later in Figure 4.15, ensures that successive iterations are less involved than the previous one. The iteration diagram illustrates the iterative design procedure in the form of a flowchart, the feedback loops indicating the return to the start of the iteration.

While the iterative process is extremely useful and necessary, the designer must realize that he cannot continually keep redesigning. The process has to end when it is judged that the goal has been reached. The danger is that the process can continue *ad infinitum*, incurring a great deal of time, effort, and cost. There must, therefore, be limitations as to the number of iterations and redesign.

The design of a bed lifting system resulted in many good ideas over which the team enthused. The lead designer reminded the rest of the team to store the ideas for later refinement after the base model had been converted into a product. If the team leader had allowed these modifications to be researched, excessive time would have been spent in refinement, resulting in a sophisticated but costly device. The main goal here was to create a basic "without-frills" product.

Ideally, redesign and reiteration must be limited to minor modifications and should be tackled at the earliest opportunity. It is easy for designers to be carried away; however, discipline and practice should be employed to keep the project moving toward the primary goal.

Good liaison between departments is a critical factor and the method of "redesign for cost" or "redesign for sustainability" should be adopted. These two concepts should be borne in mind from the very beginning of the project; however, in the later stages of the project prior to manufacture, this process should involve only minor adjustments, resulting, for example, in just switching to a different supplier for a bought-out item.

Redesign for function may be merely a simple alteration of a design sketch after tentative development tests have been conducted. For instance, this could mean changing the fixing location for a hydraulic ram.

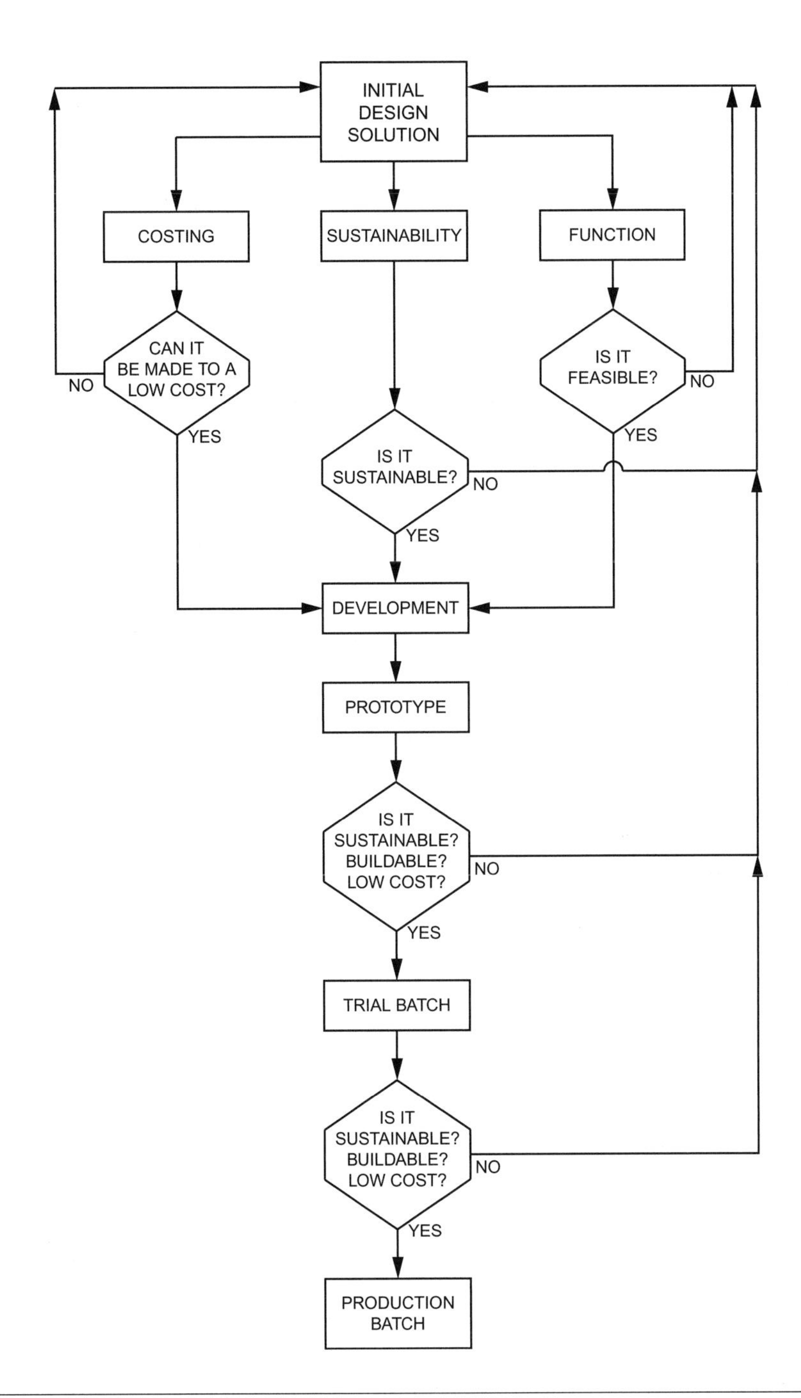

FIGURE 4.15

The Iterative Design Procedure [2].

After-sales redesign can be undertaken most effectively by considering feedback performance from a trial batch of production prototypes. Small trial batches are often supplied to a limited number of established customers for trials. The *Ford Bobcat* was a good example of this. Trial batches were sent out to trusted users who fed back useful field trial information. The *Ford Bobcat* eventually became the *Ford Fiesta*.

4.3.7 Design evolution

Design evolution takes place over many years. A product is launched in the market and immediately people try to improve it. There are hundreds of examples of design evolution. The bicycle started its life as a wooden contraption that was pushed along by the rider's feet. There were several evolutionary designs proposed such as the "Penny Farthing," but the formula of the modern bicycle emerged only when the pedal crank was connected to the drive axle by a roller chain. Modern evolution has kept the general format but improved the materials, brakes, suspension, and tires and reduced the overall weight.

The same evolutionary cycle can be cited for the motor vehicle, which started out as a horseless carriage fitted with a motor. The evolution of the vehicle has brought about improved suspension, tires, engines, and much more than can be mentioned here. Perhaps the main evolutionary successes for the motor car are improved speed and improved safety.

4.3.8 Specification (customer needs)

It is rare for a product to be designed without a purpose in mind; it usually requires an instruction to produce a solution. This instruction is really the "customer needs statement," which may be informal but quite often is likely to be set out as a specification of technical requirements. This may be supplied either by the customer or by the marketing department.

The British Standard issues recommendations for the preparation of specifications, contained in its document PD 6112. Using the British Standard, the specifications for the example of a simple electrical cable tidy may be written as follows:

1. Title

   A wall-mounted device for safely storing coiled electrical cable in a domestic environment.
2. History and Background Information

   It has been found necessary to introduce a device that can be wall mounted in a domestic environment, can safely hold up to 50 m of coiled electrical cable, and can capture the cable, holding it *in situ* until manually released. The device should be simple enough to be manufactured in tens of thousands at a low cost.
3. Scope of Specifications

   The design must incorporate a hook system with a retention device, thereby ensuring the cable will not spill out when unattended. The reason for this is that it can be expected that there may be some abuse whereby more than 50 m of cable are applied to the device.

   The device must be made from materials that are easily available and sustainable.
4. Definitions

   Cable Tidy: a hook device mounted on a wall that will enable the storage of 50 m of standard electrical cable.

Retention Device: some form of device or clip that will prevent the cable from spilling out.

5. Conditions of Use

   The device must be static when fixed to the wall and have the strength to hold over 50 m of standard electrical cable.

   There must be no environmental health hazard caused by any feature of the device during or after manufacture.

6. Characteristics

   The cable tidy should be lightweight and manufactured from readily available materials that are low cost and have a low-environmental impact. The device should be suitably finished so that it complies with the expectations of the customer when fitted within the home.

7. Reliability

   The device should have an intended life of between 5 and 10 years and should possess a safety factor for three times the intended load.

8. Servicing Features

   The device should be of a design such that it can be fitted in the domestic environment by someone with relatively low DIY expertise. Once in position, the device should need no further servicing.

4.3.9 Secondary needs

Successful design development focuses on primary needs such as function and cost. The designer is also tasked with considering the secondary needs while in the process of creating the primary needs. Primary needs ensure the design works, while secondary needs make the design comfortable. Sometimes the designer may give priority to the primary needs, leaving the secondary needs to be created afterward. Commonly, however, secondary needs are considered alongside primary needs. The following is a list of some of the typical secondary needs:

- Reliability
- Serviceability
- Ergonomics and anthropometrics
- Aesthetics
- Safety
- Economics

4.3.10 Design life cycle

A product generally goes through a life cycle, starting from the point of its initiation into the market until the need arises to redesign it in compliance with the latest market requirements. There are many reasons why a product may fall out of favor in the market. The product may become obsolete because of a change in technology or a change in methods or it simply falls out of fashion. Figure 4.16 shows a product's typical life cycle from market insertion to obsolescence.

The life cycle shown is fairly typical of many products but it fits the technology associated with still photography equipment (cameras) very closely. The whole life cycle of camera technology using light-sensitive film has taken perhaps 120 years. Originally, the picture was captured using silver halide

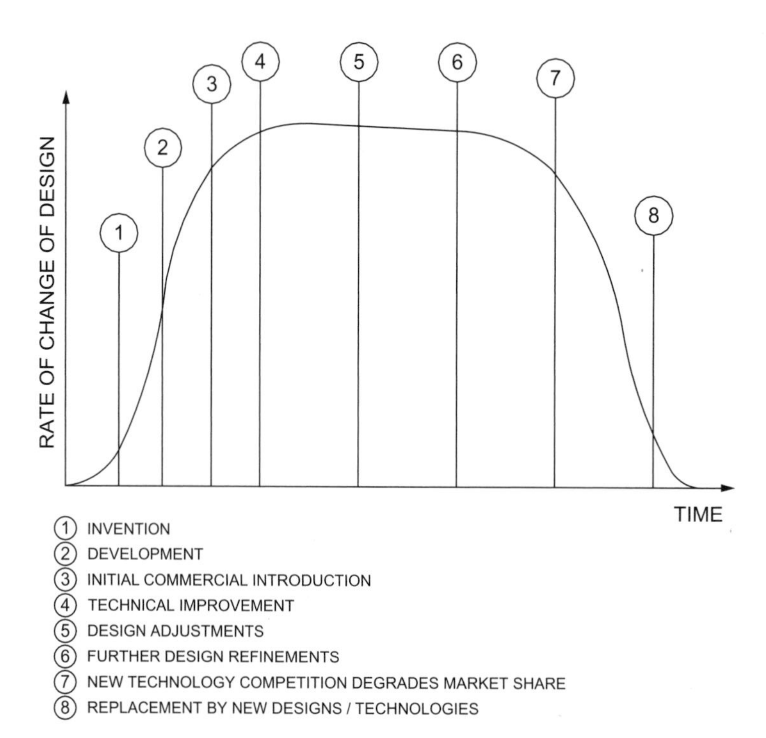

FIGURE 4.16

Typical Product Life Cycle [2].

spread in a film over a glass plate. This had to be "developed" so that the image could be seen. This took the product development to position 3 on the curve.

The next major technical development at position 4 was that of using celluloid film as the carrier for the silver halide. Improvements in camera technology, camera efficiency, lenses, etc., enhanced the technology until cameras using celluloid film reached their technological peak at position 6. At this position, digital technology became more popular, making photography much simpler for the masses.

Film camera technology thereafter began to lose its position in the marketplace to digital camera technology until at position 8, 95% of the cameras sold were digital cameras.

Improvements in technology and methods of picture capture made film camera technology obsolete.

4.3.11 Invention and lateral thinking

A new product sparked by pure inventive genius is extremely rare. Normally, an idea leading to a product is generated by some occurrence or need, which then initiates a design concept.

The concept of "necessity is the mother of invention" is very real, indeed. There are, however, many tools that stimulate the fostering of ideas and concepts once the need has been exposed. One such method is lateral thinking. There are many others, which will be covered later.

An inventor is often a freethinking individual who is not fettered by a discipline such as engineering. Thought processes are very different. The inventor will consider an idea and try and make it work. The engineer may consider the same idea and discard it after considering engineering options. It is difficult to say who is correct since both points of view have their merits. It is true, however, that inventors have greatly contributed to products that are now enjoyed worldwide.

The free-thinking inventor is usually capable of lateral thinking, often called "thinking outside the box." This usually involves considering options from many technologies rather than being limited by the technology surrounding one discipline.

4.4 DESIGN METHODS

4.4.1 The tools of creativity

A team of designers requires tools to boost creative thinking and to generate new and different ideas. Such tools also:

- Establish a creative environment
- Create a common vocabulary
- Provide new ways to understand all elements of the problem encouraging breakthrough solutions
- Incorporate the ideas and talents of the whole team
- Inject enthusiasm and energy into solving tough problems

There are several systemic tools available to teams of designers. They are discussed in the following sections.

4.4.2 Heuristic redefinition

How would one climb a mountain that has not yet been conquered? The difficulty is that there are no maps or natural pathways to the summit.

One might consider the best routes while viewing the mountain from a distance or perhaps by studying the mountain's terrain on a satellite image. This investigation would help to select the most convenient ascent to the summit. There are many different methods of ascending a mountain such as climbing on foot, parachuting from a great height, reaching the summit via helicopter, and so on. These are all *methods* of realizing the goal, but the first step is to define the goal before trying to reach it.

Often the seeds of a solution are buried in an efficient definition of the problem itself.

4.4.2.1 Example

When NASA first developed the space shuttle heat shield, they sent out a definition along the following: *"There is a requirement for a material which will withstand 15,000 degrees centigrade."*

There is no material on earth that can withstand this temperature. After several rejections, the definition of the problem was changed as follows:

There is a requirement for a system which can accommodate 15,000 degrees centigrade.

This new definition allowed design teams the freedom to develop a revolutionary heat shield that combined a heat dissipation system with specialist materials.

The *Heuristic Redefinition Tool* can often free designers from a fixed view, to propagate a free-thinking format.

4.4.2.2 Heuristic redefinition tool definition

Heuristic Redefinition is a method of looking at a system in which a problem exists and selecting an approach that will help to change and free the thought processes of the designer. Possible approaches are identified and ranked by applying criteria that are appropriate to the problem.

4.4.2.3 Use of the tool

In general, this tool can be used with any creative problem-solving process so that different perspectives can be gained as stepping stones for generating ideas. It reduces a system to the component parts and thus helps a team understand the individual elements of the design challenge in the context of the whole system.

It is especially valuable for the individual designer who is perhaps working alone and not supported by a creative team and is also helpful when a design challenge is not well defined.

4.4.2.4 Tool overview

In order to understand all the directions from which a problem can be approached, it is most advantageous to visualize the problem and all its elements in a pictorial or symbolic presentation. This could be a drawing, a flowchart, or an arrow diagram.

Each element within these symbolic representations must be analyzed with the question "How can we make sure that...?"

Each element now has the question attached to it and this represents a problem statement. The team then chooses the problem statement that has the most potential for successful problem solving.

The criteria needed to select the most promising problem statement might include, for instance, the following:

- The probability of success
- The amount of effort required
- Feasibility
- The lowest cost

The method of applying this approach is to split the problem into its component elements so that each element can be analyzed separately and then referred to in terms of the overall system. The generation of precise problem definitions in terms of statements is designed to create alternative ideas.

4.4.2.5 Applying the tool

Heuristic Redefinition forms a structured approach or algorithm and has five steps.

4.4.2.5.1 Application overview

1. *Define Goals and Opportunities*

   Objectives must be clearly stated. The goal or opportunity needs to be clearly stated in terms of the overall system, for instance,

- Is it a process specification that requires improvement?
- Is there a new function that requires integrating into the current system?
- Is there an undesirable element within the design that needs to be reduced or eliminated?

2. *Identify System Elements and Visualize Them as Parts of the System*

   Each step of the process or constituent element of the system should be visualized in the form of, for example, a process map, a functional map, a diagram, a flowchart, and parallel lanes.

3. *Label and Understand the Impact of Each Step or Element*

   Each process step or system element is analyzed in turn, asking the following questions:
   - How does this element support the system?
   - What is the role of this element?
   - What are the effects of this element, positive as well as negative?

   The answers to these questions should be expressed in fewer than five words; if more are needed, it could be an indication that step 2 is incomplete.

4. *Link Each Impact Back to Each Goal*

   Step 3 highlights the impact of each element of the system, but step 4 compares those impacts to the goals set in step 1. Step 4 questions should require the answers to the following:
   - How does the impact of this element support the goals?
   - How can the impact of this element be used to realize the opportunity?
   - How does the impact of this element contribute to the problem?

   The output is then phrased as a question that defines the new problem to solve.

5. *Organize and Consolidate Statements*

   Finally, all problem statements that are generated need to be organized. The team should particularly look for the following:
   - Recurrent themes that may point to a powerful opportunity
   - Interactions among steps or elements
   - Synergies or contradictions

   Patterns such as those listed here can be important steps on the way to solving a systemwide problem by tackling recurring anomalies at the level of individual elements.

4.4.2.5.2 Heuristic redefinition tool application overview

4.4.2.5.2.1 Step 1: Define goals and opportunities. The problem statement should be defined in positive rather than negative terms, which will tend to limit the team's thinking.

The statement should also be defined in terms of the overall goal.

- What is the end outcome?
- What has to be achieved?

The problem statement is important because it could lead the team in many different directions. A carefully written problem statement should clearly state the outcome.

This early stage in the process team should endeavor to look for directions rather than solutions.

Example: The team has been given a goal of preserving the energy of a cycle rider by improving the efficiency of the drive system, thus making available to the rider more energy at higher speeds for longer distances.

One team member suggests that the definition should be "How can we prevent the rider's energy being expended on moving the bicycle mechanism?"

This is a negative statement and restricts the team's approach.

A more positive definition would read thus:

How can we efficiently direct available rider energy to the road wheels of the bicycle?

The positive definition immediately opens up many more directions and opportunities. Clear objectives have also to be stated and for the bicycle, they are as follows:

- improving the energy efficiency of the wheels
- reducing the rider's energy expenditure on driving the mechanism
- increasing the speed of the bicycle with the same energy expenditure
- increasing travel distance with the same energy expenditure

4.4.2.5.2.2 ***Step 2: Identify system elements and visualize them as part of the system.*** The team should now describe the problem in the context of a complete system. This is often successfully achieved pictorially by means of, for example, a flip chart or previously prepared graphics. The team members are required to focus their minds on the system while allowing their creative mind to wander freely through solutions.

It is important that each major component of the system is included and illustrated so that the team members, when defining the components or parts of the system, may ask questions such as the following:

- What is it doing?
- Where does it happen?
- When does it happen to it?
- Why does it happen to it?
- How does it happen to it?
- To whom does it happen?
- Who caused it to happen?

The prepared graphic for the bicycle drive system can be seen in Figure 4.17.

4.4.2.5.2.3 ***Step 3: Label and understand the impact of each step or element.*** This step is often best achieved by creating a chart, as shown in Figure 4.18, that categorizes the responses according to each feature of the system. It is also useful at this stage to number the system elements for use later in the synthesis.

The visuals in Figure 4.17 and the composite chart shown in Figure 4.18 will help the team to consolidate and explore the results of the first and second phases. As with any design, the design team must ensure that they have captured all the key components and that they have thoroughly understood the problem and the system.

The issues now facing a team and which must be addressed are as follows:

- What is the relationship between the components of the system?
- What are the influences or relationships between the components?
- What are the rules or laws that govern the link between the components? These are often scientific or mathematical in nature.

FIGURE 4.17

Graphic Explaining the System Elements of the Bicycle Drive System. (a) Power Source, Cyclist's Legs. (b) Crank Sprocket and Pedal. (c) Chain Drive Element. (d) Rear Drive sprocket. (e) Tires. (f) Bicycle Frame.(For color version of this figure, the reader is referred to the online version of this chapter.)

The pictorial visualization of each system component in Figure 4.17 allows the team to see the element and where it is placed in the system. It then becomes easier to see the relationship between the components.

Restate the Goal. It is useful to have the goal in sight as, say, a banner, to keep reminding the team of the goal:

How can we efficiently direct available rider energy to the road wheels of the bicycle?

		How Does the Element Support the System?	What Is the Role of the Element?	What Is the Effect of the Element (+ve/–ve)
System Element	Description			
1	Cyclists Legs	Motivates the system by providing actuation energy	Provides motive power	Moves the crank
2	Crank Sprocket and Bearings	Imparts drive tension to the chain	Converts leg power into torque	Drives the chain
3	Drive Chain	Links drive sprocket to the driven sprockets	Transmits drive tension to the driven sprockets	Accepts drive sprocket force and passes it to the driven sprocket as a drive force
4	Rear Driven Sprockets and Bearings	Links the drive chain directly to the wheels	Converts drive tension to torque and links directly to the wheels	Transmits drive torque to the road/tire interface
5	Tires	Interface between driven wheel and road	Resilient member between rigid cycle components and road surface	Applies tractive force to the road
6	Cycle Frame	Supports the rider keeping all elements in position	Provides structure and strength for the system	Keeps system components in position

FIGURE 4.18

Chart Describing the Role of Each Cycle Element.

Each component should now be visited by turn. The team asks the question: What does this component do that affects the goal?

- Cyclist's legs: produce power
- Crank sprocket and pedals: transfer power to the drive chain
- Drive chain: transfers drive force to the rear sprocket
- Rear driven sprockets: transfer drive torque to the tires
- Tires: transfer tractive force to the road surface
- Cycle frame: Holds in position all the system components

A second question should now be asked:

What does each component do that affects the goal of directing available rider energy to the road wheels of the bicycle*?*

For example, what do the rider's legs do that affects the goal?

- Provide motive power
- Move the crank in a cyclic motion by pressing on the pedals

4.4.2.5.2.4 ***Step 4: State the links of the components to the goal.*** Using the data derived in step 3, the team must now examine the relationship between each component and the goal. Taking each component by turn, the question to be asked should take the following form:

How can we ensure that this component contributes to the final goal?

In particular, the question should commence thus:

How can we make sure that...

For instance, the question applied to Component 1 should read "How can we ensure that the cyclist's legs provide........?"

The answers to these questions become problem statements from which the best statement will be selected for each system component.

Example: This phase requires the graphic in Figure 4.7 to be available to the team as they work through each system component asking the question "How can we make sure that...?" This will generate problem statements for each of the system components, as shown in Figure 4.19.

4.4.2.5.2.5 ***Step 5: Construct a rating table Comparing problem statements with feasibility.*** The next phase is to practically examine each system element so that the design effort may be focused toward the most feasible aspect of the system in order to achieve the goal. A rating table should now be constructed with all the problem statements listed and compared on the basis of the rating criteria. Typical rating criteria are listed here, and a typical rating table is shown in Figure 4.20.

- The likelihood of reaching the goal
- Ease of implementation
- Expected impact on the goal

Care has to be taken when allocating scores as it can be very subjective. A more accurate score allocation can be achieved when the team members agree on a particular score. This brings all their collective experience and vision to the exercise, thus avoiding the possibility of a subconsciously biased view of an individual member.

4.4.2.5.2.6 ***Step 6: Select the most promising problem statements.*** The rating table shows that work applied to Component 5, the tire/road interface, seems to be the most promising system component on which to work. Component 2, Crank sprocket, and Component 3, drive chain, also score highly and should also be included for further consideration. Some caution should also be exercised as a high-scoring system component may not fit with the design strategy. For example, the design team may be specialists in chain drive systems, so it would be unproductive if they attempted to redesign the cycle frame. Furthermore, it is important to realize that this method may still be subject to the leanings of the design team even though the complexity of the process builds in natural safeguards.

System Element	System Component	Problem Statement: "How can we ensure that....?"
1	Cyclists Legs	Motivates the system by providing actuation energy
2	Crank Sprocket and Pedals	Converts leg power into torque to drive the chain
3	Drive Chain	Transmits drive tension to the driven sprockets
4	Rear Driven Sprockets and Bearings	Transmits drive torque to the road/tire interface
5	Tires	Applies tractive force to the road
6	Cycle Frame	Supports the rider keeping all elements in position

FIGURE 4.19

Matrix for Creating Problem Statements for Each System Element.

The team may naturally concentrate on the high-scoring components, but consideration should be given to the low-scoring elements, even if it is to eliminate them. Component 1, the cyclist's legs, represents the level of fitness and strength of the rider. At the beginning of the exercise, rider fitness was assumed to be at a particular level that could not be changed and was, therefore, outside the scope of influence of the exercise. Similarly, Component 6, the cycle frame, received a low score but it was also assumed to have a low impact on the end goal.

Scrutiny of the rating table in this way focuses the attention of the design team on the most promising development avenues.

4.5 CLASSIC BRAINSTORMING

Brainstorming has become one of the most popular forms of idea generation, probably because it is so easy and enjoyable to apply. Brainstorming is practically without rules, but it does have one rule that promotes free thinking and that is as follows:

System Element	System Component	Problem Statement	Likelihood of Reaching the Goal	Ease of Implementation	Expected Impact on the Goal	Total
1	Cyclists Legs	Motivates the system by providing actuation energy	1	1	2	4
2	Crank Sprocket and Pedals	Converts leg power into torque to drive the chain	3	2	3	8
3	Drive Chain	Transmits drive tension to the driven sprockets	3	2	3	8
4	Rear Driven Sprockets and Bearings	Transmits drive torque to the road/tire interface	2	1	2	5
5	Tires	Applies tractive force to the road	3	3	3	9
6	Cycle Frame	Supports the rider keeping all elements in position	1	1	1	3

Key:
High score = 3
Medium score = 2
Low score = 1

FIGURE 4.20

Rating Table Comparing Problem Statements Against Feasibility.

No matter how crazy an idea it must not be ridiculed.

Participants can, therefore, give free rein to their imagination. If an idea is completely impractical, it may open up another avenue of thought for someone else in the group. Brainstorming brings several major advantages to the solution generation process. These are listed as follows:

- Pools the knowledge of several people to solve a problem
- Helps to eliminate mental blocks
- Restrains people from being critical

- Allows team members the freedom to think without constraints
- Expands effective participation for the team and other externals
- Avoids discussions that are not focused on the task at hand

4.5.1 Definition

This tool allows team members to pool their knowledge and creativity in an open and uncritical environment. Brainstorming encourages "open" thinking by

- creating a host of solutions that the team members can build upon
- involving all the team members
- giving equal value to every idea
- allowing each person to be creative while focusing on the team's common purpose

4.5.2 Step 1: Identify the appropriate team

For a brainstorming session to reach its full potential, the team leader should prepare for the session in advance as follows:

- Clarify the topic by using the Heuristic Redefinition or some other tool that simplifies the best route toward the supposed solution.
- Assemble a team whose expertise is based on the problem to be solved. There should be team members who have expertise in the most likely disciplines required, but the team should also include members from a range of diverse backgrounds. These members will often provide "out of the box" ideas.
- Try to map out the various directions so that he can coach the participants if some key avenues have been missed.

4.5.2.1 Bed lifting device example

Overview: A bed lifting device is to be placed in a domestic environment. The bed should lift to the ceiling during the day converting a bedroom into a living room and lower to the floor during the evening, thus converting the living room into a bedroom. The method of actuating (lifting/lowering) the bed is required.

Requirements

- silent
- safe
- easy to operate
- rise when required, into a ceiling cavity
- lower to the floor when required

The design solution required should be defined as follows: "Determine a means of elevating to the ceiling and lowering to the floor a fully made a bed of 300 kg. And actuation, application and guidance system should be incorporated."

It is the task of the team leader to select a mix of employees, including technical experts, manufacturing personnel, marketing personnel, and others who may possess insight into the project and who may not be constrained by professional discipline.

4.5.3 Step 2: Convene the team and clarify the topic and ground rules

When the team meets for the first time, the team leader should exercise a loose control in briefing the team regarding their conduct and goal in terms of the following:

- Identifying and defining the focus topic
- Clarifying the meaning of descriptive words
- Reviewing the ground rules for participating in a brainstorming session. The ground rules are as follows:
    1. *Ground rule*: Do not criticize or judge the quality of any idea. The purpose of the session is to generate rather than evaluate ideas. This posture ensures a good flow of ideas and freedom of thought. It also prevents long-winded discussions.
    2. *Ground rule*: Combine the proposed solutions of team members to create new solutions.
    3. *Ground rule*: Allow your imagination to break free from constraints. What may seem a crazy solution may set another team member along a different path toward a winning solution.
    4. *Ground rule*: Give the team freedom to produce as many ideas as possible. The team should focus on the quantity of ideas rather than the quality. Quick snappy ideas without the encumbrance of explanations stimulate spontaneous thoughts prompt more unusual solutions from team members.

The brainstorming process is intended to be a fun and constructive experience for the whole team. During the process, each individual tem member should do the following:

- Engage himself/herself without reservation
- Not judge suggested solutions
- Look for positive elements in solutions suggested by team mates
- Be courageous
- Be carried away spontaneously
- Draw/sketch proposed solutions whenever possible

4.5.4 Step 3: Generate ideas

A typical brainstorming session should commence with several solutions being suggested. The intensity should build to a peak and then reduce to a pause in the solution flow. This is the first "dead point" where the team members need time to think.

After a short period of time, the flow of solutions should "self-start." This does not always happen, in which case the team leader may be required to "jump start" the session using several techniques. One method would be to review the list of solutions and to request the group to expand on one or more of the interesting ideas.

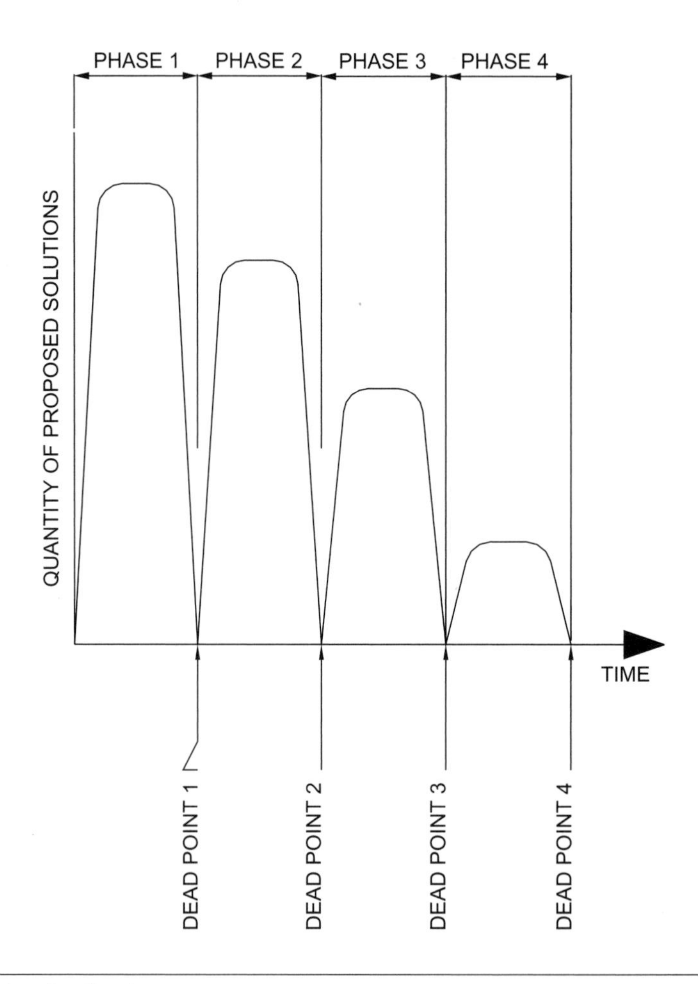

FIGURE 4.21

Phases of a Brainstorming Session.

The second phase usually reveals a small number of ideas. Since most of the more obvious solutions have been put forward in the first phase, the second phase reveals the more creative solutions. The second phase will also end in a dead point.

Sometimes it is useful to extend the session to a third and perhaps a fourth phase, each phase gradually progressing toward a more useful conceptual design. Figure 4.21 illustrates the phases of a brainstorming session.

4.5.4.1 Example: bed lifting device

The team leader sets the scene by explaining the ground rules and the requirements of the device as follows:

Overview: A bed lifting device is to be fitted in a domestic environment. Methods of lifting and lowering the bed are required.

Requirements: The bed lifting system must

- Lift into the ceiling cavity, under control
- Lower to the floor, under control
- Be easy to operate by one person
- Be safe to use
- Be low cost

The problem statement is "It is required that a means of elevating to the ceiling and lowering to the floor a fully made up bed of approximately 300 kg. The solution also requires a guidance system."

The definition and requirements have been deliberately left vague (open-ended) to avoid directing the team in a particular direction.

During the three sessions, the team generated the following solutions:

4.5.4.2 Phase 1-generated solutions

1. Apply lift cables from all four corners of the bed
2. Lift using a hydraulic ram from underneath the bed
3. Apply a gas bag underneath the bed
4. Apply a tractor beam (as used in Star Trek)
5. Apply screw jacks at each corner
6. For guidance, use rollers on a track
7. For guidance, use slides fixed to the wall
8. Use a counterbalance system enabling hand lifting
9. Use a lever mechanism to provide lift
10. Apply an electrically powered drive

Phase 1-generated solutions are usually adventurous and risky but, as can be seen here, many are impractical.

4.5.4.3 Phase 2-generated solutions

1. Treat the bed as a cantilever beam with the cantilever base at the wall
2. Apply rollers in a vertical channel at the base of the cantilever
3. Apply pneumatic rams to lift and lower the bed
4. Apply electric linear actuators
5. Apply a cable lift with a wind-up pulley-and-motor system

Phase 2-generated solutions are much more practical and begin to move away from conceptual solutions to more practical and achievable solutions.

4.5.4.4 Phase 3-generated solutions

1. Apply a roller chain lift with sprockets
2. Create the same lift from both sides of the bed using a pulley at each end of a shaft. The shaft is keyed to the pulleys so that they rotate at the same pace
3. Actuate the system using an electric motor with reduction box
4. Wind up the cable using a helix-grooved wind-up pulley
5. Reduce motor speed to an appropriate pace using a worm-wheel reduction box.

Phase 1-generated solutions show the list of solutions that emerged during the first phase of the brainstorming session. There were a large number of solutions contributed, several of which were not practical, but as the session progressed, the solutions became more usable.

In order to kick-start the next phase in a practical direction, the team leader asked the group to expand on the roller track system (from solution 6). Phase 2 solutions show that the team's response was more practical and usable.

At the second dead point, the solutions became more technical as the team began to consider how the roller-and-channel system could be applied. The solutions in this phase also included a guidance system.

In order to start off the team contributions, the team leader then asked the team to expand on and incorporate solutions 1, 4, and 5. Phase 3 solutions show that the team's response was very practical and came very close to the final solution.

Solutions generated during the third phase tended toward practical solutions, using known technology. The team leader decided that the appropriate direction had been isolated and now required specific design work.

Before the design work commenced, the solution set was reviewed by a nonattendee who dispassionately considered the solutions to

- Check the validity of the design route
- Ensure avenues had not been missed

4.5.5 Step 4: Clarify ideas and conclude the session

The list of brainstorming solutions should now be reviewed to check the design direction. This review should not be done during the session but later, by someone who was not at the session. The review should also clarify the next actions and the method of reporting the review to the team. This could be formal or informal, but it is necessary to focus the team on the next design phase.

The review for the bed-lift example made the following suggestions for the next design phase:

- Cantilever the bed from a structure incorporating rollers and channels
- Apply an electric motor/reduction drive system
- Develop a cable/pulley lift system alongside a chain/sprocket lift system

Efficiencies and costs of the chain lift and cable lift would then be compared.

4.6 BRAIN WRITING 6-3-5

The previous method of generating ideas, brainstorming, is a highly verbal process allowing a group of members to share their ideas. Brain writing also allows sharing of ideas, but the team members have to write their solutions down on paper. The most popular version of this technique is *Brain writing 6-3-5*: this has six people in a group, three ideas for each round, and 5 min of idea generation for each round. It has distinct advantages in situations where there may be dominant group members or, conversely, members who have little confidence in a group.

There are additional advantages to brain writing, which at first may not seem obvious. They are as follows:

- Participation number in brain writing is unlimited. You need to just add another group of six
- Participants can think in more depth without distraction of discussion
- There is no criticism of ideas because there is no discussion
- No one person can dominate the discussion
- Lower-ranking participants may share their ideas more freely
- Shy people may feel more encouraged to participate

The disadvantages, however, are as follows:

- There is no spontaneity
- Crazy ideas are not immediately usable to create alternative trains of thought
- There is no team leader to guide the discussion
- Clarification of solutions is not immediate

4.7 IMAGINARY BRAINSTORMING

This is a variation on Classic Brainstorming, where the introduction of a *hoax e*lement of the solution requirements tends to shift a person's thinking in a different direction.

The *hoax* involves replacing one element of a problem by a completely new element. The *hoax* change may be any aspect of the session. Even the wearing of underpants on the outside or pretending to snowboard in the office may be the change required.

The hoax changes the problem to something strange though it is still strongly related to the original problem. The effect on a group of members when they are solving the problem out of context is that it directs thoughts in new directions where old perceptions are put aside.

4.7.1 Tool overview

The start of an imaginary brainstorming process usually involves a classic brainstorming session where initial possible solutions are generated, listed, and clarified to the team members. The next step is to change one of the elements so that new ideas can be generated.

Once these essential problem elements have been listed, one of the elements may be changed to generate new ideas.

4.7.2 Example: mini cement mixer

The design problem given to a group of design students was to "design a mini cement mixer to mix 10 kg of aggregate."

Key elements were defined as follows:

- Mixing a loose dry substance of 10 kg
- Mini size cement mixer

The team were concentrating on the word "Cement," which led them down certain design paths. In order to get the team to consider alternatives, the word "cement" was changed to "Animal Feed." This changed the team's perspective and allowed them to create a nonpowered drum type, pull-along mixing device.

4.8 WORD-PICTURE ASSOCIATIONS AND ANALOGIES

An expert in a particular field tends to apply techniques he has applied previously. It is safe but perhaps unimaginative. When a new person enters the team he/she will have a different perspective simply because he/she is not "hide bound" by current or past methods. New technologies or new materials may be introduced that the "old" team have not considered.

A forward-thinking design team may realize that a fresh injection of ideas is required. Without the introduction of a "new" person, the same effect can be achieved by using *associations*.

Associations are really stirrings of memory that make mental connections to another place or time. A particular song may trigger the memory of a girlfriend/boyfriend or the smell of freshly baked bread may stir memories of a visit to a Paris boulangerie.

Analogies are a more direct comparison of a system or device.

Perhaps the best known analogy is that observed by the Swiss inventor George De Mestral who investigated how burdock burs stuck to his trousers after a country walk. Investigation using a microscope showed that each bur possessed hundreds of "hooks" that caught on to anything with a loop, such as clothing or animal fur. These direct analogies led De Mestral to develop a new type of fastener, Velcro. A microscopic image of Velcro can be seen in Plate 4.1.

Another example is provided by the designer RJ Mitchell, who designed the Spitfire and the Schneider Trophy aircraft. In the 1920s and early 1930s, they were cutting-edge float plane designs with a single wing, when most aircraft at the time were biplanes. Mitchell analyzed the flight and structure of birds' wings in order to formulate the streamlined nature of his aircraft and the design of the wings. The Supermarine S6, winner of the 1929 Trophy, can be seen in Plate 4.2.

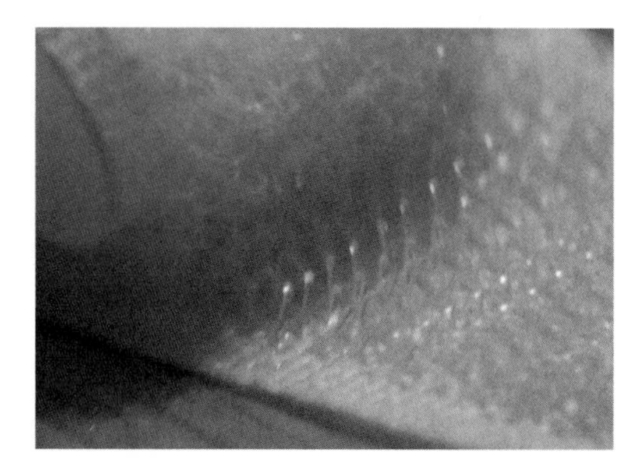

PLATE 4.1

Microscopic View of "Velcro" [3].

PLATE 4.2

Supermarine S6, Schneider Trophy Winner [3].

PLATE 4.3

Supermarine Spitfire Designed by RJ Mitchell [3]. (For color version of this figure, the reader is referred to the online version of this chapter.)

RJ Mitchell famously went on to design the Supermarine Spitfire, the WWII British fighter plane. Plate 4.3 shows a Spitfire. Design similarities between the Spitfire and the Supermarine S6 can be clearly seen.

4.9 METHODS OF GENERATING ASSOCIATIONS AND ANALOGIES

4.9.1 Random words

There are several methods of generating associations and analogies that can influence a design. One method is to introduce random words into a problem. The association and probable analogies generated allow a team to examine the unique characteristics that the word might provide to the problem. During the process, the team should be asking questions such as:

- What associations are brought to mind?
- What are the connections that the random word generates?
- What comparisons does the random word stimulate?

4.9.2 Pictures

The human condition uses visual aids almost 80% of the time. It makes sense, therefore, to use visual aids (pictures/diagrams) to stimulate associations and other analogies. The team members may review a random picture (magazine article, technical brochure, etc.) in order to stimulate associations that might apply to the design problem.

In this way, the use of visuals creates thought processes that allow individuals to replace insufficient or incomplete information in their imaginations with an image.

For example, during the design of a sports car, the specification requires wide tires with alloy wheels, but the memory carries only a skeletal outline. The insertion of a picture of wide tires with alloy wheels serves to indicate a more complete design than that in the imagination of the team members.

4.9.3 Biotechniques (biomechanics)

This is the process where a naturally occurring device or structure is analyzed to understand its function. George De Mestral did exactly this when inventing "Velcro." The modern aircraft wing is basically a cantilever beam. Cantilevers abound in nature in the form of tree branches. Naturally formed arches, perhaps at the entrance to caves, are naturally occurring lintels, which are used extensively in buildings in the form of support beams above doorways or windows.

This technique makes use of the enormous number of solutions nature has developed. The analysis of natural phenomena is always an excellent beginning for idea generation.

4.9.3.1 Biotechnique tools: overview

The use of an analogy or an association is a fairly high-level method of solution generation and requires an initial input of possible solutions derived from Classic Brainstorming or Brain Writing.

The team leader then has a choice as to whether to use random words, pictures, or biotechniques/ biomechanics as the primary feed to generate associations or analogies. Depending on the problem, a single element or a combination of elements may be used. The team leader guides the team through the generation of associations or analogies. These solutions are then linked to the solving of the problem.

4.9.3.2 Example: bed lifting device

A team has been assembled to explore solutions to the problem: "It is required that a means of elevating to the ceiling and lowering to the floor a fully made up bed of approximately 300 kg. The solution also requires a guidance system."

During the initial phase 1 brainstorming session, the team generated the following solutions:

Phase 1-Generated Solutions

6. Apply lift cables from all four corners of the bed
7. Lift using a hydraulic ram from underneath the bed

8. Apply a gas bag underneath the bed
9. Apply a tractor beam (as used in Star Trek)
10. Apply screw jacks at each corner
11. For guidance, use rollers on a track
12. For guidance, use slides to the wall
13. Have a counterbalance system for hand lifting
14. Use a lever mechanism to provide lift
15. Apply a four-bar link system

It is now necessary for the team to define the technique to use: word association, picture association, or biotechniques/biomechanics.

4.9.3.3 Word associations

The team leader would need to plan ahead so that he would have collections of words that are independent from the problem. The words could be derived from several sources, such as a thesaurus, dictionary, newspaper, magazine, or a mail order catalog.

The choice of words should be random, but general themes could be applied so that for technical problems, such as the design of a vehicle power plant, the random words could relate generally to, say, aircraft or other similar material objects.

4.9.3.4 Picture associations

A randomly chosen set of pictures can be shown to the team, perhaps projected onto a screen. The picture should portray a variety of situations and activities that are unrelated to the problem. In the case of a bed lifting device, pictures could be shown of elevated buildings or escalators, or perhaps scissor lifts for cargo.

Pictures can be sourced from almost anywhere. Popular sources are as follows: magazines, books, travel magazines, industry magazines, and the Internet. It is often difficult to judge the response of the team, but the team leader should endeavor to select pictures that may lead the team to give an upbeat and positive response.

4.9.3.5 Biotechniques (biomechanics)

The application of biotechniques is perhaps the most difficult as the team will need to familiarize itself with models, books, specimens, and other research materials. In order to do this and to carry out appropriate research, the team would need to first analyze the actual design problem. The research will remove the randomness of the activity and help in a positive fashion as the team will become better informed.

4.9.3.6 Stimulating the team

Whichever element is selected (word, picture, or biomechanics/biotechnique), the team leader should stimulate the team by asking appropriate questions while it is considering one of the elements. Difficult questions with which to stimulate the team can be seen in Figure 4.22.

	Team Stimulus Questions
Technique	**Stimulus Questions**
Random words	What images does it generate? What is the meaning of the word? Connections? What is the structure? How does it function? Shape? Attributes? How is it used? What are its effects?
Random pictures	All the questions above—plus: What is happening in the picture? Who/what is happening—how? where? when? why?
Biomechanics	Motivations? How does it function? Does it breathe? What other natural functions? Does it have problems? Does it have unique features?

FIGURE 4.22

Stimulus Questions for Analogy and Association Generation Techniques.

For the clockwork mechanism shown in plate 4.4, the team worked through the stimulus questions outlined in Figure 4.22 and derived 23 associations, some of them, however, were overlapping associations. The team leader then sifted through the results, eventually creating three categories as follows:

Attributes

- Controlled mechanism
- Controlled spring energy
- Precision
- Wind-up and release system
- Instrumentation
- Movement of analog fingers for humans to read the time

Method of Operation

- Small gears
- Shafts rotate on bearings
- Uses spring energy

PLATE 4.4

Picture of a Clockwork Mechanism [3]. (For color version of this figure, the reader is referred to the online version of this chapter.)

Uses

- Telling the time
- Enabling the calculation of longitude
- Timing of races
- Moving with precision

4.9.3.7 Apply the generated ideas to the problem

Guided by the team leader, the team now considered each idea by turn, creating associations and analogies linked to the bed lifting device. The team leader then created a chart that listed the new ideas next to the associative trigger. These associations and analogies can be seen in Figure 4.21.

After considering the first trigger the team moves on to the next trigger creating further associations and analogies. The process continues until a sufficient number of ideas have been generated (Figure 4.23).

4.10 TILMAG

The methods already discussed are extremely prolific in their idea generation; however, for really tough and stubborn problems for which there is a scarcity of ideas, the structured TILMAG approach is most helpful.

TILMAG is a solution generation technique created by Dr. Helmut Schlicksupp. The word TILMAG is an acronym for several German words, which loosely translated means “the transformation of ideal solution elements in an association matrix.”

Associations	Application to the design problem
Controlled mechanism	Computer control Control mechanism Height and bottom stops
Precise	Accurate mechanism Machined tracks Machined rollers
Wind up and let go	Powered lift, power-controlled fall
Intrumentation	PLC controller
Small gears	Reduction box fitted to motor, worm/wheel/gears/pulleys/belts
Shafts rotate on bearings	Engineering plastic bearings, slow speed, zero maintenance
Uses spring energy	Electric motor power
Precise movement	Controlled from PLC

FIGURE 4.23

Associations and Analogies Related to the Bed Lifting Device.

4.10.1 Technique operation

The general operation of this technique is with a matrix that allows associations to be made; pairs of possible solutions are combined so that they can be explored, often providing unusual breakthrough solutions.

The team first generates several desirable solutions using a basic technique, such as brainstorming. The solution is then inserted into the matrix for evaluation. A step-by-step procedure is provided here.

Step 1. Generate solutions using a basic technique such as brainstorming
Step 2. Define ideal solutions
Step 3. Construct a TILMAG Matrix incorporating the ideal solutions
Step 4. Consider each empty cell by turn, making associations by pairing ideal solutions and the associations in the matrix boxes
Step 5. Combine the new associations derived in step 4 with the ideal solutions defined in step 2

4.10.2 TILMAG matrix

The basic TILMAG Matrix can be seen in Figure 4.24. It shows how seemingly different solutions can be matched with other solutions. In practice, this combination technique often leads to new and useful solutions.

4.10.3 Example: bed lifting device

Consider the bed lifting device, which was first considered as an example of brainstorming. The problem definition is as follows:

> *It is required that a means of elevating to the ceiling and lowering to the floor a fully made up bed of approximately 300 kg. The solution also requires a guidance system.*

4.10.4 TILMAG phase 1

The first phase in any complex solution generation method is to revert to a basic solution generation technique such as brainstorming. The team-generated initial solutions from the "Phase 1" brainstorming session are listed below:

Phase 1-Generated Solutions for the Bed Lifting Device

1. Apply lift cables from all four corners of the bed
2. Lift using a hydraulic ram from underneath the bed
3. Apply a gas bag underneath the bed

	Ideal Solution 1	Ideal Solution 2	Ideal Solution 3	Ideal Solution 4	Ideal Solution 5
Ideal Solution 6					
Ideal Solution 5					
Ideal Solution 4					
Ideal Solution 3					
Ideal Solution 2					

FIGURE 4.24

The Basic TILMAG Matrix.

4. Apply a tractor beam (as used in Star Trek)
5. Apply screw jacks at each corner
6. For guidance, use rollers on a track
7. For guidance, use slides fixed to the wall
8. Use a counterbalance system enabling hand lifting
9. Use a lever mechanism to provide lift
10. Apply an electrically powered drive

4.10.5 TILMAG phase 2

Sometimes these generated solutions may require some refinement. The most promising "ideal solutions" were selected by the team for insertion into the TILMAG Matrix. They are listed here:

1. Apply lift cables from all four corners of the bed
5. Apply screw jacks at each corner
6. For guidance, use rollers on a track
8. Use a counterbalance system enabling hand lifting
9. Use a lever mechanism to provide lift
10. Apply an electrically powered drive

4.10.6 TILMAG phase 3

After having isolated the most promising "ideal solutions," insert them into the TILMAG Matrix as shown in Figure 4.25.

4.10.7 TILMAG phase 4

Consider each empty cell by turn, making associations by pairing ideal solutions and the associations in the matrix boxes. As can be seen in Figure 4.25, the ideal solutions have been combined in such a way as to create new solutions.

4.10.8 TILMAG phase 5

Combine the new associations derived in step 4 with the ideal solutions defined in step 2.

The team can now use its experience and expertise to consider the solution displayed in each box. The scrutiny of each solution may involve elements such as

- cost
- feasibility
- fitness for purpose
- ease of manufacture
- noise
- maintenance
- power source

	IS 1 Lift Cables	IS 2 Screw Jacks	IS 3 Rollers in a Track	IS 4 Counter balance system	IS 5 Lever Mechanism
IS 6 Electric Drive	Motor-driven pulleys	Direct drive-through gearbox	Individually powered rollers	Linear actuator	Linear actuator
IS 5 Lever Mechanism	Cables attached to a lever	Screw jacks operate lever mechanism	Lever mechanism actuates bed guided by rollers	Counterbalance actuated by lever	
IS 4 Counter balance System	Classic counter balance using cables and weights	Counterbalance system reduces size of screw jacks	Counterbalance actuates bed guided by rollers		
IS 3 Rollers in a Track	Lift cables actuate bed lift using rollers as guidance	Screw jacks actuate bed lift using guidance rollers			
IS 2 Screw Jacks	Screw jacks provide actuation for cable lift				

FIGURE 4.25

TILMAG Matrix Applied to the Bed Lifting Device.

The team may also consider combining boxes. In practice, the bed-lift device employed a combination of the shaded boxes in Figure 4.25. The system employed motor-driven pulleys, which wound lift cables, thus actuating the lift of the bed using rollers running on a track for guidance. This was the most cost-effective and practical method.

4.11 THE MORPHOLOGICAL BOX

The application of morphology to a problem is essentially trying to change current parameters into something else. The morphological box is useful when considering whole systems rather than individual elements. It is important, therefore, that the team or, indeed, an individual designer needs to understand the whole system and its fundamental problem structure.

The first stage of the process is to list all the parameters that can be included or might affect the outcome of the design. This could be motor type, guidance system, actuation method, etc. For each parameter, multiple options are defined and listed within the morphological box. Once the tool has been created, it is a fairly easy exercise to combine the various options from other parameters. The result of the exercise is a larger number of alternative solutions.

4.11.1 Tool overview

The first step in the process is to identify and define the characteristics that are essential to a solution.

In a technical example, a wheelbarrow could be the focus of attention; the parameters would be as follows:

- Can carry a weight of 50 kg
- Can carry loads of any shape
- Can be operated by one person
- Can be operated over rough terrain
- Can be used outdoors

In setting up the morphological box as shown in Figure 4.26, it can be seen that by combining parameters by joining options with lines, several different and unique solutions can quickly be shown. A more complex morphological box will naturally lead to many more options.

This is an excellent method to view the whole system and can be used by a team or by an individual designer.

Perhaps the most difficult time for a designer is right at the beginning of the project when he/she is faced with a blank sheet of paper. In devising the system, the designer will, in his mind's eye, go over

Parameters	Options					
Transport 50 kg	Sack	Rucksack	Drum	Tray	Bucket	Sled
Any Shape of Load	Ropes	Elastic hooks	Bucket	Large bag	Box	Tray
One Person Operation	Sled	Rollers	Wheels	Lift it	Lever	Light in weight
Operation over Rough Terrain	Sled	Rails	Wheels	Tracks	Motor	
Used Outdoors	Waterproof	Washable	Robust			

FIGURE 4.26

Morphological Box for Wheelbarrow Alternatives [2].(For color version of this figure, the reader is referred to the online version of this chapter.)

Morphological Box for the Bed Lifting Device						
Parameters			**Options**			
Actuator	Electric Motor	Pneumatics	Hydraulics	Manual	Air Motor	Manual
Mechanical Control	Slides	Rollers	Linear Bearings	Mechanism	Constrained	Free
Actuator Control	PLC	Manual	Pneumatic	Mechanical	Semiautomatic	Fully Automatic
Lift Method	Cables	Screw Jacks	Pneumatic Rams	Hydraulic Rams	Air Bag	Counterbalance Mechanism
Lift Position	From Each Corner	Centrally Underneath	Centrally from the Top	Bed Head	Bed Foot	Bed Side
Structure	Cantilever	Beam Structure	Steel	Wood	Plastic	

FIGURE 4.27

Morphological Box for the Bed Lifting Device [2]. (For color version of this figure, the reader is referred to the online version of this chapter.)

the various parameters and review the options for each parameter. The morphological box represents this process as a visual aid, which can be added to and used many times for other similar projects.

Figure 4.27 shows the morphological box devised for the bed lifting device. The left-hand column shows all the important parameters such as the actuation, the control systems, and the lift method. The columns to the right represent the options available for each parameter. For instance, the actuator for the system could be an electric motor, pneumatics, or hydraulics. It is now a fairly simple task for the designer to select the options and merely join them with straight lines. This method can quickly simplify the process of devising a whole system.

The morphological box shown in Figure 4.27 offers two alternatives.

The first method suggests that the bed lifting operation can be powered by an electric motor driven lifting rollers controlled by a Programmable Logic Controller (PLC). The control is deemed to be semi-automatic so that human input can at least start it on the lift cycle or descent cycle. The structure is made from steel and takes the form of a cantilever secured at the bed head end where lift cables are able to raise and lower the bed.

The second method is powered by pneumatics and is constrained to move using pneumatic brands with pneumatic control. The actuation is from the bed head because the structure is based on a cantilever manufactured from steel.

4.12 DESIGN AND PLANNING METHODS

The practice of design is often viewed by nonpractitioners as an open-ended and largely artistic discipline. In fact, engineering design combines a creative aptitude with practical discipline and is often time-dependent so that designers are often working round the clock to meet tight deadlines. The achievement of good designs and their timely conclusion to projects is usually the result of good organization and scheduling.

To ensure that a design project is completed on time, the design manager must prepare a suitable plan, which must take account of such things as the creative process, team aptitude and experience, component selection, materials selection, detail design, and checking.

Two common techniques used in the project planning stage are

- Gantt charts
- Network planning charts (network analysis)

4.12.1 The Gantt chart (activity sequence chart)

This was the first formal method used to establish the start and finishing times for various phases of a product. It was devised by Henry L Gantt at the turn of the twentieth century.

A typical though fairly simplistic Gantt chart can be seen in Figure 4.28. Generally, the activities are displayed in a column on the left-hand side while a time scale is placed horizontally along the top of the chart. The activity column in Figure 4.28 shows a typical design process, which can be fairly detailed. The time scale is shown in weeks though any convenient time scale can be used depending on the project.

The strength of the Gantt chart lies in its simplicity. It can be understood at a glance without a great deal of training. Figure 4.28 shows the planned time allocated to each duty as a blue bar. As the project progresses, the actual time spent on each activity can be planned. This is shown as a red bar. This kind of chart is extremely useful for the project manager, because he can see exactly where time overruns and bottlenecks are occurring.

It should also be noted that some activities can be carried on simultaneously with other activities. For instance, the generation of solution possibilities can occur even while research into the brief is ongoing. These kinds of multiple activities are quite common in most projects.

The Gantt chart is a very useful project planning tool and can be as simplistic as that shown in Figure 4.28 or hold much more detail such as the example shown in Figure 4.29.

There are many excellent software packages that allow Gantt Charts to be easily contrived. Some offer only Gantt charts while others offer other planning tools such as network analysis and activity listings. The Gantt Chart shown in Figure 4.29 has been created in *Microsoft Project*.

4.12.2 Network planning (critical path analysis)

The Gantt Chart covers some useful aspects of project planning, but it does miss some vital project planning elements. The Network Planning technique lends itself to more complex activities and is often used on large projects as a single planning tool. Its main advantage is that it gives planners the opportunity to plan a "critical path" that allows for longer lead time items to be integrated into the program.

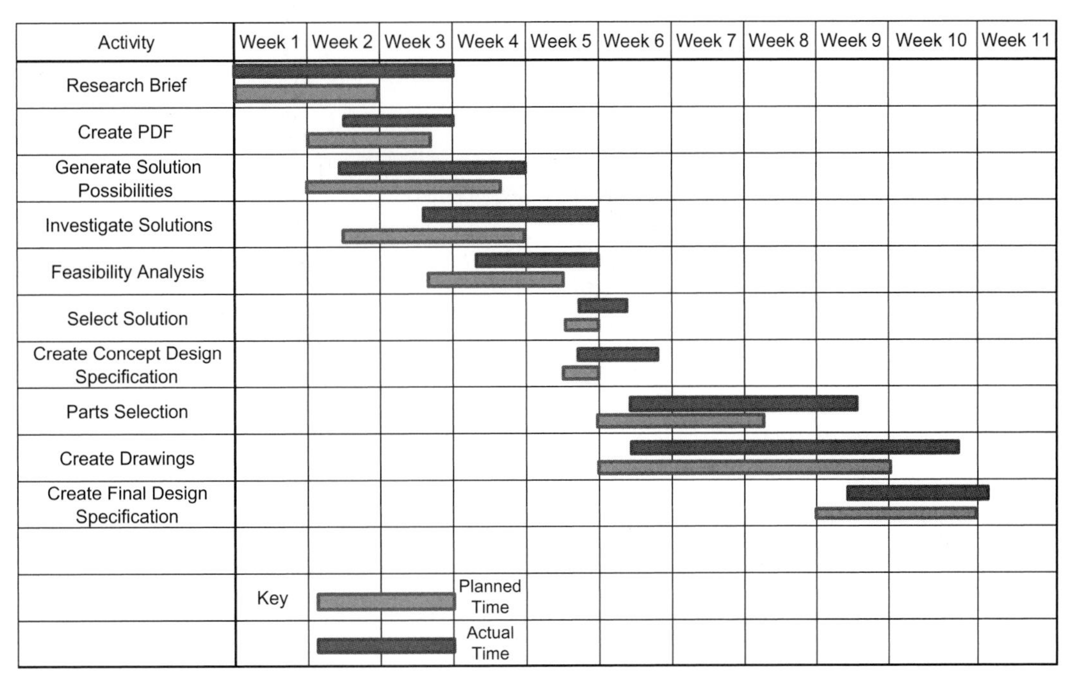

FIGURE 4.28

A Typical Gantt Chart for a Design Activity. (For interpretation of the references to color in this figure legend, the reader is referred to the online version of this chapter.)

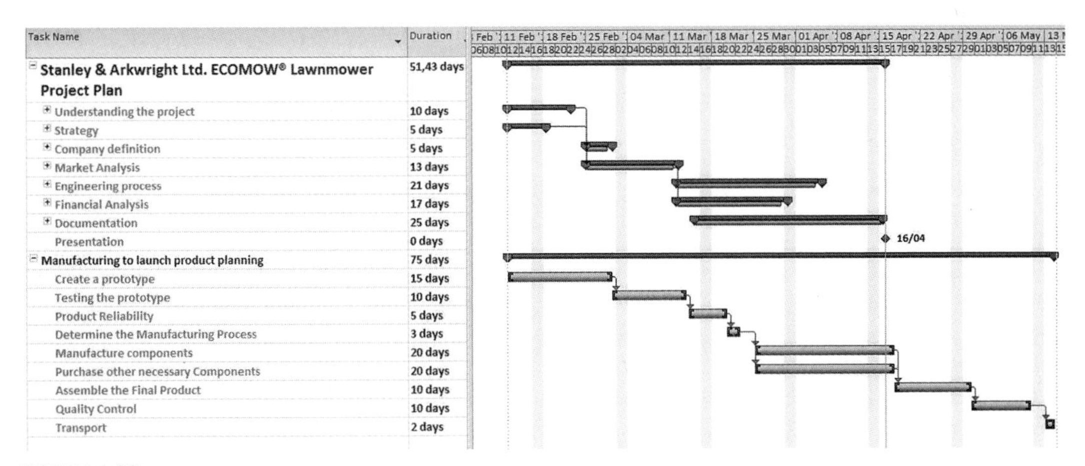

FIGURE 4.29

Typical *Microsoft Project* Gantt Chart [4]. (For color version of this figure, the reader is referred to the online version of this chapter.)

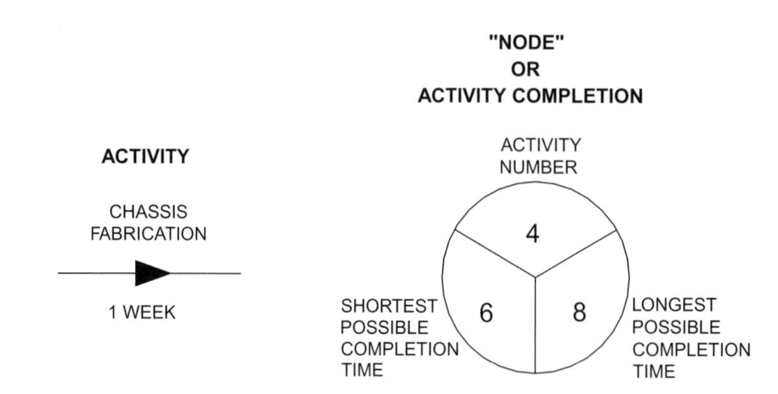

FIGURE 4.30

Elements of a Network Diagram.

Network planning, however, lacks the immediacy of a Gantt Chart and for this reason, it is commonly used alongside Gantt charts when large projects are being planned.

The main advantage of network planning is that activities can be allocated a length of time for completion. The supply of a critical component such as a screw compressor may take, say, 3 months from the order date. If the project is to be completed in 4 months, then this critical component will need to be ordered very early in the project time scale. Other activities may include chassis fabrication, which will take only 1 week. This is clearly not a critical length of time and the activity can be planned within the overall time plan for the part to be ready in time for the assembly of the compressor.

The various tasks within the project can therefore be classified into high-priority or low-priority activities, as shown here:

- Low-priority activities are those that can be inserted into the time scale flexibly without affecting the overall completion date
- High-priority activities, however, cannot be delayed without extending the project duration. High-priority activities are said to lie on a *Critical Path.*

A network analysis consists of two major building-block elements: *Activities* and *Nodes* (*Completion of Activities*). Examples of these are shown in Figure 4.30.

- *Activities*: They are denoted by straight lines complete with arrowheads. The description of the activity is usually placed above the line, with the activity time allowance placed below the line.
- *Nodes*: They are denoted by circles and include
  - An activity number
  - The earliest possible completion time
  - The latest possible completion time.

In order to illustrate the way to construct a network diagram, let us analyze the simple task of making a mug of coffee. Figure 4.31 shows the list of activities required. On the network, activities 1, 2, 3, and 4 do not lie on the critical path and can take secondary routes from the starting point.

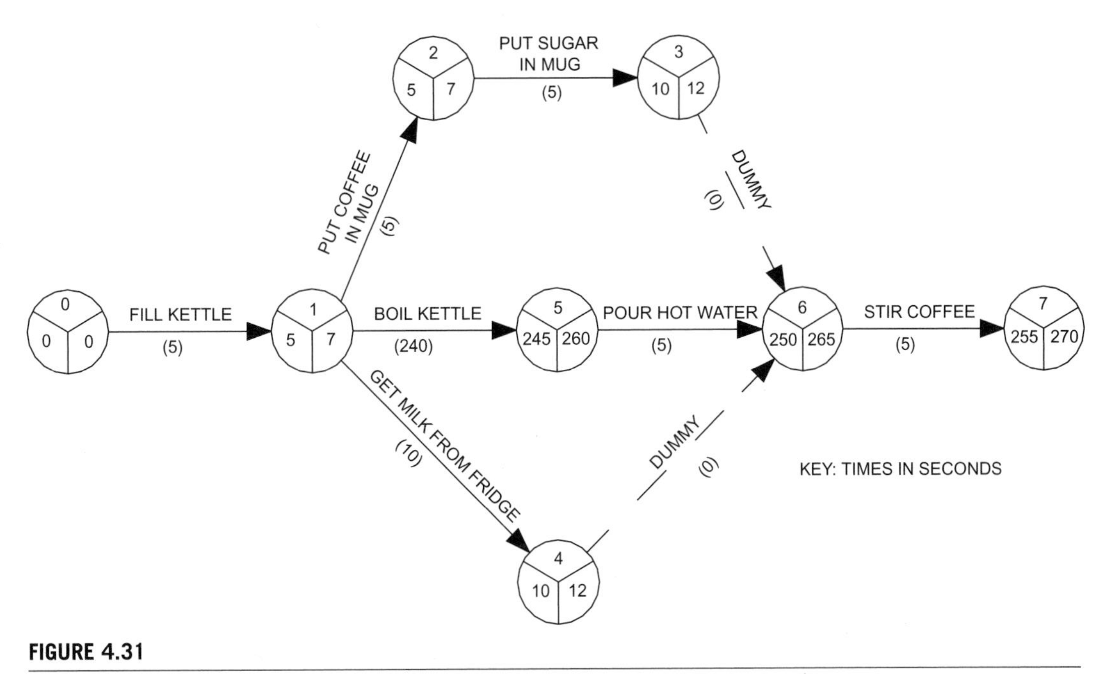

FIGURE 4.31

Network Analysis for Making a Cup of Coffee.

Activity List	
1. Fill the kettle	5 s
2. Put coffee in the mug	5 s
3. Put sugar in the mug	5 s
4. Get milk from the fridge	10 s
5. Boil the kettle	240 s
6. Pour hot water into the mug	5 s
7. Stir the coffee	5 s
Total expended time	275 s
Critical path time minimum	255 s
Critical path time maximum	270 s

4.12.2.1 Time allocation

Time allocation is noted underneath the activity line. This is added at the node. Also at the node can be a time tolerance value. In node 1, the allocated time is 5 s but the time tolerance value is 7 s. Estimated tolerances such as this can be inserted into any node should the activity require it.

The left-hand quadrant of each node displays the shortest time and should continually add up to the shortest time as the diagram progresses from left to right. The right hand quadrant displays the longest time and should similarly add up to the longest estimates as the diagram progresses through the project.

4.12.2.2 Critical path

The critical path is the line that shows the earliest and latest completion times. The time allocation for the longest critical activities lies on this line. Activity 5 is allocated 240 s and is the longest of all the activities. Activity 6 is dependent on activity 5 and, therefore, has to follow it. Though activity 5 takes only 5 s, it lies on the critical path simply because it cannot be actioned until the kettle has boiled.

Several activities may converge on a single node. In this case, insert the longest time values. These nodes should normally lie on the critical path, but in complex networks, there may also be a minor critical path, which may possess convergent nodes. Purists may argue that there should be only one critical path, but some complex projects may demand several network routes.

The critical path can be recognized as the line that equals the earliest and latest completion times. The critical path from Figure 4.31 is as follows:

Critical path = 0-1-5-6-7.

The activities that lie on the critical path determine the value of the overall completion time and are high priority. Any delays in these activities will result in an extended project.

Activities 2, 3, and 4 do not lie on the critical path and are therefore not critical activities. They can be executed at any time during the period between activities 1 and 6 but must not take longer or they will lengthen the project time.

4.12.2.3 Dummy activities

Dummy activities show no process and have no time allocation. They facilitate the return of a low-priority activity to the critical path through a convergent node.

4.12.3 The complete planning exercise (PERT)

A technique that brings together all the planning activities associated with a project is known as PERT (Programme Evaluation and Review Technique).

PERT is a management method that analyzes the tasks involved in completing a given project. In particular, the method closely considers the time needed to complete each task, and identifies the minimum time needed to complete the total project.

PERT was developed for large complex projects where planning had to be simplified. PERT was originally developed for the U.S. Navy Special Projects Office in 1957 in support of the U.S. Navy's Polaris nuclear submarine project. PERT is a technique that uses events where other techniques use beginning and completion strategies and is often used in projects where time rather than cost is important. PERT is very useful in many diverse projects, but it is particularly advantageous when used to plan large-scale, one-time, complex, nonroutine projects. The technique was used to great success in the planning of the Grenoble Winter Olympics in 1968.

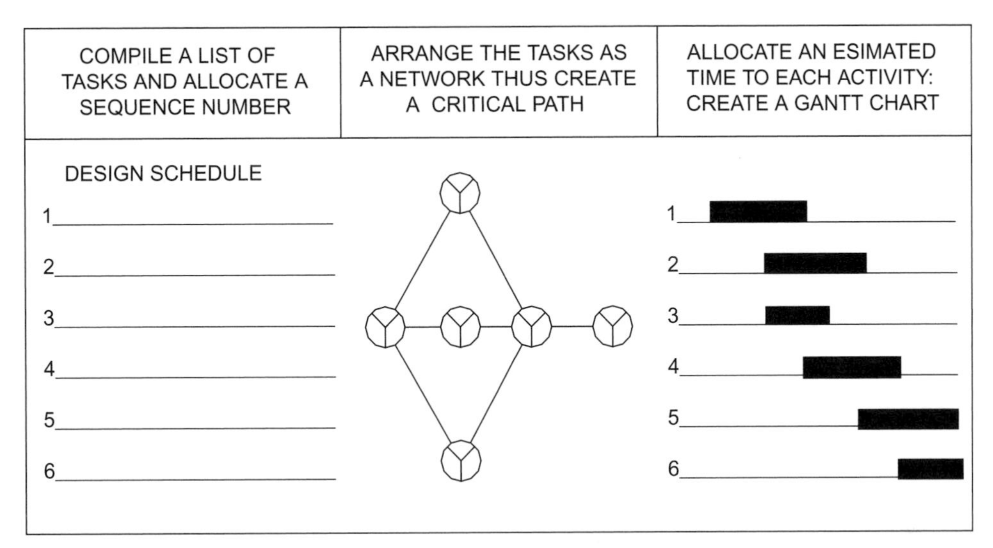

FIGURE 4.32

Tasks Involved in PERT Analysis.

There are three main activities associated with PERT and they are listed as follows:

1. Complete a list of tasks and allocate a sequence number
2. Arrange the tasks in a network and define a critical path
3. Allocate an estimated time to each activity, thus creating a Gantt Chart

An example of PERT tasks can be seen in Figure 4.32.

4.12.4 Case study: flywheel energy storage system

A design manager has created a design team that has been tasked to design a large flywheel energy storage system. Figure 4.33 shows the conceptual design of the flywheel energy storage system. The schematic shown is merely a concept and requires the design of the manufactured components and the specification and selection of bought-out components.

4.12.4.1 Complete a list of tasks and allocate a sequence number

This is perhaps the hardest part of the planning exercise, where the manager has to visualize the activities required to complete the project and list the tasks as closely as possible to the natural sequence of events. When he/she is satisfied with the list, the design manager can allocate personnel with appropriate expertise to each task. Figure 4.34 shows the task list for the flywheel design.

4.12.4.2 Arrange the tasks in a network and define a critical path

Armed with the list of tasks, the design manager can now create a network analysis. This will show the critical time-dependent activities that have to be completed within the allocated time in order to keep the project on track. The network analysis for the flywheel energy storage device can be seen in Figure 4.35. It should be noted that several activities, such as 6, 14, and 15, do not take as long as those

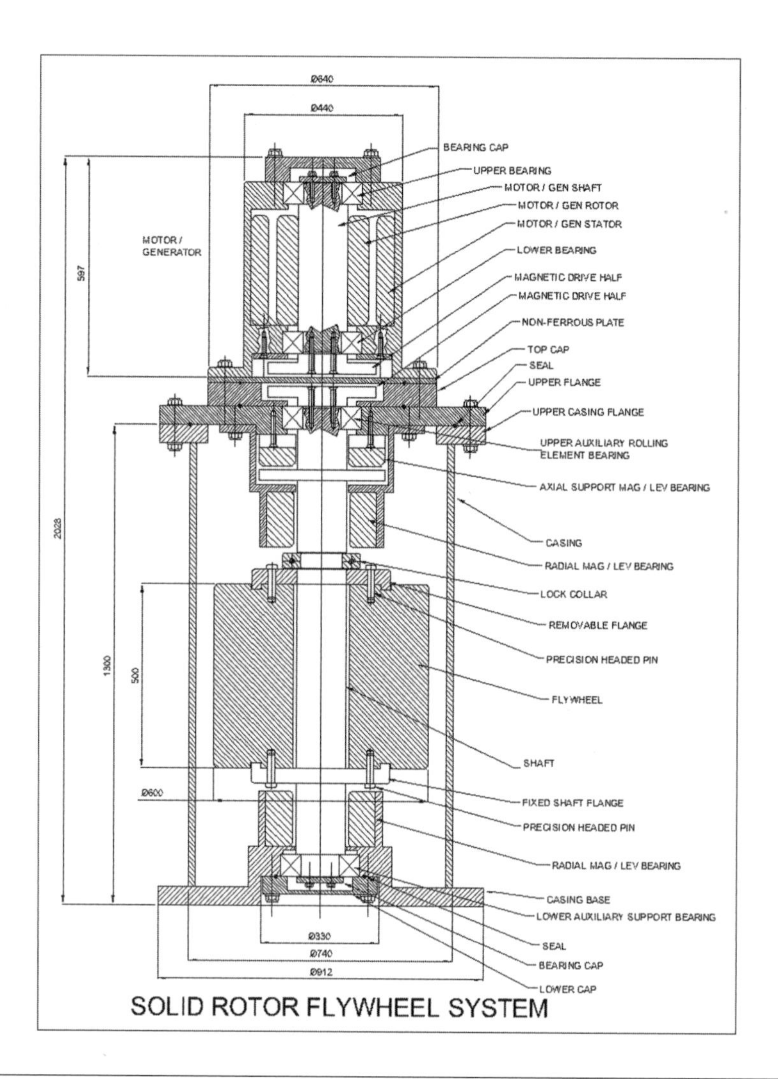

FIGURE 4.33

Conceptual Design of a Solid Rotor Flywheel Energy Storage System. (For color version of this figure, the reader is referred to the online version of this chapter.)

on the critical path, which is: 0-1-2-5-11-12-13-17-19. The critical path activities require full attention if the project is to meet the projected completion time of 70 days.

4.12.4.3 Allocate an estimated time to each activity, thus creating a Gantt Chart

The main tools of the task list and the network analysis chart are now in place. The Gantt Chart (activity sequence chart) can now be assembled using the data from the network analysis and the task list. The Gantt Chart represents the whole project and allocates tasks to team members who have particular skills. The Gantt Chart fulfills other more subtle duties in that it allocates responsibility for a task completion to a particular team member.

Activity	Task	Team Member	Estimated Time (Days)	Actual Time
0	Begin project			
1	Specify rotor dimensions	ADJ/SM	5	
2	Primary analysis stresses in rotor	SM	20	
3	Rolling element brgs radial design	ADJ	5	
4	Rolling element brgs axial design	ADJ	5	
5	Rotor/shaft manufacturing drawings	ADJ	10	
6	Magnetic coupling spec/design	ADJ	5	
7	Motor generator specification	ADJ/MD	15	
8	Motor generator design	ADJ/MD	15	
9	Supplementary magnetic thrust bearing selection	ADJ/MD	8	
10	Layout rotor control systems	ADJ/MD	5	
11	Motor generator casing design	ADJ/MD	5	
12	Rotor housing design	ADJ	5	
13	Manufacturing drawings	ADJ	15	
14	Vacuum analysis	EU/RM	5	
15	Vacuum equipment spec/design	EU	5	
16	3-D digital modeling	EU	15	
17	Design review/assembly issues	Team	5	
18	Specification documentation	ADJ/EU	5	
19	Visualization/publicity	EU/ADJ	5	
	Total days		158	
	Critical-path days		70	
Key	Anthony Johnson	ADJ	Design/dynamics	
	Steve Mumford	SM	FE analysis	
	Mike Dolent	MD	Electrical	
	Eric Umberto	EU	Vacuum/graphics	
	Rob Matterson	RM	Fluid dynamics	

FIGURE 4.34

Task List for the Flywheel Energy Storage Design [2].

The Gantt Chart for the flywheel storage device design can be seen in Figure 4.36.

The Gantt Chart allows a design manager to manage his team. Inevitably, a design team will be working on several projects at once, and personnel may be required on the critical elements of other projects as well. The PERT procedure and, in particular, the Gantt chart allow the design manager to move his staff between projects so that the critical path times can be met.

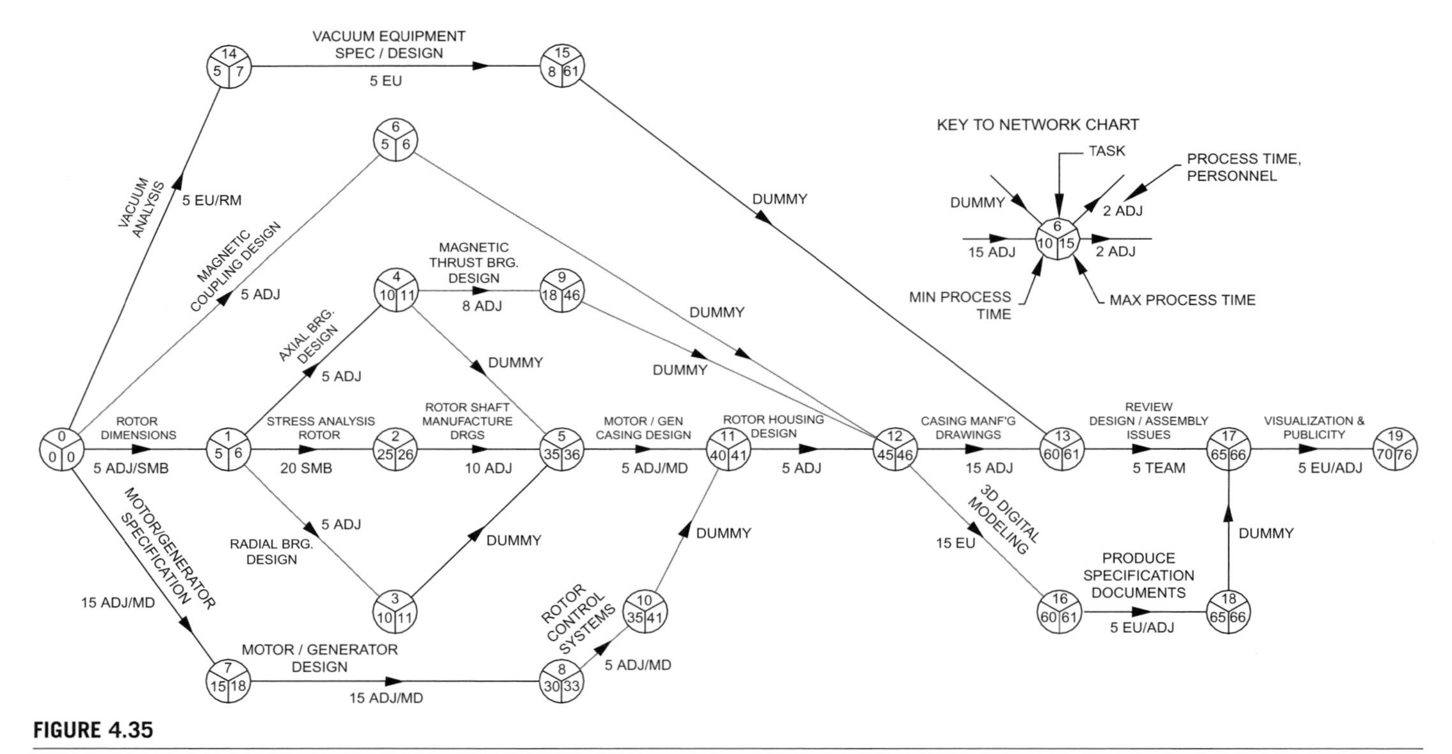

FIGURE 4.35

The Network Diagram for the Flywheel Energy Storage Device Design [2].

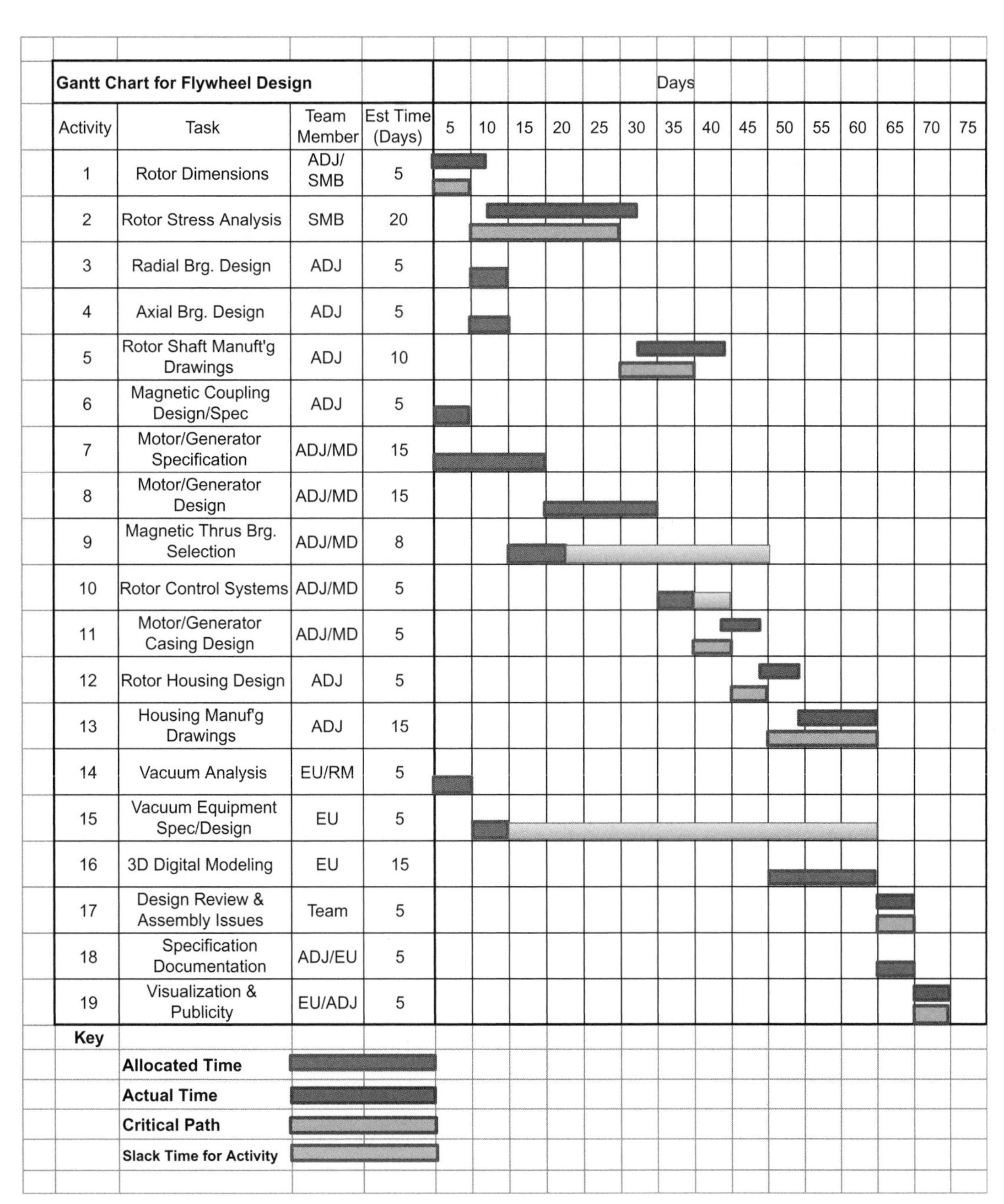

Gantt Chart for Flywheel Design

Activity	Task	Team Member	Est Time (Days)
1	Rotor Dimensions	ADJ/ SMB	5
2	Rotor Stress Analysis	SMB	20
3	Radial Brg. Design	ADJ	5
4	Axial Brg. Design	ADJ	5
5	Rotor Shaft Manuft'g Drawings	ADJ	10
6	Magnetic Coupling Design/Spec	ADJ	5
7	Motor/Generator Specification	ADJ/MD	15
8	Motor/Generator Design	ADJ/MD	15
9	Magnetic Thrus Brg. Selection	ADJ/MD	8
10	Rotor Control Systems	ADJ/MD	5
11	Motor/Generator Casing Design	ADJ/MD	5
12	Rotor Housing Design	ADJ	5
13	Housing Manuf'g Drawings	ADJ	15
14	Vacuum Analysis	EU/RM	5
15	Vacuum Equipment Spec/Design	EU	5
16	3D Digital Modeling	EU	15
17	Design Review & Assembly Issues	Team	5
18	Specification Documentation	ADJ/EU	5
19	Visualization & Publicity	EU/ADJ	5

FIGURE 4.36

Gantt Chart for the Design Process of the Flywheel Energy Storage Device [2]. (For color version of this figure, the reader is referred to the online version of this chapter.)

References

[1] B. King, H. Schlicksupp, The Idea Edge, Goal/QPC 2010.
[2] B. Hawkes, R. Abinett, The Engineering Design Process, Longman 1998.
[3] K.T. Ulrich, S.D. Eppinger, Product Design and Development, McGraw-Hill 2003.
[4] Courtesy of Justina Vaitkute, University of Huddersfield, UK 2013.

CHAPTER

Communication for Engineers

5

5.1 COMMUNICATION OVERVIEW

"I am an engineer, not a communicator." To the contrary—to be an engineer *is* to be a communicator. To be a design engineer is to be an *exceptional* communicator. Clear communication leads to effective solutions. Design engineers solve problems in situations where the solution very often lies in collaborating with other experts, personnel, designers, analysts, and marketers. Communication with those professionals is absolutely necessary.

It is important that design engineers recognize that they must communicate their ideas and developments to other parties. This can be done in several ways, depending on the kind of information that needs to be imparted. Information relating to manufacturing may usually be disseminated through the use of technical drawings and models. This kind of information is delivered toward the end of the design process. Prior to that point, there is an enormous requirement for precise and clear communication to other interested parties. This communication may be verbal, visual, written, or physical, depending on the information that needs to be imparted.

Designers may also need to liaise with a variety of interested parties, such as clients, manufacturers, suppliers, lawyers, patent agents, marketing specialists, distribution specialists, and packaging providers. Each of these individuals and groups will require information in a particular format and delivered in a particular style so that they can fulfill their work function. For instance, marketing specialists will require a description of the product's benefits. Packaging people will themselves liaise with distribution specialists but will require information relating to the envelope of the product and the necessity for protection during shipping.

Whatever the information that needs to be imparted, several methods can be used to communicate it, depending on the audience. Generally design engineers need to be serious about the methods and accuracy of their communication techniques. To become an adept communicator, it is inevitable that basic tools of communication need to be assembled and used. These tools could include:

- Dictionary
- Thesaurus
- Reference books on grammar and writing styles
- Standard presentation techniques
- Computers with appropriate software

Depending on the audience, there are several methods of communication that, in one form or another, engineers should master. Some of these methods are discussed in the following sections.

Sustainability in Engineering Design. http://dx.doi.org/10.1016/B978-0-08-099369-0.00005-4

5.2 WRITTEN COMMUNICATION

5.2.1 Accuracy

As a matter of course, people often treat what they read as the final word. There is a great responsibility, therefore, for an author to write and publish data that is not only accurate but is backed up by appropriate analysis or information sources. In a technical report, the author should never offer opinions but should offer sound reasoning based on facts and other hard data. Furthermore, it is important that the author names the source, which has the additional advantage of avoiding plagiarism.

5.2.2 Presentation

Written communication in an official document needs to look official. Presentation is paramount in this case. A tatty, unstructured, unbound, unprofessionally presented document to get a quote for a project will give a poor impression to a client. On the other hand, a well-presented, bound, clear, concise, professionally presented written document would immediately create a good impression. This professional presentation indicates that the presenter cares for his work and would probably treat the project in the same meticulous way. Presentation can make or break a project, since it is a clear indicator of the presenter's level of professionalism. It is useful here to remember the adage "There is only one opportunity to create a first impression."

5.2.3 Research

During the first stages of a project, research needs to be done. That research generally takes two forms:

1. Information that allows an understanding of the relevant issues, facts, problems, and possible solutions
2. Information that will be presented to the audience

Often these two information elements overlap and vary greatly from project to project. The first type of information allows the designer to engage with the project and formulate a solution. The second type will probably overlap with the first type but might need to be greatly simplified so as not to confuse the audience with too many facts.

It is very easy for a researcher, or indeed a designer, to become so involved with and enthusiastic about the project that he imparts every detail to the audience. The reaction of the audience is likely to be that of boredom and eventual disinterest at the overwhelming level of tedious information.

It is important to realize, however, that research is an absolute necessity from the design point of view as well as to enable the designer to correctly respond to the audience. For instance, it would not be appropriate to impart technical information such as stresses, vibrations, and materials to an accountant. The information would serve no purpose and probably would not be understood because the discipline of the accountant is not tuned to understanding such technical information. In relating a project to said accountant, you must present the information in terms of costs, timescales, and budgets.

This example shows that information imparted to an audience needs to be presented in a particular way. The required research enables the designer to understand the audience and gives them the opportunity to tune the presentation material to attain the best reception from the audience.

5.2.4 Guidelines for effective writing

Effective writing really means planning ahead. A writing project needs a beginning, a middle portion, and an end. Planning is therefore required to understand what information needs to be put into each section. When you're first beginning a project, this might seem a daunting task, but with appropriate research, appropriate design application, and effective planning, the task should become straightforward.

5.2.5 Planning

Planning is an essential component of writing effectively. Just as the designer may see the end result in her mind's eye, so should the writer. The writer should see the shape of his narrative and be able to break it down into individual elements. In terms of a design report, the process should start with a brief, which can then be broken down and explored in smaller elements. The middle portion of the report should be split into smaller elements, each of which explores a single idea. The end of the report is a summary of these ideas leading to a conclusion or recommendation.

The preparation of an outline of the proposed work is highly desirable. This exercise helps the writer think through the composition and improve its internal order before actually commencing writing. In preparing an outline, an orderly framework is created with additional benefits of productivity and efficiency.

During the course of a long report the titles may change as the script develops. This may be due to recognition of a more efficient way out. Often the change in the script layout is due to changes in the design or new revelations brought about through ongoing research. An orderly framework allows the writer to easily change the format of the report.

5.2.6 Report structure

Writing the prose is always done in sentences that are elemental components of paragraphs. Typically, readers find it daunting to be faced with a page of pure, unbroken text. The text is therefore broken up into smaller units, called paragraphs. A paragraph is really the unit of composition and is grouped together in chapters, sections, and subsections, perhaps with appropriate headings or titles. If the end goal was given to the reader in one huge portion, he would become confused and disinterested. The chapter, subsection, and paragraph approach to structure leads the reader toward the end goal in a building-block approach, building on the story as the paragraphs progress. This leads to an interesting read for the audience, who should want to know more and more, rather like reading a good novel.

5.2.7 Crafting an interesting narrative

If the script is not interesting, the audience will not read it. The writing has to be tuned to the audience in order to make it interesting. An interesting read to a chief designer may prove to be completely boring to the company accountant. This point has been made previously, but it is so important to aim the writing toward the intended audience.

The narrative requires structure so that the reader is almost "led by the hand" through the report. Each chapter needs a start, beginning, and an end, which is exactly what people like to see. Chapters written in this format are effectively short stories.

It is always discouraging for a reader to be faced with lots of descriptive text. The authors should consider inserting pictures, diagrams, tables, and the like to improve readers' understanding.

5.2.8 Graphics

"A picture is worth a thousand words." This adage is very old but nevertheless is still very true. It is said that 80% of human learning occurs through visual input. Newspapers use pictures on their front pages to entice readers. In terms of report writing, this revelation indicates that pictures, diagrams, charts, graphs, and other visual representations can reduce the number of words in the report and grab the reader's attention.

To emphasize this point, try describing a wheelbarrow to someone who has never seen one. Write your description as a letter or perhaps as an e-mail. The exercise will probably take at least two sides of A4 paper. Isn't it easier to just sketch a drawing of the wheelbarrow?

Using graphical aids, we can reduce the number of words in a report, keeping the reader interested and better informed and improving understanding of the project.

There are several rules for inserting graphics into text, some of which are indicated here:

- Give each graphical element a number, such as figure number, table number, graph number, or plate number.
- Always refer to the graphics element from the main body of the text.
- Annotate sketches for improved clarity.
- If the report is to be photocopied, present graphics in black-and-white to ensure a sharper reproduction.

5.2.9 Brevity and clarity

The secret of good writing is often to strip every sentence to its minimal components. Short sentences are generally preferred over long sentences; short words are usually better than long words. A few examples follow:

Near is preferred over *in close proximity*
Scarce is preferred over *in short supply*
Now is preferred over *at this point in time*
Used car is preferred over *previously owned vehicle*
Library is preferred over *communications resource center*
Garbage collector is preferred over *sanitation engineer*

During an interview on a radio news bulletin, a subject specialist was heard to say, "It has a subnormal moisture content." What he really meant was, "It's dry"! The main rule here is "Keep it simple."

Technical writing requires a certain professional standard whereby words such as *okay, tremendous, terrific* are not really suitable. They are not descriptive and merely convey opinions. Other faddish words and expressions such as *prioritize, finalize,* and *the bottom line* may also be viewed as unprofessional in a technical report.

5.2.10 Adapting the writing style to the intended audience

This element has been alluded to, but it is very important to understand how the audience will receive and understand your writing. The author must consider the audience's educational background, socio-economic level, age, and interests. For instance, writing intended for technical journals may contain scientific language, formulae, chemical symbols, and analysis. The target audience for this kind of publication would normally be educated to a level where the message could be easily understood. Articles or reports destined for a general audience require a much different style that uses plain language and simple illustrations. Personal and practical implications may also be stressed in such a report.

In presenting engineering subjects to young teenagers, the presenter must consider their educational level. In that age group, reading ability in technical areas could be expected to be almost nil, and to keep the interest of a normally "short attention span" audience, the presentation would need to be almost all graphical. Employing such tactics in the presentation is likely to keep the audience interested.

5.2.11 Things to avoid

In writing for any audience, try to avoid the following:

- Long words
- Jargon
- Abbreviations
- Euphemisms
- Calculations in the main body of text (put them in the appendix)
- Poor grammar
- Poor spelling

5.2.12 Rewriting and proofreading

In working on a project, one of the most difficult parts is starting with a blank piece of paper. In writing, it is creating the first draft—essentially creating something out of nothing. In the first draft, the author synthesizes the research, the requirements, and the perceptive needs of the audience into a readable piece of writing.

Once the first draft is completed, the next phase is rewriting. Rewriting is the molding of the first draft in an effort strives to make the piece readable and meaningful and to accurately reflect the work and thoughts developed during the project. When the rewrite is finished, the meaning should shine through, enabling the reader to absorb the essence from a seemingly effortless piece of writing.

Rewriting may take several revisions and should be proofread before publication. Rewriting may also be termed *revising and editing*. The writer must realize that a first draft will contain factual errors, spelling mistakes, and grammatical mistakes and should never be published without rewriting it at least once. Proofreading is usually the final step in preparing a document and is often applied to journal papers, books, and other published works.

5.2.13 Guidelines for revising, editing, and proofreading

The revision of the first draft should be done in a structured way. Guidelines are as follows:

- Structure and content:
  - Ensure that the structure (report, proposal, etc.) is what the audience expects.
  - Ensure that the data is complete.
  - Are any omissions justified?

    - Ensure that information gleaned from external sources is fully referenced. Information from open sources may be simply referenced by web site and access date.
    - Check that section headings match the contents page.
- Check the logic and flow:
    - Is the logic based on the reader's knowledge base or the author's?
    - Ensure that each passage follows a logical theme from the previous passage.
- Edit line by line:
    - Perform a line-by-line check for grammar and style.
    - Check sentences for sense, clarity, and precision.
- Punctuation, spelling, grammar:
    - Check for correct punctuation, spelling and grammar.
    - Check for words left out.
    - Check for consistency of headings.
    - Read the document aloud to "hear" any remaining errors.

5.2.14 Logs and notebooks

Taking notes and minutes at meetings is normal practice because it creates a record of the details of the communication. It also allows names to be assigned to points to be actioned. Many engineers also maintain an informal record of their work by keeping a day-to-day diary, log, or notebook. Diary entries may be made from short conversations with colleagues, but more serious logs or notebooks should always be made when performing laboratory experiments or pursuing a particular part of a design, such as the selection of a power plant or completing some analysis. Carefully made and maintained records of this type provide a permanent and ready source of information for memoranda, letters, technical reports, and product complaints.

In one particular case, an extending handle for a golf bag was designed as two tubes, one sliding into the other. The safety factor was calculated at 4:1, which was deemed high for the duty. A subsequent product complaint was received, saying that the handle had bent, the inference being that it was not strong enough for the duty. Design analysis notes were quickly found and used to refute the complaint.

5.2.15 Memoranda, business letters, and e-mail

Before the e-mail age, memoranda and business letters were often used to communicate between departments within the same company and between businesses or to converse with clients. Memoranda were generally used to communicate between departments within the same company and fulfilled a similar function to that of e-mails today. Notably, today the use of memos has almost ceased with the advance and ease of use of e-mail systems.

Business letters, which used to be the norm, have also been overtaken by the ease of use of e-mail. Business letters are still frequently used, but for particular purposes. When it hits your desk, a business letter has a great deal of impact simply because today it is not the norm to receive paper communications. Business letters therefore tend to be used for specific reasons, such as delivering a contract, delivering an affidavit or legal document, or making the recipient notice the message by giving it paper impact.

In the modern business and design engineering environment, e-mail is one of the most common means of communication. E-mail is a message device that allows almost conversational communication but is clinical in that e-mails often do not follow the niceties of grammar rules, which make the message flow. E-mail's main advantage is that it is instantaneous and global. A message sent from the United Kingdom can be received in the United States in seconds. Business letters cannot do this unless they are faxed, which incurs extra cost and low-quality reception. E-mails are so common that important messages can be overlooked. In such eventualities, the sender may choose to send a business letter, giving the message impact and garnering more attention.

5.3 PROJECT REPORTS/TECHNICAL REPORTS

5.3.1 General approach to project reports

A technical report will follow a particular format that contains the necessary building blocks to create a complete information source. The report should be written in the third person, which means negating the use of personal pronouns such as *I, we,* and *they*. An example of the use of third person is: "The team decided that . . ." instead of "We decided that. . . ." It is important to give the report a professional feel, and the use of the third person goes some way toward accomplishing this requirement.

The aim is to reduce written matter to a minimum. This can easily be accomplished by incorporating annotated sketches, diagrams, and 3-D models in preference to lengthy written descriptions. A credible report will also possess statements that have clear backing, either in calculation or by reference to an information source. This technique then avoids statements that are really opinions, such as, "It is heavier" or "It is cheaper." It is necessary for the writer to qualify statements and suggest *why* it is heavier or cheaper and on what premise the author has based that statement.

Presentation can make or break a report. The adage "There is only one opportunity to create a first impression" should be paramount in the author's mind. Sloppy, untidy reports may lead to readers' rejection of the product idea, no matter how good the proposed design.

Every language has its preferred spelling and grammar, and it is important that the author create a document that is perceived to be grammatically correct and with correct spelling. Without this correct presentation, the report and its contents will be judged unprofessional.

It is necessary to adopt a reference method so that information elements can be stored in separate areas of the report. Items such as book titles, research articles, analysis, graphical information, bought-out parts, and company data are often placed in an appendix, bibliography, or reference sections and are referred to from the main body of text. One typical reference system is the Harvard Referencing System.

Several key elements constitute a formal report. These are listed as follows:

- Title
- Cover sheet (author's name and perhaps contact details)
- Contents page
- Terms of reference
- Summary
- Introduction
- Main body of text

- Conclusions and recommendations
- References
- Bibliography
- Appendix

5.3.2 Title

A good descriptive title often sets the tone for the report and introduces the subject to the reader. A title is sometimes unavoidably lengthy in order to be comprehensively acceptable for an information retrieval system. An example of an effective project title could be: *The Specification and Design of an Automatic Clamping Device for the Location and Clamping of a Family of 30 Components Within a Flexible Manufacturing Cell.*

5.3.3 Cover sheet

As with a published book, this is the first page after the front leaf of the report and does not need to be numbered. The cover sheet should contain:

- Title
- Author's name
- Contact details
- Project designation number

The cover sheet is normally signed and dated by the author.

5.3.4 Contents page

The contents page is the guide to the contents of the report and should list section and subsection headings, along with their page numbers. Depending on the size of the report, the contents "page" may extend to several pages. The inclusion of a contents page also means that the report should have numbered pages. This might seem obvious, but an incomplete contents page with no page numbers may be deemed unprofessionally presented. The contents page should therefore contain:

- A hierarchical numbering system for each section and subsection
- The titles of the main sections and subsections of the report
- Tables, figures, and plates detailing their number, full title, and page number
- Appendices with appropriate numbered sections and subsections

5.3.5 Terms of reference (generally the project brief)

Terms of reference are instigated by some commissioning body or person who is usually responsible for providing the brief. Any project requires a reason, but more important, who instigated or commissioned the project and subsequent report. Within a company the commissioners may be the board of directors. In government the commissioners may be a governmental subcommittee.

The brief is only one element of the terms of reference. Full terms of reference can also include measurable elements of the project and how those elements will be measured by the commissioners.

Terms of reference may be fairly complex, since they give the commissioning body the means to control the outcome and, most often, the financial input. The creation of detailed terms of reference is therefore often critical, since they define:

- Vision, objectives, scope, and deliverables (i.e., what has to be achieved)
- Stakeholders, roles, and responsibilities (i.e., who will take part in the project)
- Resource, financial, and quality plans (i.e., how it will be achieved)
- Work breakdown structure and schedule (i.e., when it will be achieved)
- Success factors, risks and constraints

5.3.6 Summary (executive summary or abstract)

A summary should be brief and to the point. It will give the busy reader an insight into *what* the report contains without going into the detail of reading the whole report. The summary should be a précis of the report contents, culminating in a brief design specification and a list of product benefits. The summary should therefore contain:

- Objectives and purpose of the report
- Background information
- *What* the report has achieved
- The main findings
- Brief conclusions
- Specification (if appropriate)

The summary is important because it is used as the basis of decision making. After reading the summary, alongside the conclusions and recommendations sections, the technical manager or managing director should be able to make a decision as to the next stage in development.

5.3.7 Introduction

An effective introduction familiarizes the reader with the design challenge and provides background information. Essentially the introduction explains *why* the project is being undertaken. The introduction provides:

- Context: A cursory historical review
- Scope: An explanation of the challenge to be solved
- Purpose: An explanation of the basic need
- The reason for carrying out the work
- Procedure and methods

A typical introduction might be along the following lines:

> *There is a need to design a new form of rear-view mirror for the pedal cyclist. In the past, rear-view mirrors have taken the form of a small mirror mounted on a stem that protrudes sideways from the pedal bike frame or handlebars. When the rider dismounts and leans his bike against a wall, the mirror is often caught on the wall and bent out of position or out of shape.*

It is required to design a mirror-mounting system that is not affected when the cycle is parked. The mirror-mounting system needs to be lightweight, robust, and easily viewed by the rider and must hold the mirror in a position for the rider see to the rear when cycling.

The reason for carrying out this work is to improve cyclist safety and to provide a solution to a perceived problem that is not catered for by the current cyclist mirror market.

Such an approach enables the reader to be gradually introduced to the design challenge. Should the reader want to glean more meaningful information, he or she can then read the report's contents.

5.3.8 Main body of text

The main body of the report contains all the details of the method of thinking and design decisions. It is an explanation of the actions taken during the design process and the reason those actions were carried out.

This section should be explanatory but as concise as possible. Full use should be made of annotated sketches, 3-D models, charts and graphs, photographs, and indeed any other visual representations that reduce the amount of text. The aim here is to impart the information to the reader in an easy but accurate fashion. All visual items should be labeled in a numerical sequence preceded by labels: *Figure 1, Figure 2, Figure 3*, and so on for sketches, graphs, and diagrams; *Plate 1, Plate 2, Plate 3* for photographs, and *Table 1, Table 2, Table 3* for charts and tabular data.

Always qualify statements using data that has been derived by analysis or by research. A statement such as "It is cheaper" or "It is too heavy" is merely an opinion.

It is often necessary to include the results of analysis or research, but do not be tempted to include large swaths of analysis in the main report body. These results should be available to the reader in the appendix, suitably referenced in the main body of the report. Similarly, company brochures may be referred to from the main report body, but only selected pages should be available in a section of the appendix.

The technical report should always be written in the third person. Avoid the use of personal pronouns such as *I, we, me, us,* and *they*. A typical example here would be: "It was considered that a 28-tooth spur gear should be used because . . ." rather than "I decided that a 28-tooth spur gear should be used because...."

Presentation can make or break a report. A scruffy presentation leaves the reader with an impression of sloppy work, whether or not the research and background work has been carefully executed. Consider the method of referencing, formatting, and eventual binding of the report. You have only one opportunity to create a first impression.

5.3.9 Conclusions and recommendations

Any piece of work that requires a written report is not complete without appropriate conclusions and recommendations. Often, conclusions and recommendations are the whole point of performing the work.

Conclusions and recommendations should be included at the end of the main report and should be short and to the point and should clearly indicate the major findings of the work.

Recommendations are the logical continuation of the work and outline what should be done in the future to continue the work that has just been completed.

5.3.10 Bibliography

Usually the bibliography takes the form of a list of books and papers and useful background reading. These are often related to a particular subject or a particular author and may form part of the references.

5.3.11 References

The reference section should contain a list of books and papers from which specific items have been used. Whenever any outside source has been consulted and subsequently included in the report, it must be included in the reference section. Any work referred to must include the title, author, location, and a reference number. This information is required so that readers can find the work themselves. For example, "It has been shown by Becker (1) that a rectangular orifice should tend to prevent bridging, whereas...." In the reference section, this item would be referred to as follows:

1. Becker, P., "Bulk Flow of Powders," Journal for Flour Graders, *vol. 40, pp 981–994.*

The intention of a reference is to direct the reader to a specific section of a book or article rather than for the book to be read completely.

5.3.12 Appendix

The appendix is often the final element of the report and may be large enough to warrant a separate volume. Information that appears in the appendix should be directly relevant to the design and should enhance the report by giving a clear understanding of the product being designed. Specific company brochure pages may be included rather than the whole brochure. The appendix may also give specific information to the purchasing department for components that have been specified as part of the design.

The appendix should contain:

- Information on standard bought-out parts
- Mathematical proofs
- Worked examples
- Tables
- Charts
- Information that can be used as evidence to substantiate the statements within the report body

Analysis should be presented here, with an appropriate reference number so that it can be referred to from the main body. The report must list the meaning of the symbols used in any formulas the report includes. It is usually necessary to quote the source of your equations, but it is not necessary to prove the formulae unless several equations have been combined.

5.3.13 Project layout

The same attributes you would expect to see in a published book should also be seen in a professional report. The layout should be neat, tidy, and formal. It should be easy to read in a structured manner, following a logical progression. A large part of the project layout aspect of the report is visual. The professional nature of a report is primarily visually conveyed (what people see) and secondarily by what people read. The visual and nonvisual layout of the report may be achieved with subtleties discussed here.

5.3.13.1 Logical progression

No matter what the subject, the report should convey a story and should unfold that story in a logical, step-by-step progression. In a design report, the beginning should consider a brief with the initial requirements and convey elements of research on which design decisions have been based. This first section should conclude in the concept design. The detail design section should then tell the story of components and materials selected, with reasons, and conclude with the final product specification.

During the process of conveying this information, the logical progression should take the reader from one aspect to the next logical step so that there is no reader confusion as to the route by which the design has arrived at a particular point.

5.3.13.2 General layout guidelines

5.3.13.2.1 Single page of text

In a formal report, it is normal convention use A4 paper and to place any text and graphics on the right margin of an open document. This leaves the left margin for the reader to make notes. See Figure 5.1. This style is very different to that of a published book, where the reader is not expected to make notes.

5.3.13.2.2 Viewing position

When bound, the report should be opened so that the spine is vertical, which allows the reader to view the text in portrait format, that is, from the top of the right-hand page to the bottom. See Figure 5.1.

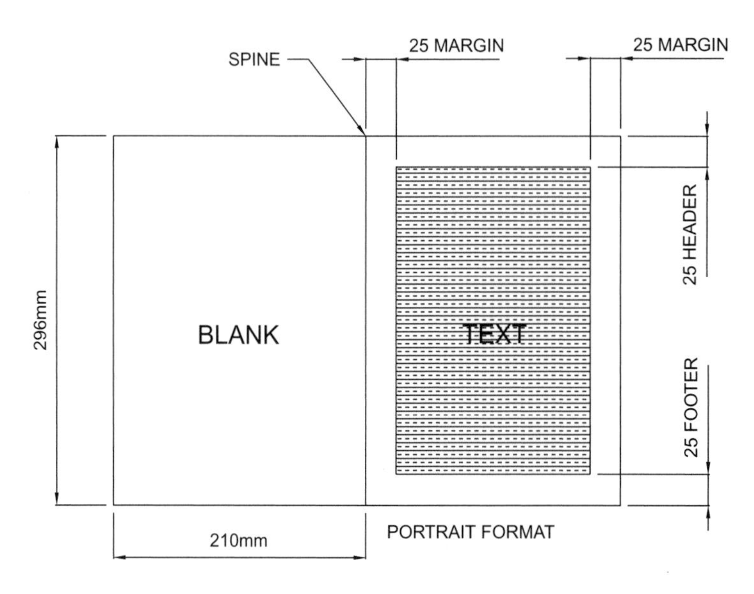

FIGURE 5.1

Portrait Viewing Position for a Formal Report Using A4-Sized Paper [1]

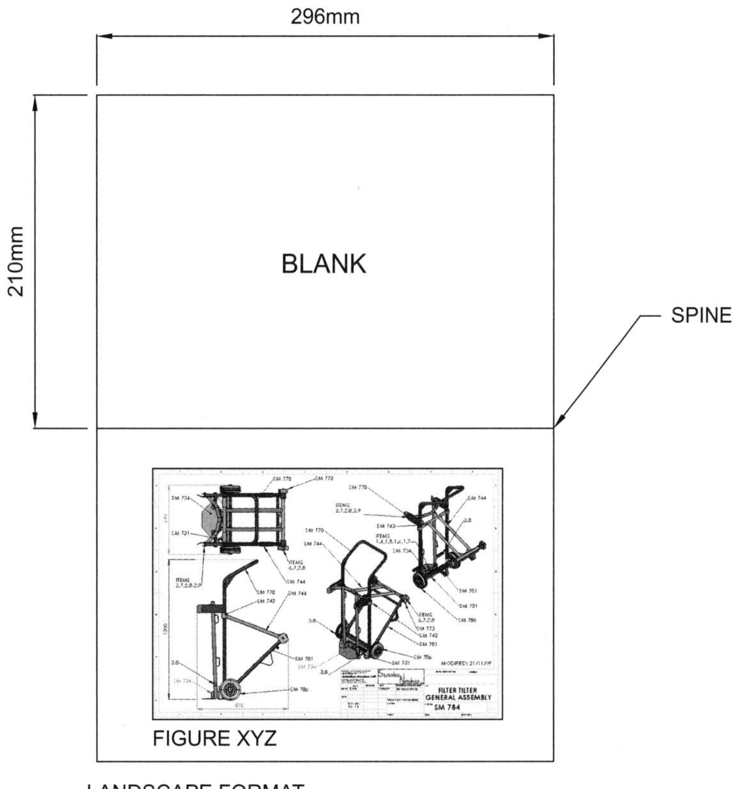

FIGURE 5.2

Landscape Viewing Position for a Formal Report Using A4-Sized Paper [1]

Should wider pictures and diagrams be required to run in landscape format, it is usual to hold the spine horizontally and view the graphic from the bottom leaf of the report. See Figure 5.2.

5.3.13.2.3 Margins

A margin is the area between the main content of the page to the page edges, as shown in Figure 5.1. It helps to define where a line of text begins and ends. The margins' purpose is to hold the text in a particular area on the page, subtly telling the reader that there is no text that has overspilled the page.

There are several formats for margins, depending on the document. A letter may have very wide margins in contrast to a technical report, which may have narrow margins. It must also be remembered that the left margin may be larger than the right margin to accommodate binding.

Word processing software tends to default to 25 mm margins, but this can be changed in the software setup.

5.3.13.2.4 Grammar and spelling

Grammar and spelling contribute greatly toward creating professional documents. Good grammar and perfect spelling create good impressions and promote clarity and flow. With practice, attention to detail, and diligence, the author can achieve a very professional presentation.

5.3.13.2.5 Pictures, diagrams, and tables

Any graphics placed within the text must be referred to from the text using figure numbers, table numbers, or plate numbers. An unreferenced graphic in the text will leave the reader confused. Any such graphics must also be source-referenced if they have been gleaned from other works.

5.4 ACADEMIC PUBLISHING (TECHNICAL OR JOURNAL PAPERS)

Academic publishing typically involves publishing technical or journal papers or textbooks. It describes the subfield of publishing that distributes academic research and scholarship and usually refers to journal article, book, or thesis formats. Most established academic disciplines have their own journals and other outlets for publication, although many academic journals are somewhat interdisciplinary and publish work from several distinct fields. There is also a tendency for existing journals to divide into specialized sections as the field itself becomes more specialized. The Institution of Mechanical Engineers, for example, represents the engineering field very broadly and specializes its activities in 18 specialist journals in areas such as automotive, rail, manufacturing, control engineering, and engine research.

In academic publishing, a paper contains original research results or reviews existing results. Such a paper will be considered valid only if it undergoes a process of peer review by one or more *referees* (academics or specialists in the same field) who check that the content of the paper is suitable for publication in the journal. A paper may undergo a series of reviews, revisions, and resubmissions before finally being accepted or rejected for publication. This process typically takes several months. There is often a delay of many months before an accepted manuscript appears.

Journal papers require a very rigorous approach and may be submitted not only to journals, as mentioned, but also as conference papers. An author may submit a paper to a particular conference in the appropriate field and then present it in person on the day of the conference. Conferences and journals require authors to submit technical works in a particular format, including specified font type and size, general format in terms of columns, heading styles, and so on. Conference papers and journals are quite different in style from a formal report. They tend to be less dry, reflecting a slightly more journalistic style aimed at a broader audience.

5.5 GRAPHICAL COMMUNICATIONS

Graphical communications in one form or another may be considered perhaps the most important communication method engineers employ. Graphical communications in an engineering sense often relate to detail drawings, frequently still called *blueprints* due to the ultraviolet printing process used in the past. The sophistication of printers in the modern computing environment means that drawings may be printed very clearly using black or colored ink on a white paper background. For many years there have been initiatives to reduce the number of paper drawings in favor of electronic display systems, but it seems that people in many cases prefer traditional paper methods of displaying technical details.

Whatever the level of engineer, it is important that the engineer is able to read drawings and possesses some skill in the preparation of drawings. Since drawing practice has evolved into a very precise scientific communication method, it is often the province of specialists to prepare drawings. As technology evolves, it has become possible for drawing preparation to become semiautomatic after we first prepare a 3-D model.

As the design project evolves, the level of graphical communication becomes more sophisticated and precise. Low-level graphics such as hand sketches are excellent for conveying ideas and solutions at the beginning of a project, when precision is less important than idea generation. As the project progresses and solutions become more probable, the level of precision increases, along with the graphical communication method. The progression of graphical communication from low- to high-level precision is as follows:

- Hand-drawn sketches
- Layout drawings
- 3-D models
- 2-D drawings
  a Detail drawings
  b. Subassembly drawings
  c. General assembly drawings
- Digital data transfer

All industries do not move at the same pace. Many high-tech manufacturing companies may start a new design with hand-drawn sketches but might not utilize all the processes mentioned in our list. In a highly technological manufacturing environment, it may be more efficient to progress to 3-D models, also bypassing 2-D drawings so that digital data transfer can be used directly to machine tools.

In other branches of industry, companies may not need 3-D models. A company specializing in fabrication of steel plate may only require 2-D drawings, which can then be directly downloaded through digital data transfer onto a laser cutter, thus quickly cutting out the steel shapes to be fabricated.

5.5.1 Sketches

There is more to graphical communications than detail drawings, however. In following a new project through the design route from the first sight of the brief, the designer will sketch various ideas, possible solutions, and options. This is a means of assisting the visualization of ideas and possibilities. It is also a method of sharing fledgling solutions with colleagues and clients. Sketching is a useful skill and can be used to convey a reasonable amount of detail, but care must be taken, since sketches are often hand-drawn and not to scale. Building components directly from sketches may lead to dimensional errors and errors in perception.

During the design of a drill rig, a rotation gearbox sketch was first made of the major components such as seals, bearings, gears, spacers, and, of course, the main shaft. The original sketch is set out in Figure 5.3.

Sketching is a fairly low-level method of graphical communications and is extremely useful for assisting in the visualization and formulation of possible solutions to design problem. The use of

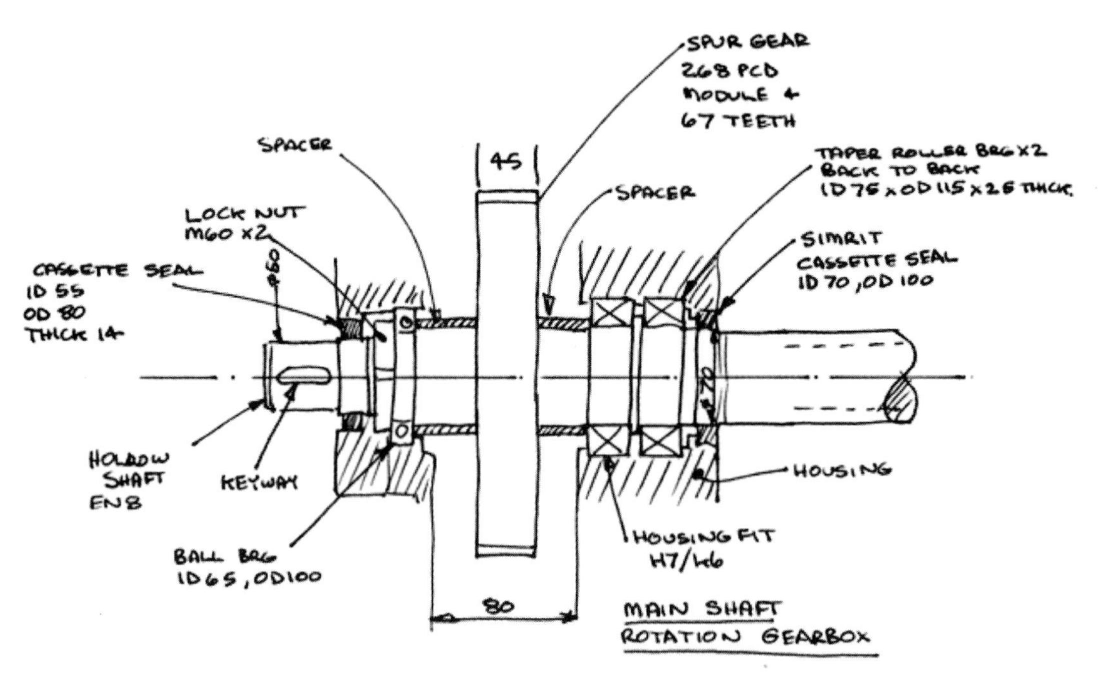

FIGURE 5.3

Initial Sketch for Major Parts of a Rotation Gearbox [1]

sketches is not restricted to design problems, however; visual representation may be very useful in other areas—for example, as a precursor to analytical problems. When working out stresses in beams, the situation can be clarified by a neat sketch. In some cases the development of beam analysis requests neat sketches so that shear-force diagrams and bending-moment diagrams can be created. These are just some of the instances where sketches can be used to enhance the development of a solution or communicate to another party.

Sketching is a low-level communication device and its use should be avoided for direct inclusion in professional reports, since inclusion often reduces the professional impact, perhaps giving the reader the wrong impression.

5.5.2 Layout drawings

Layout drawings are created after the sketch stage. Layout drawings are prepared at the design stage by the designer, who has calculated elements such as bearing sizes, seal sizes, and gears. These items are then applied to scale to the overall assembly. These drawings lay out the item mechanism or assembly so that each item can be annotated with component sizes, designation, tolerances, sizes, materials, and other information required to enable draftspeople to prepare detailed drawings.

When designs were prepared and drawn by hand on a large drawing board, these layout drawings were an important step in the process. Often a designer would prepare a layout drawing and pass it to a draftsperson. Commonly the designer would complete the detail drawings personally.

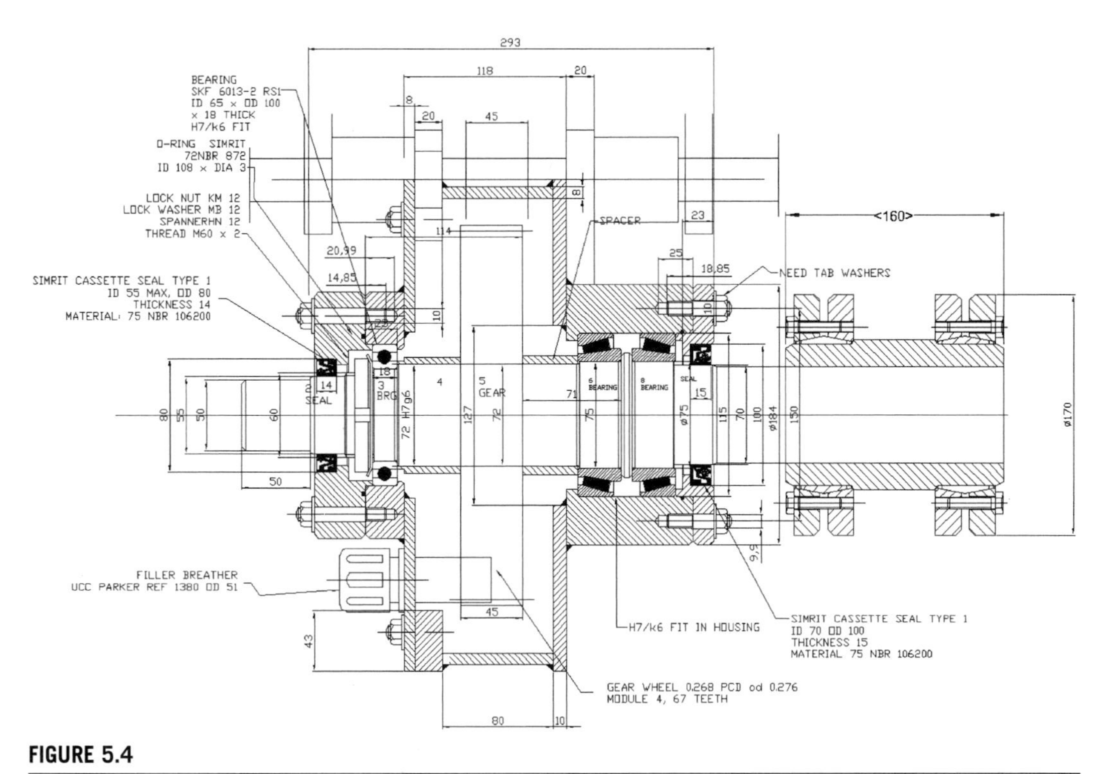

FIGURE 5.4

Part of a Layout Drawing for a Rotation Gearbox Showing Major Components [1]

The application of digital drawing packages has changed the process and improved the speed of drawing preparation. In many cases the layout drawing can be bypassed altogether, but some complex designs still require the clarification offered by layout drawings. Figure 5.4 shows a part of the layout drawing created from the sketch in Figure 5.3.

Figure 5.4 shows the major gearbox components in context and to scale. Annotations show tolerances, specifications for the bought-out components, and all manner of detail that would be required to create detail drawings.

The advent of 3-D modeling has changed the process. Figure 5.4 was created on a 2-D drafting system, but the individual component outlines were copied to be recreated as 3-D models. This may be considered a long-winded approach, since many 3-D systems allow direct creation of components without referring to the layout stage of the design process. The choice of process and method of creating the models depends entirely on individual preferences and the complexity of the design.

5.5.3 3D modeling

Humans tend to see objects and components in three dimensions. Three-dimensional modeling has allowed engineers to communicate much more efficiently by creating accurate 3-D images. Three-D

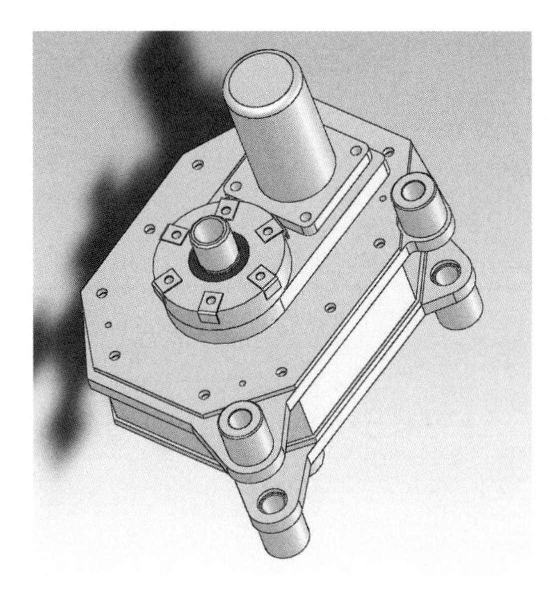

FIGURE 5.5

Three-Dimensional Model of a Rotation Gearbox Case [1]

models are able to be expanded, reduced, turned around on the computer screen, sectioned, and colored, as well as many more options that enable the human designer to better understand the shape and function of the component. The gearbox shown in Figure 5.4 was assembled with a group of 3-D components. The gearbox casing is shown in Figure 5.5.

Three-D components and assemblies digitized as models can be forwarded to clients via e-mail; then, using suitable viewing software, the clients can view the proposed design. The sharing of designs and other information has now become an almost instantaneous paperless activity, with the great advantage that such communication can now be global. Designers are no longer restricted to designing in a location next to the manufacturing facility; they can now design in, say, the United States and e-mail manufacturing information, in the form of digitized drawings and models, to a manufacturing operation in South Korea.

The digitization of a proposed component or assembly allows it to be stored as a computer file. These files can be loaded into specialist software to perform complex analytical operations such as finite element stress analysis, dynamics analysis, thermal analysis, and fluid flow prediction using computational fluid dynamics (CFD) techniques.

These powerful tools have been developed principally as a result of the digitization of components and can perform multiple calculations per second that a human designer would find impossible to complete. The result is to give the designer a powerful means of performing accurate calculations and analysis and predicting the performance of components to a level and accuracy hitherto unknown. These tools also speed up the design process, since components can be assembled within the computer environment, saving time and cost. A prototype may be assembled digitally before any metal has been cut. It has been said that designing assemblies in this fashion can reach between 80% and 95% of the complete prototype before any metal has been cut.

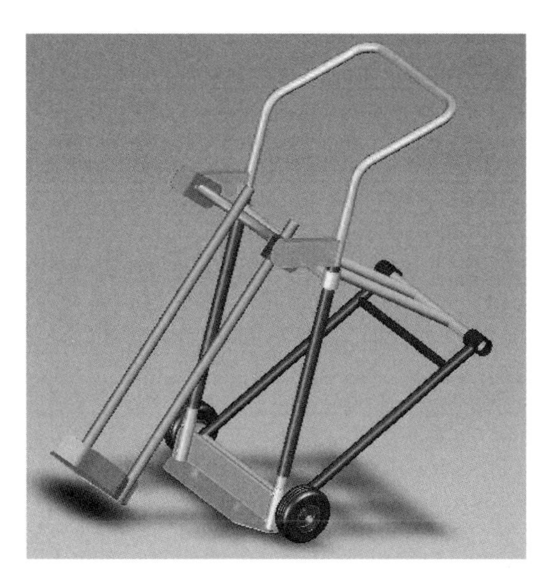

FIGURE 5.6

Digital Assembly 3-D Model of the Filter Tilter [1]

Compare this to the design process that created the Avro Lancaster WWII bomber, which required drawings to be produced on paper so that a team could build the prototype aircraft. The completion of the design prior to the prototype build was probably as low as 50%, thus creating an enormous cost in terms of redrawing, redesigning, and manufacture of new parts.

Three-D modeling, along with the associated tools, has expanded the design function so that time to market and therefore cost have been dramatically reduced. For instance, the time required for a vehicle to be manufactured from the concept stage in the 1960s was about five years. Digital design techniques have much reduced that time to market to perhaps 18 months to two years.

The Filter Tilter shown in Figure 5.6 was designed and digitally modeled prior to the manufacture of any components. It was estimated that the design was 95% complete prior to manufacture.

5.5.4 Two-dimensional modeling

Two-D modeling is a modern term given to the production of two-dimensional drawings. Historically, 2-D drawings represented every component, every subassembly, and the general assembly. A 2-D detail drawing has to be as technically accurate as the component it represents. The technique of drafting has grown, by necessity, to be a very precise method of describing the shape and size of components. It is subject to precise standards such as British Standards BS8888 or the American Society of Mechanical Engineers (ASME) Y14.2, Y14.3, and Y14.5.

5.5.4.1 Detail drawings

Detail drawings, sometimes called *manufacturing drawings,* usually show a single component and give all the information necessary for the manufacture of that component. The items these drawings should specify are:

- The form or shape of the component
- Full dimensions, tolerances, and surface finish
- Specific material designation, heat treatment, and the like
- Machining instructions and processes

5.5.4.2 Assembly and subassembly drawings

Assembly and subassembly drawings are essentially the same thing. However, an assembly drawing is of a discrete product, such as a pneumatic bollard. A subassembly is an assembly of parts, such as those in an internal combustion engine, which are destined to be part of a much larger assembly, such as a passenger vehicle.

An assembly drawing shows an assembly of parts and specifies how the whole component is assembled. The drawing should show the finished product with all the parts assembled in their correct relative positions. This is done by producing the drawing in such a way as to show how adjacent parts fit together. Sometimes an assembly procedure is stated. Certainly assembly instructions such as "Grease shaft before applying seals" should be stated on the drawing, as should other similar instructions. It is beneficial to add overall size dimensions and perhaps specify the position of parts relative to some datum.

5.5.4.3 General assembly drawings

A complicated product such as a passenger vehicle is made up of several subassemblies, such as engine, the vehicle body, seats, suspension, and the like. The general assembly drawing is an assembly of all the subassemblies, showing their location within the overall product.

5.6 GENERAL DRAWING APPLICATION

There is a great deal of responsibility laid on the draftsman or the designer. A lack of concentration on someone's part may mean that the wrong size is set down on the drawing. This could translate down the line to the component not fitting and needing be scrapped, which of course is costly. A grave mistake could be made, perhaps of inserting a weaker specification of material on the drawing than that which is actually required. For example, the chain anchor on a rock drill mast lifts the rotation gearbox out of the ground, along with tonnes of drill string. On one occasion this chain anchor was specified on a drawing as 070 M20 (mild steel). The designer had requested a much stronger alloy steel. The consequences were that the chain anchor was weak and snapped in service. The 3,000 kg of gearbox and drill string crashed down the drill rig mast, narrowly missing the driller. This was a close call and could have ended in a fatality.

Two-dimensional drawings can be generated using a 3-D CAD package. Once the component is modeled in 3-D, the software can semiautomatically generate the drawing. The drawing usually needs attention from the designer, who will arrange the views, dimensions, and tolerances so that the drawing is a true and accurate representation.

An example of a complete detail drawing is shown in Figure 5.7. This is the rotation gearbox shaft, part of the gearbox shown in Figures 5.3, 5.4, and 5.5. We can see that it possesses all the detail necessary to manufacture the shaft. Attention to detail is absolutely paramount for this application, since a

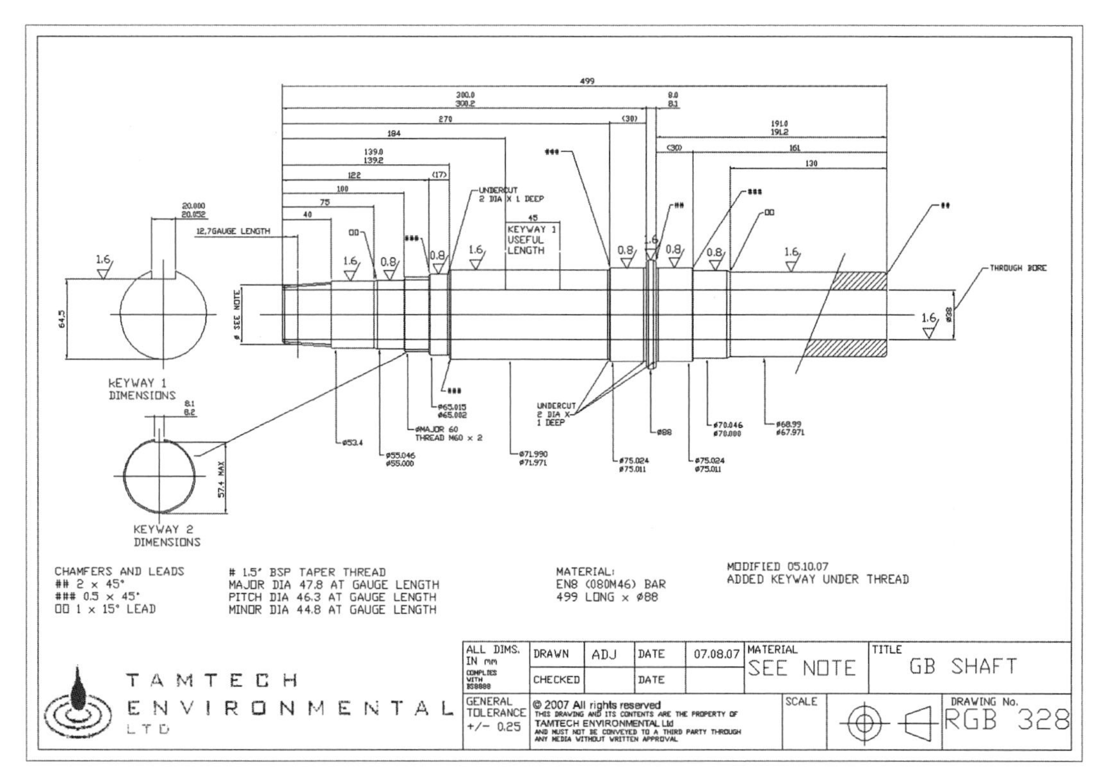

FIGURE 5.7

Detail Drawing of a Rotation Gearbox Shaft [1]

missing dimension or perhaps an omission of material specification, surface finish, tolerances, or the like could lead to at best the item being scrapped or at worst a component failure leading to a fatality.

Our discussion of 2-D drafting has thus far covered detail drawings, but other types of drawings, such as assemblies, can also be represented. Figure 5.8 shows the general assembly of the Filter Tilter, showing each labeled component and its location and interaction with other components. The assembly drawing is often perceived as a picture that in reality is a pictorial assembly instruction. Often assembly procedures are specified on the drawing along with components required to assist in assembly. Such assembly-assist items could include grease to assist the assembly of bearings and seals, specialist tools, jigs, and other small items such as end plugs to prevent dirt ingress into hydraulic pipes.

5.6.1 Drawing practice basics

5.6.1.1 Information clarity

The whole point of creating a drawing is to impart precise, technical information to a third party. With this in mind, the draftsperson must ensure that the drawing is clear and unambiguous. U.K. and U.S. standards have been devised around this principle. The skill of the draftsperson is to create a drawing to a drawing standard so that it can be read and understood by another person who is also familiar with that

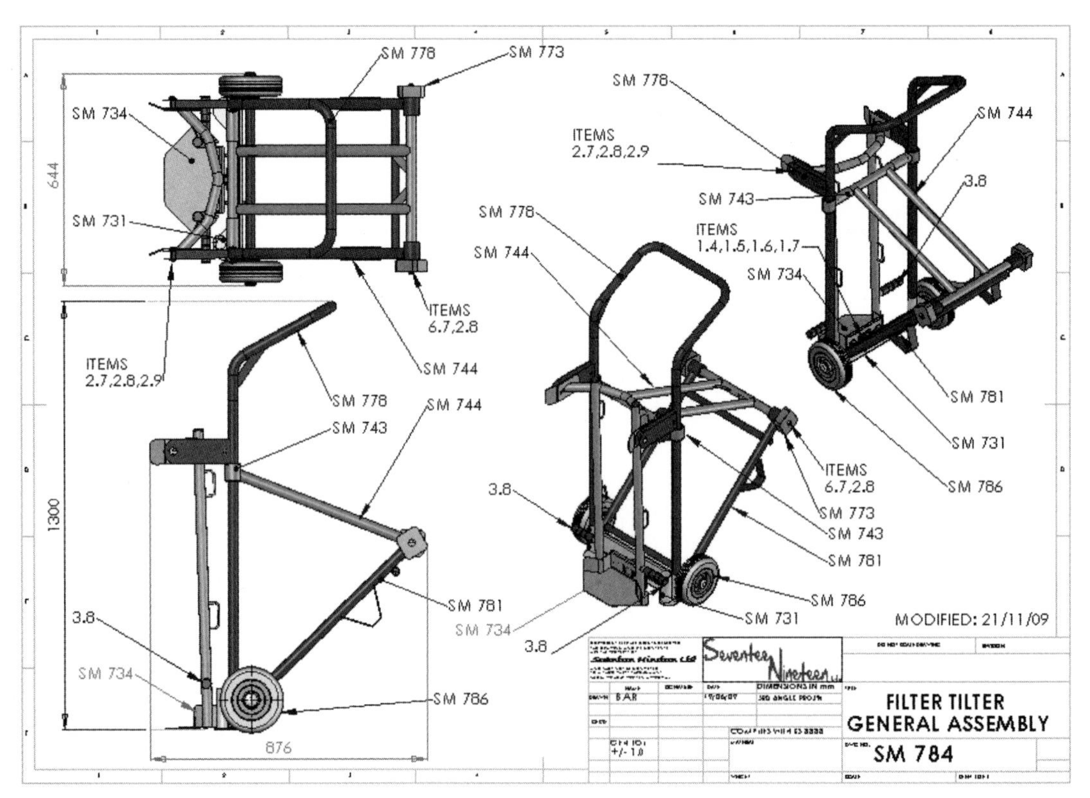

FIGURE 5.8

General Assembly of the Filter Tilter [1]

drawing standard. Producing a drawing to a drawing standard can be likened to speaking the same language. It should be noted, however, that within the drawing standards there is scope for individual styles by different draftspeople. A drawing of a component produced to the same standard by one individual will convey the same information but will be a very different drawing than that done by another individual.

5.6.1.2 Orthographic projection

Orthographic projections are 2-D representations of 3-D objects. Features of 3-D parts are presented on a flat plane, as though the outlines of the part had been projected onto a screen. Figure 5.9 illustrates the principle.

The viewing angles are split into quadrants of first angle, second angle, third angle, and fourth angle, as shown in Figure 5.10.

In practice only first angle quadrant and the third angle quadrant are used since these are pure views. The second angle quadrant and fourth angle quadrant are combination of first and third angle projections and cause confusion to the reader.

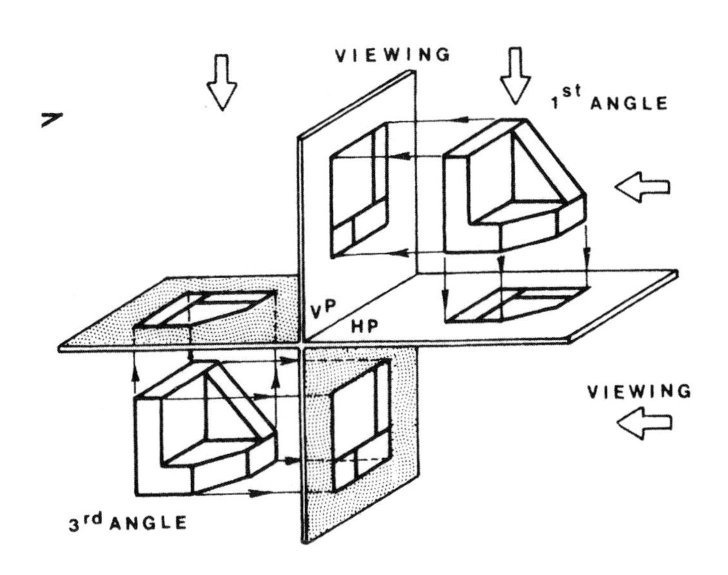

FIGURE 5.9

The Principle of Orthographic Projection [2]

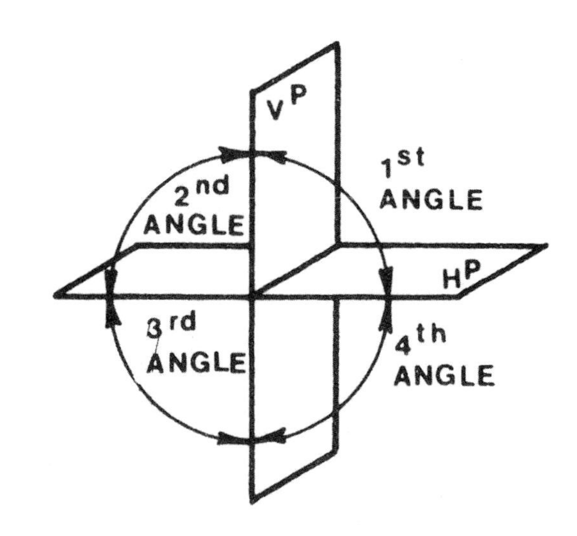

FIGURE 5.10

The Quadrants Used in Orthographic Projection [2]

The principle of first-angle projection is shown in Figure 5.11. Here the viewer looks at the 3-D shape and projects the image onto a screen *behind* the component.

The principle of third-angle projection is shown in Figure 5.12. Here the viewer looks at the 3-D shape and projects the image onto a screen *in front of* the component.

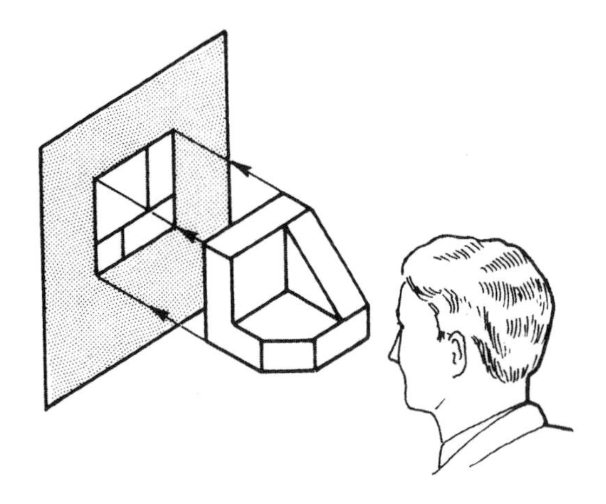

FIGURE 5.11

The Principle of First-Angle Projection [2]

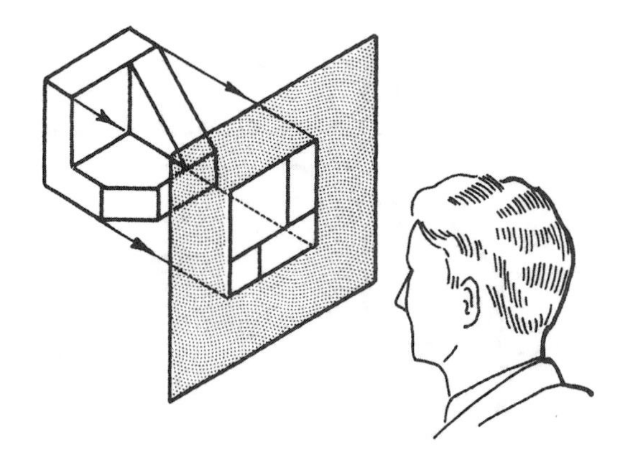

FIGURE 5.12

The Principle of Third-Angle Projection [2]

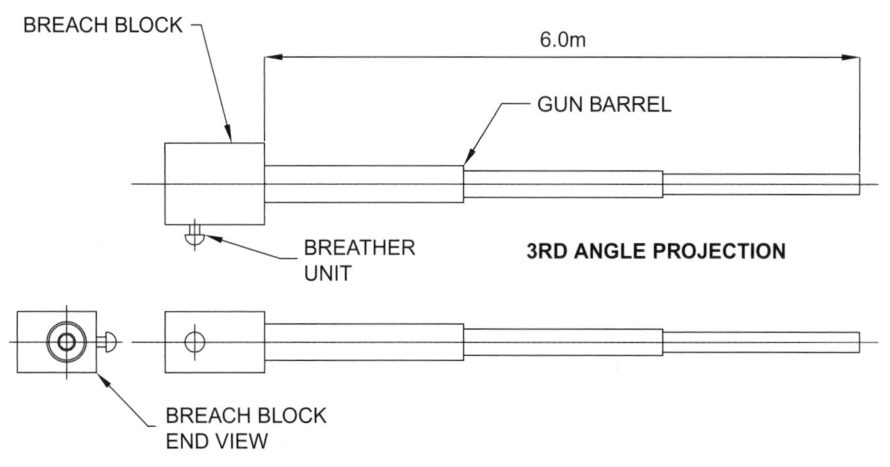

FIGURE 5.13

A Gun Barrel Drawn to Third-Angle Projection [1]

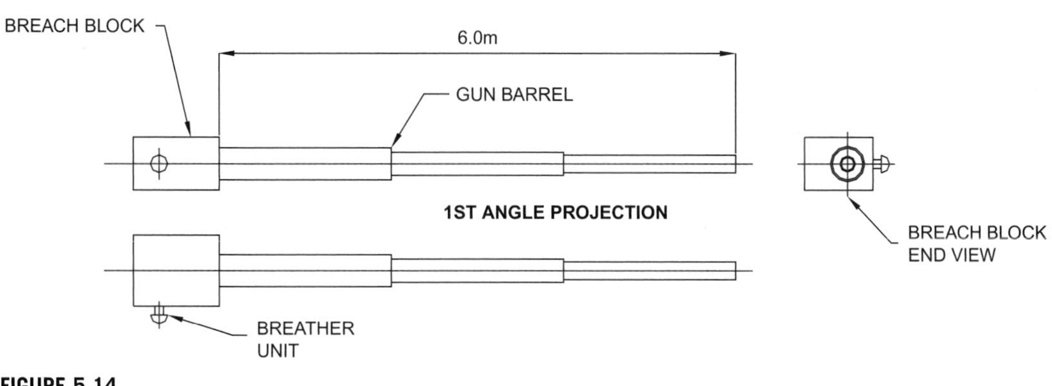

FIGURE 5.14

A Gun Barrel Drawn to First-Angle Projection [1]

The difference between the two projections might seem trivial. However, there are great advantages in using third-angle over first-angle projection. In third-angle projection, the view of a component is drawn next to where the view was taken. In first-angle projection, the view is drawn on the other end of the component, at the opposite end from where the view was taken. Figures 5.13 and 5.14 show the differences using an example of a large ship's gun barrel.

In the case of 3rd angle projection the breach block end view is drawn at the same end as the breach block. Conversely in the case of first angle projection the breach block end view is drawn at the opposite end of the gun barrel to the breach block.

In the case of small components, first-angle projection might not present any difficulty, but small components tend to be fitted to larger components and subassemblies, which can require a third-angle projection approach. Convention therefore dictates that draftspeople do not mix projection angles in the same subassembly. The normal projection tends to be third-angle projection for all drawings, from subparts through general assemblies. Though first-angle projection is still included in the standards and is still used in some industries, it is the least popular of the two preferred projections.

5.6.1.3 Views and sections

It is normal for the component to be represented in at least three elevations. These are usually:

- Front elevation
- Side elevation
- Plan (overhead view)

Depending on the complexity of the component, these views may be expanded to include others as required to achieve clarity. The selection of the other views depends on the skill of the draftsperson, who selects the most appropriate views that most efficiently describe the component. Sometimes a simple component can be described with just two views and perhaps a section, as shown in Figure 5.15.

Components can be very complex, requiring an inside view for a full explanation. Sections are often taken to reveal this view. One of the conventions the drawing standards require is that dimensions must always be placed on an outline. The only method with which to lay bare internal structure of a component is to take a section. This may be a complete section of the component or it may be a scrap section

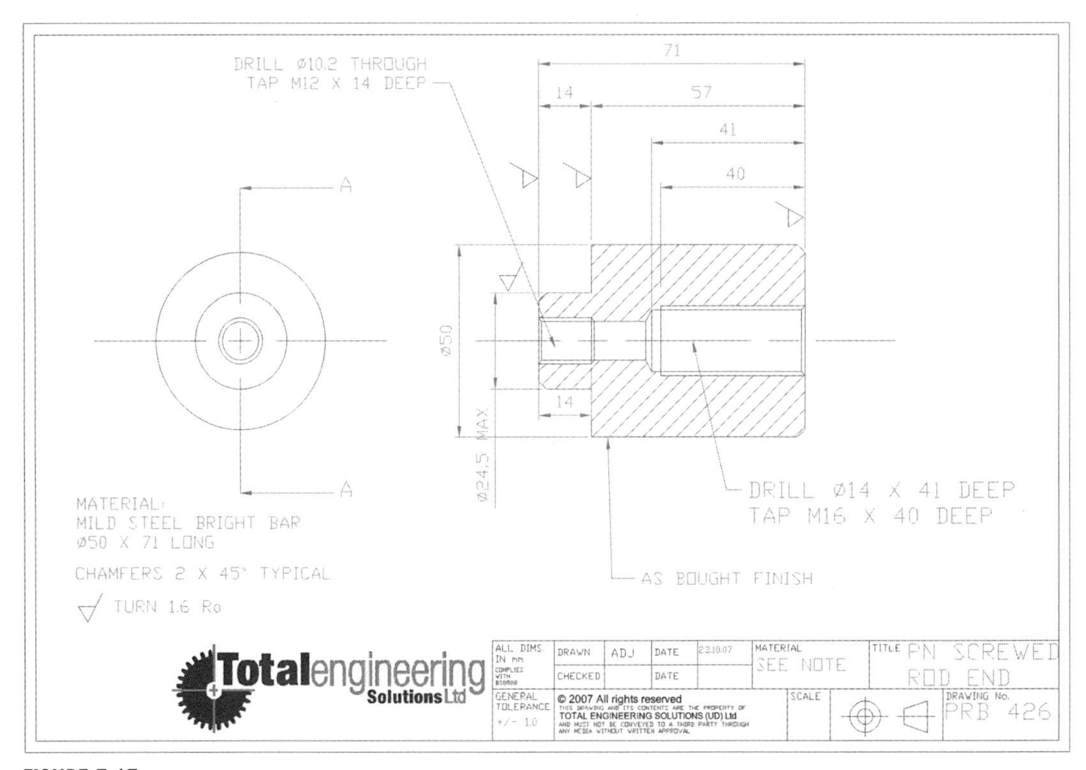

FIGURE 5.15

A Detailed Drawing of a Rod End Showing Two Views, Incorporating a Sectional View [1]

or an auxiliary view. Figure 5.15 shows a simple component requiring only two views for its full explanation. This is a classic example of a component in which the internal shape is more complex than the external shape and therefore requires a section for its full explanation. Figure 5.16 shows an assembly drawing of a pneumatically actuated bollard. This is a full section, which allows the fitter to "see" the internal components and their positions.

5.6.1.4 Hidden detail

Hidden detail is very useful to show internal features of a component but should be used sparingly, since it does not show the depth of a feature. Furthermore, if several features are shown with hidden detail, the view becomes very complex and the reader is unable to decipher which feature is which. In such a case, it is better to take several sections so that each feature is shown individually and can be referred to using section planes shown on other elevations of the drawing. The confusion brought about using hidden detail to show multiple features is the main reason the drawing standards prefer dimensions to be drawn from solid outlines, that is, sections.

5.6.1.5 Dimensions

A drawing without dimensions is merely a picture. The outline of the component shows the "shape," but the dimensions define its size and convert the picture into a technical drawing. There are two types of dimension:

- Size dimensions
- Positional dimensions

5.6.1.5.1 Size

Every feature on a component needs to have its shape defined by *size dimensions*. It is important to note that these size dimensions need to be applied in all three dimensions, that is, in terms of:

1. Height of the feature
2. Width of the feature
3. Depth of the feature

5.6.1.5.2 Position

Every feature a component must have its position defined by *position dimensions*. It is important to note that these position dimensions need to be applied in all three dimensions from a datum, that is, in terms of:

1. Position in the depth of the component
2. Position along the width of the component
3. Position within the height of the component

Figure 5.17 shows a simple cube in which two holes have been machined. In dimensioning such an item, we should treat the component feature by feature.

5.6.1.6 Size dimensions

External component size: Height, depth, and width entered at 150 mm for each element stop
Circular hole size: Height, 50 mm; diameter, 30 mm
Square hole size: Height, 50 mm; width, 30 mm; and depth, 30 mm

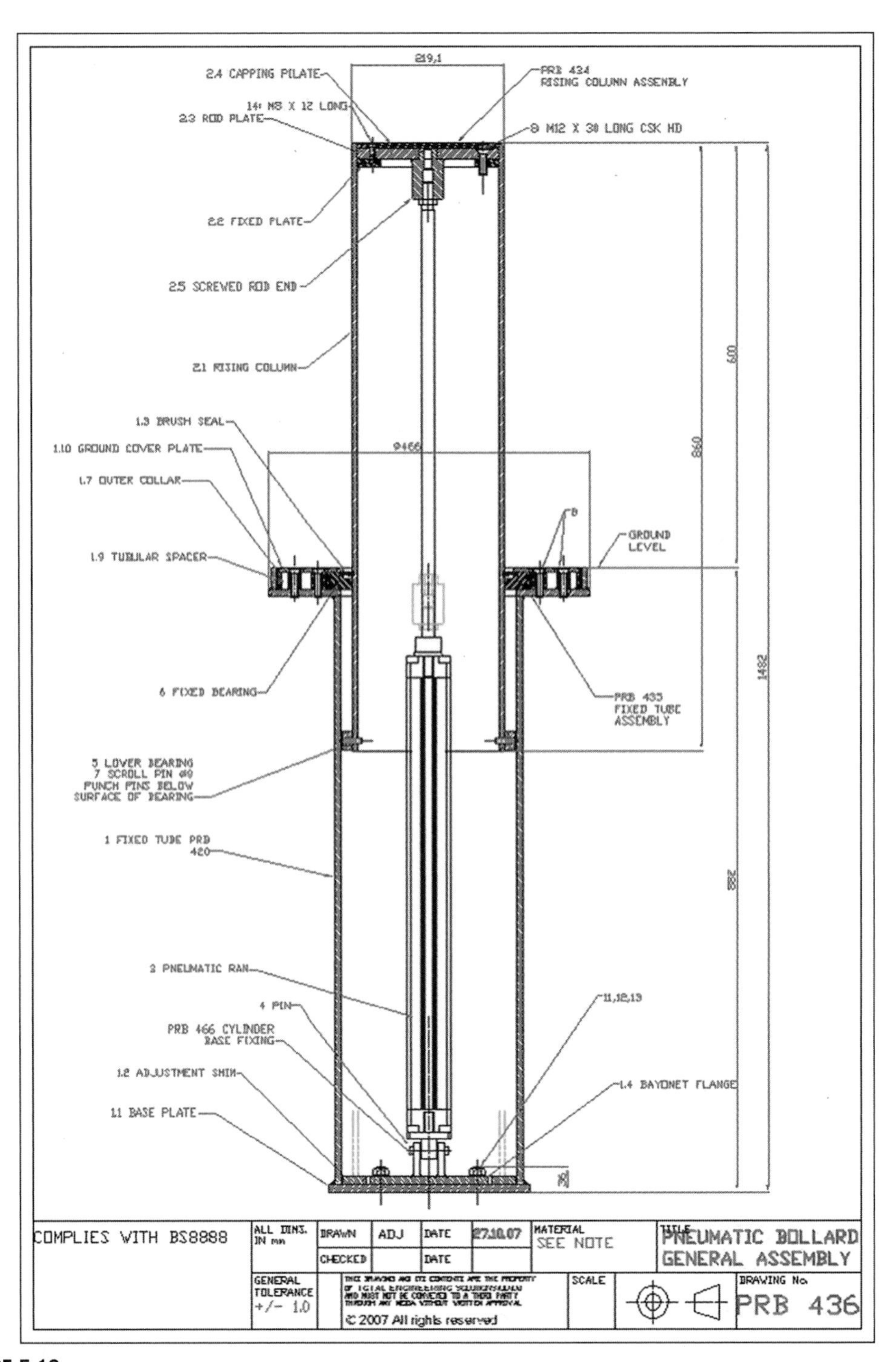

FIGURE 5.16

Sectional View of a Pneumatically Actuated Bollard Assembly [1]

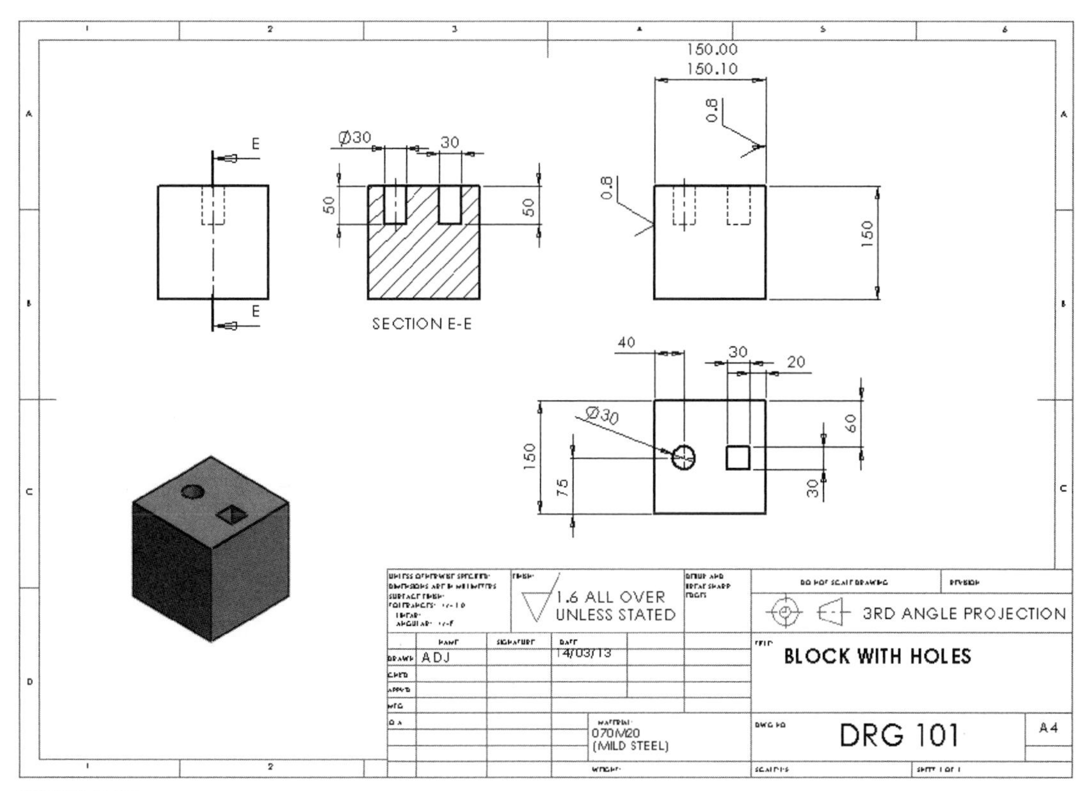

FIGURE 5.17

Cube with Machined Holes

5.6.1.7 Positional dimensions

There are two features, the circular hole and the square hole:

1. *Circular hole position.* Two positional dimensions, from the edge of the block to the center of the hole, position the hole in two dimensions. The third dimension is given by the depth of the hole.
2. *Square hole position.* Two positional dimensions are shown, from the edge of the block to the edge of the hole. The third dimension is given by the depth of the hole.

We can see that sometimes positional dimensions and size dimensions become one and the same. The depth of each hole, for instance, becomes a combined size dimension and positional dimension.

5.6.1.8 Dimensioning strategy

As with writing reports and other documents, dimensioning strategy needs to consider who will be reading the technical information displayed on the drawing. Dimensioning strategy will change from industry to industry, but nevertheless, dimensioning should be tuned to manufacturing personnel who directly translate those sizes and positions to the component.

Sheet metal workers will require not only cut sizes but also position of bend lines. Foundry personnel (pattern makers in particular) will require component sizes plus an extra thickness to account for casting tolerances, draw angles, and so on. The dimensions for a machined component will need to take into account how that component is to be held in the vice or chuck. Figure 5.7 showed a shaft that requires machining.

We see from Figure 5.7 that the dimensioning strategy has been developed so that the machinist (turner) can turn the shaft from one end while holding the other end in the lathe chuck. This has been achieved by creating a datum at each end of the shaft. The turning tool will cut from right to left, finishing at the flange. Some measurements will be taken from the right-hand shaft end along the length of the shaft toward the chuck. When one end of the shaft has been completed, it is merely turned round so that the machined end can be inserted into the chuck, allowing the machining process to be completed on the unturned end.

Attention is also drawn to the method of dimensioning the keyway. U.K. and U.S. keyway standards indicate the depth of keyway is taken from the upper surface of the shaft. In practice this surface is machined away and is an impractical dimension for the machinist to measure, since there is no material from which to measure. Instead the machinist will use a vernier gauge or micrometer to measure the distance between the bottom of the keyway and the diametrically opposite surface of the shaft. It is therefore incumbent on the draftsperson to calculate this dimension and insert it into the drawing.

The approach to dimensioning is to give the artisans numerical dimensions that can be directly applied to the component. The artisans are often in a dirty environment and are ill equipped to perform calculations and conversions while at their machine tools. It is the draftsperson who has the means and is in an environment that is conducive to performing these calculations and manipulations. The aim is to allow the components to be manufactured efficiently while reducing mistakes due to poor information.

5.6.2 Other information for complete drawings

5.6.2.1 Tolerances and surface finish

Sizes are only part of the story of making the drawing into a true technical representation of a component. Dimensional accuracy requires tolerances, and with tighter and tighter tolerances come the requirement for smoother and smoother surface finishes. The shaft in Figure 5.7 and the cube in Figure 5.17 both possess toleranced dimensions with the incumbent surface finish specification.

It is important to realize that tolerance and surface finish indicate a machining process and should be carefully applied. A tight tolerance cannot be achieved with a very rough surface finish where the surface roughness could even be larger than the tolerance. Specific tolerances are usually inserted on critical dimensions such as diameters, which house bearings or seals. The title block of a drawing normally has a general tolerance. This is a tolerance that applies to all the dimensions on the drawing that are not specifically toleranced. It should be as generous as possible but is specified to keep the overall dimensions within reasonable bounds.

The quest for accuracy generally means tighter tolerances and smoother surface finish. Unfortunately, tighter tolerances incur exponentially, increasing machining costs. For example, a shaft turned to a diameter of 50 mm has a tolerance of:

ϕ50.00 / 50.11 mm,
Surface finish 1.6 Ra (1.6 roughness average)

These sizes specified to two decimal places generally mean that the machining process is to be a turning operation.

If the tolerance is now specified at:

ϕ50.000 / 50.111 mm,
Surface finish 0.8 Ra

the specification to three decimal places means that the machining process is now specified to be a cylindrical grinding operation. This also means that the cost has quadrupled compared to a turning operation.

The lesson here is that tolerance accuracy and its application on the drawing may lead to enormous increases in component cost. The designer or draftsperson must continually ask the question, "Do the tolerances really need to be so tight?"

Golden rule: Apply tolerances that are as wide as is practical.

5.6.2.2 Center lines

Center lines denote a circular feature such as a shaft or a hole. A rectangular feature seen on an elevation of a drawing could be identified either as a circular feature or a rectangular feature. The center line is the method of quickly identifying the shape.

The sectional view of the cube in Figure 5.17 shows the two hole features. In that particular view, the only way the circular hole can be identified is by its center line. If the center line is removed, the two features are identical. The shaft shown in Figure 5.7 is also shown with a center line. If the center line were not present, convention suggests that it should be considered to be a rectangular cross-section.

In the reading the drawing, clarity dictates that center lines should always be entered on circular features.

5.6.2.3 Material specification

There are many other features that should be included on a drawing because they are essential for manufacture. One of those is material specification. This must be a specific material designation, such as:

EN (070 M20): Low-carbon steel (a version of mild steel) used for general-purpose steelwork
EN8 (080 M46) (AISI 1040): Medium-carbon steel, usually used for shafts
EN24 (817 M40) (ASTM A1011): High-strength steel, used where strength and low weight are important

The correct designation of material on the drawing specifies that material exactly. If general terms such as *mild steel* or *aluminum alloy* are inserted, it leaves the choice of material open to guesswork by well-intentioned manufacturing staff who probably do not have a grasp of the end use of the component and are therefore ill equipped to select appropriate material.

5.6.2.4 The title block

The title block houses information such as drawing numbers, title, scale, author, date, revisions, and other data essential to making the drawing complete. This element of the drawing is just as important as the outline of the component, since it gives background information and source information. The drawing number is the key to the designation of parts and its position in the overall assembly. Without the drawing number, the component would be lost in a jumble of parts. Examples of title blocks are shown in Figures 5.7 and 5.17.

5.6.2.5 Text and formal lettering

Drawing standards set out the preferred standards for text and numbering. They take into account the size of text compared to the size of the drawing sheet, the style of the text and its clarity, and whether the text is handwritten. In the days before CAD, when many drawings were hand-drawn on large drawing boards, the text was also handwritten; however, most CAD systems now offer a variety of fonts and the ability to change font size and style.

On occasion it is still necessary to write on drawings by hand. On such occasions it is useful to adhere to the appropriate standards. Figure 5.18 is an extract from the American National Standards Institute (ANSI) that indicates the style and size of handwritten lettering.

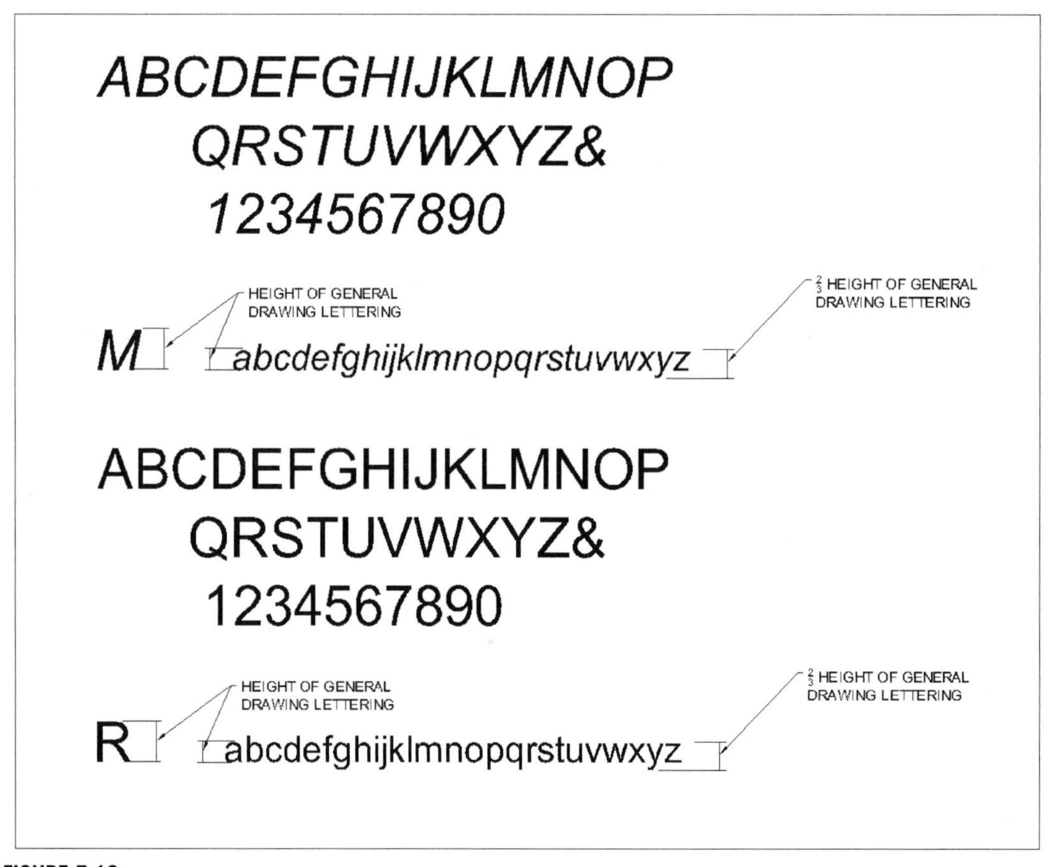

FIGURE 5.18

Drawing Standards Applied Lettering Convention [3]

5.6.3 Drawings checklist and common errors

It is useful to check through the drawing to make sure all the data is present. To this end, here we include a general checklist.

5.6.3.1 Title block and border

- Border
- Component title
- Drawing number
- Date
- Author
- Material specification
- General tolerance
- Scale
- Dimensions (in mm/inches)
- Revision/modification number
- Drawing angle 3rd/1st

5.6.3.2 Overall drawing

- Dimension to suit the machining method
- Dimension: Critical dimensions
- Dimension: General dimensions
- Dimension: Features
  - Size of each feature in three dimensions
  - Position of each feature in three dimensions
- Angle dimensions
- Specific tolerance: usually located with a critical dimension
- General tolerance
- Geometric tolerances
- Annotations: Always use capital letters

5.6.3.3 General drawing instructions

- ϕ diameter symbol: Designation of diameter dimensions
- Center lines: Designates a round feature
- Surface finish marks: Usually located near a critical dimension
- Cross-hatching: Designated sectioned area
- Scrap sections
- Section reference planes: Position of cut line for sections
- Welding marks: Designates weld type/style/position/method/size

5.6.3.4 General remarks often found on drawings

- References to another drawing if a view is on a separate drawing; e.g., another view on a different drawing
- Ballooned parts numbers should specify components on assembly drawings
- Remove sharp edges

- Unless specified, all surfaces to have the following surface finish: _____
- Unless specified, all welds to be as follows: _____
- Unless specified, _____

5.6.3.5 Oft-repeated convention errors

Please don't make these mistakes!

- Cross-hatch webs
- Cross-hatch round objects
- Cross-hatch bought-out components
- Dimension to hidden detail: Use sections or scrap sections
- Dimension to the end of shafts; confusing
- Dimension diameters as a radius; machinists work to diameters

5.6.4 Drawings: common errors

5.6.4.1 Dimensioning

- Keyways are depicted in a particular way as shown in Figure 5.19. Dimensions depicted in this way will allow the machinist to measure the feature and compare a direct measurement to the dimension.
- Avoid tolerance buildup by dimensioning from another dimension, as cumulative tolerances will lead to dimensional errors.
- Dimensioning to three places of decimals, e.g., 24.987, forces the machinist to work to those fine limits. Give the machinist the largest tolerances possible. Ask yourself, "Do I really need the dimension this tight?"
- Apply dimensions to features that the machinist can directly measure.
- Dimension diameters, not the radius of a shaft. The machinist can directly measure diameters.
- *Do not* dimension to end views of shafts. It is confusing and does not help the reader understand the information. See Figure 5.20.
- Dimension 38.97. Round it up!
- Ensure that *all* relevant dimensions are on the drawing—PCDs and sizes and positions of features.
- Check component size in three dimensions and feature position in three.
- Tapped holes: Ensure tapping drill size and depth are given as well as tap size and depth.

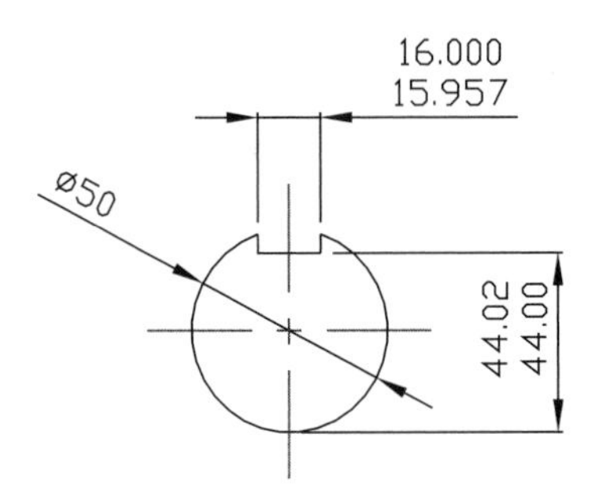

FIGURE 5.19

Dimensioning and Measurement of a Key Way [1]

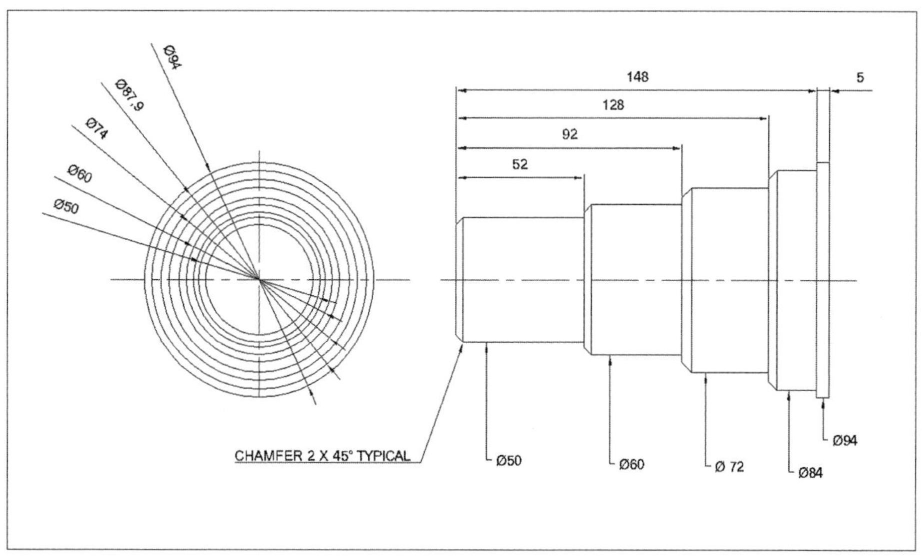

FIGURE 5.20

Dimensioning to the End of Shafts Leads to Confusion [1]

- Dimension to suit the manufacturing process—e.g., turn a shaft from both ends because the shaft has to be held in a chuck on the end that is not being machined.
- Dimensions referring to diameters should be preceded by the diameter symbol ϕ. Dimensions without this symbol will be treated as rectangular.
- Assembly drawings: Include overall sizes and/or scale. Without these items the size of the component is impossible to determine.
- Ensure that arrowheads can be seen. Match the size of the arrowhead to the size of the text.
- If you draw arrowheads by hand, they should be blocked in with a ratio of 1:1/3. See Figure 5.21.

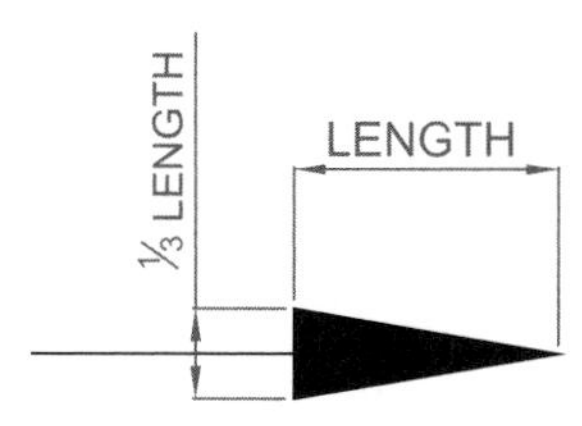

FIGURE 5.21

Preferred Arrowhead Ratio [1]

5.6.4.2 Drawings: general

- Circular features require a center line. This line must be visible; therefore, AutoCAD yellow is not sufficient because it cannot be seen when printed.
- Each component must be drawn on a separate drawing, complete with a unique drawing number.

- Heat treatment has to be specified on the drawings. It can be specified to a particular hardness or a particular stress level. This needs to be applied in conjunction with steel type used.
- Do not cross-hatch shafts or webs.
- Avoid drawing small outlines on a large sheet of paper. *Fill* the paper, either by reducing the paper size or increasing the scale of the outline.
- Ensure that scales are correct as specified on the drawing. Sometimes CAD systems automatically apply a scale.
- Title blocks: *Use them.* They include important and useful data.
- Use standard sizes of screws, e.g., M8, M10... but *not* M9. There is no such size.

5.6.4.3 Materials

- Specify materials exactly using the appropriate designation! A general material description of "Steel" or "Brass" is too broad.
- Often cast iron is specified. Gray cast iron is brittle in tension and cracks in cold weather. Select an appropriate cast iron, such as spheroidal graphite cast iron, which has the strength of mild steel. UTS $320 MN/m^2$.
- Materials: Use medium-carbon steel EN8 (080 M46) for the shaft and mild steel (070 M20) for general steelwork applications.
- Use hidden detail sparingly; otherwise, it becomes confusing.
- Use scrap sections in preference to hidden detail. These are clearer because the depth can be seen.

5.6.4.4 Parts lists

A parts lists must include the following:

- Part number
- Quantity
- Description
- Material and cut-off sizes
- Drawing numbers
- Bought-out part numbers
- If bought-out, the supplier

5.6.5 Allocation of drawing numbers

Drawing number systems vary from project to project and from company to company. Every drawing should be allocated with a unique drawing number. The drawing number may be a pure number, perhaps computer generated or a combination of alphanumeric characters. The possibilities are so vast that only a general overview can be given here.

The drawing number for a totally computerized system may be of the following order:

01-12345-02

This number may be unintelligible to someone outside the company, but fed into a computer it can be very precise. Often drawing numbers possess a sequence that has meaning for practitioners who work with the drawings. The number may reflect a client or a general assembly or a subassembly and finally a part and subpart, as shown in Figure 5.22.

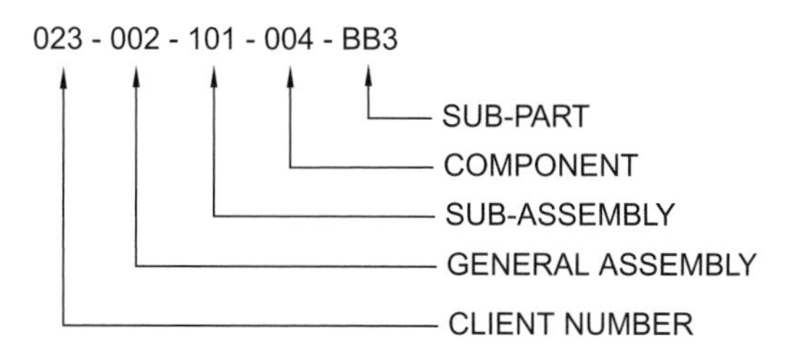

FIGURE 5.22

Example of a Drawing Number [1]

- Subpart: This is an item such as a web that will be part of a fabrication, creating a full component.
- Component: This could be a welded fabrication assembly. In the case of an engine, it could be the cylinder head block or the crank shaft.
- Subassembly: This is an assembly of parts that form a discreet product, such as an internal combustion engine. This will eventually be fitted to a much larger assembly.
- General assembly: This is a "map" that shows where the subassemblies fit on a large assembly. The whole assembly of a product such as a passenger vehicle is split into convenient elements, which are the subassemblies. A subassembly may be a suspension system, a rear seat, a wheel hub, or an internal combustion engine. The general assembly brings the whole of the subassemblies together.

On a smaller design, the drawing number may be much simpler, such as that shown in Figure 5.23. The drawing number shown in Figure 5.23 is much simpler in that it shows the designation for the project, the particular subassembly, and the component.

Drawing numbers may be so simple that they may be merely a number. This depends on the complexity of the assembly. An uncomplicated drawing numbering system is shown in the example parts list shown in Figure 5.24. These drawing numbers are typically of the style:

PRB 435

PRB represents the project, which is a pneumatic rising bollard (PNB), whereas the number *435* is merely the next drawing number in the sequence.

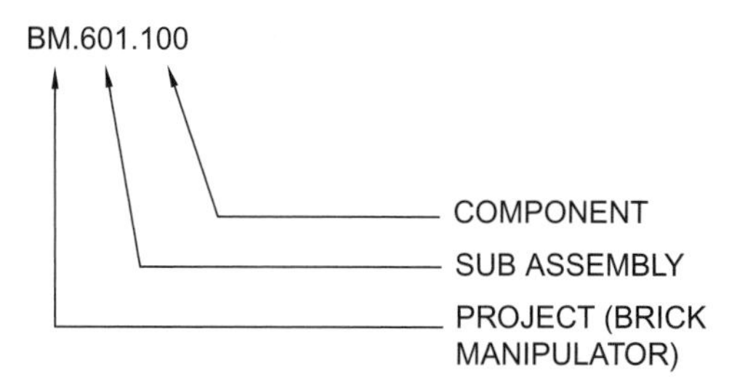

FIGURE 5.23

Alternative Drawing Number [1]

Part No	Quant	Description	Part/Drg no	1 0ff Cost Each	Total	
		Pneumatic Rising Bollard 220 dia	PRB 436			
		General Arrangement				
		Fixed Tube Assembly	PRB 435			
1	1	Inner Fixed Tube	PRB 420	50.00	50.00	
		Mild Steel CHS:				
		OD 273 x ID 253 x 829 long				
1.1	1	Base Plate	PRB 421	32.00	32.00	
		BMS Plate OD 293 x 10 thick				
1.2	1	Adjustment Shim BMS Plate or Nylon 6	PRB 423			
1.2.1	1	BMS Plate OD 249 x 10 thick		15.85	15.85	
1.2.2	1	BMS Plate OD 249 x 8 thick		15.85	15.85	
1.2.3	1	BMS Plate OD 249 x 2 thick		15.85	15.85	
1.3	1	Brush Seal		7.00	7.00	
		Formseal Ltd 810 long base on 257 diameter	CS 2/20			
1.4	1	Bayonet Flange	PRB 422	15.00	15.00	
		BMS Plate OD 170 x 10 thick				
1.5	1	PN Ground Plate	PRB 429	20.00	20.00	
		BMS Plate OD 466 x 10 thick				
1.6	1	PN Inner Collar	PRB 430	1.00	1.00	
		BMS Bar 861 long x 25 x 6				
1.7	1	PN Outer Collar	PRB 431	1.00	1.00	
		BMS Bar 1315 long x 25 x 6				
1.8	1	PN Containment Collar	PRB 432	1.00	1.00	
		BMS Bar 1365 long x 32 x 6				
1.9	8	Tubular Spacer	No Drg	0.50	4.00	
		MS Tube OD 20 x ID 14 x 25 long				
1.10	1	Ground Cover Plate	PRB 433	22.00	22.00	
		OD 440 x 6 thick				
		Customer Specified BMS Plate, Stainless Steel Plate				

FIGURE 5.24

A Typical Parts List [1]

Sometimes the efficiency of the build is dependent on how the drawing number is designed. The drawing number should possess certain elements that are known to the manufacturing and design personnel. Ideally the drawing number should have the attributes described in the following sections.

5.6.5.1 Must-have attributes

- The drawing number should be unique.
- It should normally follow an intelligible sequence.

5.6.5.2 Desirable attributes

- Indicate the project type
- Indicate the customer
- Indicate the subassembly
- Indicate the general assembly

5.6.6 Parts lists

Parts lists have many names. They may be called bills of materials, component lists, or inventory lists. A parts list is created by the design department as the designs are created and parts and components are specified. The most obvious function of the parts list is to itemize all the parts required to build up an assembly, but its real task is to act as a build itinerary and a guide to the various functions (e.g., purchasing department, planning department, manufacturing, accounting department) that are required to manufacture the product.

The parts list is the key with which to identify components and cross-reference drawing numbers, parts numbers, and material specifications. Of all the documents that make up an assembly of parts, the parts list is probably the most important. Without the parts list, the cross-referencing of materials and components cannot take place. The parts list provides the background information to key together all the disciplines required to manufacture the assembly.

A parts list must contain:

- A complete list of parts
- A list of subassemblies
- A physical description of parts and raw materials
- Drawing number designations
- Links of part numbers to drawing numbers
- Links of parts numbers to bought-out components

Throughout a manufacturing company, various disciplines contribute to the final creation of the product. Each will use the parts list in a specific way, as itemized in the following sections.

5.6.6.1 Purchasing department

Purchasing uses the parts list as a shopping list, sourcing all the bought-out parts and raw materials that will eventually be manufactured.

5.6.6.2 Planning department

Planning uses the drawings and parts list in combination to organize the progress of the components through the works. After considering all the machining processes required for each component, the planning department allocates manufacturing time for each station and plans the progressive manufacture of each component through the final stage of assembly.

5.6.6.3 Manufacturing department

Manufacturing uses the parts list to identify subassemblies and link raw materials to components while planning work schedules for manufacturing personnel.

5.6.6.4 Accounting department

Accounting allocates material costs and labor costs against each part and eventually completes a cost analysis against each subassembly. In this way a costing model can be compiled.

5.6.6.5 Contents of the parts list

As we've shown, the parts list is used by several departments that will apply their own information. It should be noted that the design, drawings, and eventually the parts list are generated in the design office. It is incumbent on the designers to apply certain information that the other departments can use. This general information is as follows:

- The part number
- The quantity required
- The name or description of the part
- The material from which the part is to be made
- The overall size of the raw material
- The designation of bought-out parts such as bearings, seals, motors, and the like
- The contact details for suppliers of bought-out parts

A typical parts list is shown in Figure 5.24. Many of these details are present here, but in this case costs have been added so that a total cost can be tallied.

5.7 THE CLIENT

Depending on the type of project, the client could be a large institution, such as a government or a multinational company. The client may also be an individual for whom the designer is acting as a consultant. Most designers work for institutions between these two extremes.

5.7.1 Large institutions

Large institutions tend to issue large, high-value projects that are bound by rules and penalties set out in a contract. It is normal practice for the client and the design team to name liaison officers who routinely meet and share regular updates, ensuring that checks are in place and that stage deadlines are being met.

This system of checks is important because clients on the contract often offer stage payments. For example, the contract might suggest that once the concept has been presented and accepted by the client, a payment will be made to further the development process. In such a case, the liaison between designer and client is likely to be formal in terms of shared reports and weekly or monthly meetings. It is very likely that appointed liaison officers will be in informal, regular contact through e-mail or telephone.

5.7.2 The individual client

Depending on the type of project, the communication process between client and designer could be either very formal or very informal. It is likely that communication between client and designer would occur through e-mail, phone calls, and personal contact. Progress reports are unlikely to be as formal as with the large institution; however, regular contact is absolutely essential. It is likely that between an individual client and the designer or design team, there must be a great deal of trust. If communication fails, human nature suggests that the client will naturally assume that the project has faltered and his project funds are being squandered. This type of client requires constant reassurance.

5.7.3 The medium-sized client

Most designers will be involved in projects that are supported by their own company. Communication therefore is often between colleagues within the same design team and on similar levels based on the client's team. Often the task of formal communication with the client is left to the design team leader and may be organized on a formal basis. In such projects, however, it is important that communication is established at lower levels, between members of the client team and the design team, so that technical information can be shared.

5.7.4 Client requirements

No matter what the project, the basic requirements for the client are a willingness to fund the design and development of a product or service. Projects vary in size and require different approaches. The institutional client will require a great deal of formality and formal reporting; the individual client may want to have input into the design process. The detailed involvement of the client can either be very helpful or counter-productive.

For example, in the design of a bed-lifting device, discussed in earlier chapters, the design of the folding leg at the foot of the bed became a major design issue simply because the client was never satisfied with the eight different options that were put forward. The design team was heavily constrained by the lack of fixing points and the visual impact of the leg when raised to the ceiling. The problem was solved when a fresh approach was applied by external designers responsible for the aesthetics of the device.

In other areas of the design, the client was extremely helpful, but this particular point illustrates the problems that can arise when a client becomes closely involved in a project. In the longer term, client involvement may cost more than if the client simply accepted the design team's suggestions.

5.7.5 The Client's wishes

One of the most difficult parts of a project is always the beginning, when the designer is attempting to understand the requirements and constraints. The client could actually make this process more difficult, since he will have his own ideas as to how to solve the design issues, even though he might not be a specialist. In such a case the designer has to develop an attitude whereby he will absorb all information but then step back and take account of what is *required*. It is easy to be led by the client to a solution that might not work. In such instances it is useful to write down a list of requirements and constraints.

For example, an individual client (a small company) requested that a designer design a simple pneumatic press that would cut out double-sided adhesive tape into heart shapes. The double-sided adhesive tape was to be used to build up a soft three-dimensional heart to be placed on the front of a Valentine's card. Several layers of fabric, card, and foam were to be combined using the double-sided sticky-tape heart shapes.

The client described the type of machine required and much of the technical details, such as airline pressure, physical size, controls, and the like, and had even allocated a space for the machine once built. It would have been a simple task for the designer to design the required press; however, the designer took a "mental step back" and considered what was actually required. The designer set out a list of requirements and constraints as follows:

Requirements:

- Throughput speed: 10 units per minute
- Low cost
- Combine parts using adhesive
- Assembly performed within the factory environment
- Assembly performed by one person

Constraints:

- Personnel safety (adhesive vapor inhalation)
- Dexterity of personnel in assembly
- Perform assembly within the factory environment
- Cost

It can be seen that in writing down the requirements and constraints, there is no obvious requirement for a pneumatic press. The client was focusing on what he could apply to solve his problem rather than focusing on the real problem, which was, "How do I glue these components together?"

The key to the solution was the third requirement, "Combine parts using adhesive." The correct isolation of the problem led to the use of a pressurized spray adhesive. This evolved into a solution that was much lower cost than the pneumatic press and had more efficient throughput of over 20 items per minute.

In conclusion, it is true to say that sometimes the client does not know what he needs to solve his problem. The designer's skill and self-discipline allow him to step back from the problem, possibly ignoring the client's suggestions so that he can identify the real requirement.

5.7.6 Project requirements

Whatever project is undertaken, the client has employed the designer or the design team as specialists who can solve a particular design problem. The designer must evaluate each project by considering many conflicting requirements. Some of these requirements are:

- Technical requirements
- Cost requirements

- Market requirements
- Production requirements
- Sustainable requirements and possibilities

The contradicting nature of some of the requirements may require constant feedback to the client so that the direction of the project is as the client wishes. For instance, say that a design team has been tasked with designing a lean-burn internal combustion engine with very low emissions. The sustainability requirements are such that the cost increases. Liaison is required with the client to ensure that the increased cost direction of the project can be sustained by the client.

As a design progresses, the use of new materials and processes increases the likelihood of unseen costs. Communication liaison channels are therefore required to keep the client informed of new developments on the project.

5.7.7 Presentation of designs

The presentation of a design depends on the stage of the project and the formality required by the client. An institutional client will probably require stage feedback, but at certain points during the project, such as at concept stage, the client might want to have an independent body scrutinize the design before signing the approval document, thus allowing the project to continue. In other less formal arrangements, the client might merely want to see the concept design in terms of 3-D models for drawings or perhaps as a product specification. At the end of the design-and-build cycle, the commissioning of large items such as a ship is dependent on a formal report from the shipping line. This would normally happen before the final payment.

5.7.8 Styles and methods of presentation

There are many and varied styles of presentation, as this chapter has explained. The presentation needs to be tuned to the understanding of a particular client. The client may have a team of specialists, in which case the presentation should cater for the main points of the design as well as its more technical and cost elements, to which the experts on the team will be receptive.

The presentation could take many forms, contingent on the client's requirements. Any presentation will most likely require a verbal presence accompanied by a visual slide presentation or even posters, reports, or the like. On another level, the presentation may also involve the prototype.

Whatever the design or the presentation process, the designer needs to possess multiple skills, not only in engineering design but also as a diplomat and an expert communicator.

References

[1] Author supplied: designs and drawings from the authors' commercial and teaching experience.
[2] O. Ostrowski, Engineering Drawing with CAD Applications, Butterworth Heinemann, 1989.
[3] American National Standards Institute (ANSI) cross referred to BSENISO5457 (1999).
[4] Open source.
[5] H. Heart, Engineering Communication, (2009), Pearson.
[6] P.H. Right, Introduction to Engineering, Wiley, 1999.

CHAPTER

Performance Prediction

6

6.1 WHY PERFORMANCE PREDICTION IS NECESSARY

When a product is designed and built, the designer needs to specify what the product is capable of achieving and how it may perform. Without performance prediction techniques, the specification of performance would be guesswork. Some designers might estimate materials and strengths by sheer experience. This is a dangerous game, as it could result in a structure that is too weak or one that is heavily overdesigned. Traditionally, designers have made structures heavier than necessary to be "on the safe side."

To optimize material usage and potentially drive down the cost of products, product performance prediction needs to be as accurate as possible within specified limits. A Formula One racing car engine is designed for high performance but to last for only one race, after which it is refurbished and any sacrificial or worn components replaced. This level of accurate performance prediction can only be achieved using sophisticated mathematical techniques in various fields, the most common of which are listed as follows:

- Stress analysis
- Vibration analysis (dynamics)
- Mechanical analysis (velocity motion, inertia, energy, power, etc.)
- Fluid dynamics analysis
- Thermodynamic analysis

Performance prediction can generally be split into two categories:

- Performance prediction to ensure the product is safe
- Performance prediction to obtain a precise working specification

Engineering analysis combines a fundamental understanding of natural phenomena, the laws of physics, and a mathematical ability with which the designer can arm himself in order to perform a particular analysis for a particular product for a particular situation.

6.2 HISTORICAL ASPECTS OF ANALYSIS

Modern man has always manufactured tools as aids to living. Early *Homo sapiens* designed hunting tools such as spears and clubs. If the tools did not work or broke during use, they were modified until they worked, creating a trial and error approach to design. As time went on, society required more sophisticated products such as carts, bridges, advanced weaponry, and so on. Up to the late Middle

Sustainability in Engineering Design. http://dx.doi.org/10.1016/B978-0-08-099369-0.00006-6

Ages, the devices so built were not designed in the modern sense but were manufactured and tried out. If the devices did not work, they were modified.

During the Industrial Revolution, engineers began to quantify some of the parameters that would assist in the design of various devices and products, but still, much of the work was based on experience. Many engineers gained their "experience" from one or more failures: learning from their mistakes. Famous engineers such as Brunel, Telford, Arkwright, and Stephenson were gifted people who were able to quantify some of the parameters of their designs and who had a wealth of experience to guide them on the remainder.

Modern design began to emerge when drawings and plans were created so that builders, engineers, and other artisans could understand what was in the mind of the lead engineer. True design came of age when science and mathematics were applied to the design of structures and other artifacts. Engineers and mathematicians such as Newton, Bernoulli, Reynolds, Euler, Boyle, and Faraday, among many others, experimented in the physics of their field to define laws and to apply the mathematics to model physical phenomena. This assisted later engineers, who applied these mathematical principles on future projects and individual designs.

Perhaps the greatest and earliest breakthrough of modern engineering mathematics was by a Ukrainian *Stephen P. Tymoshenko* (1878-1972), who is known as the father of modern engineering mechanics.

Tymoshenko was instrumental in developing "the theory of elasticity," which quantifies and sets out analytical techniques for calculating stresses and strains in materials under varying loads. He expanded his theorems and techniques to include general mechanics and buckling plus many other areas of stress analysis, which engineers now use as an everyday tool in their analytical armory. It is significant that *Tymoshenko* also used the *Rayleigh* method of Finite Element Analysis as early as 1911. After moving to the United States in 1922, *Tymoshenko* became a professor, first at the University of Michigan and later at Stanford University where he and his research teams greatly expanded stress analytical techniques and methods.

With the advent of quantifiable stress analysis techniques, it was possible to analyze components using computing power. The finite element method of stress analysis had been known for decades, but its use only became widespread with the development of more powerful and sophisticated computers.

6.3 MATERIALS TESTING

The analytical techniques put forward by *Tymoshenko* and others were excellent tools for predicting the stresses within a component, but these calculated stresses had to be weighed against the strength of the material. Alongside the expansion of analytical techniques was an expansion in testing methods, which quantified the strength of the material plus many other parameters.

Materials such as steel behave in different ways under different conditions. Perhaps the most useful test is the tensile test where a standard-sized specimen is put into tension (pulled apart) until it breaks. The result of such a test provides information about the following:

- elasticity (Hook's Law)
- yield point of the material

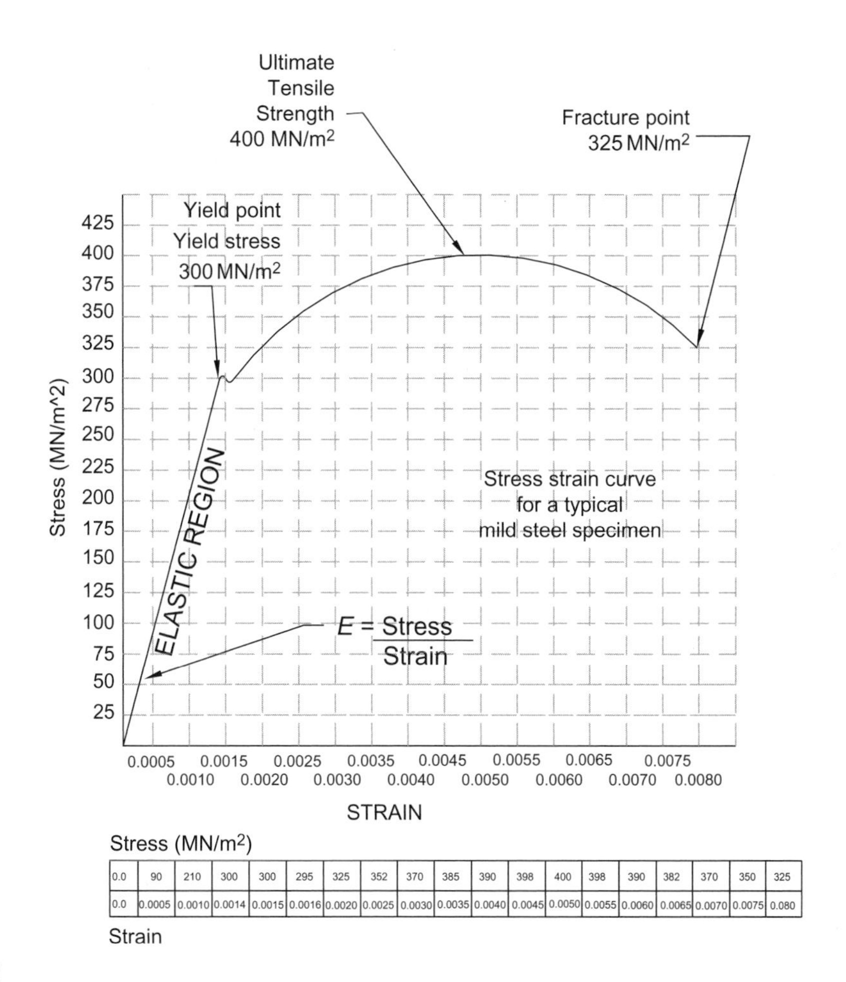

Stress (MN/m^2)	0.0	90	210	300	300	295	325	352	370	385	390	398	400	398	390	382	370	350	325
Strain	0.0	0.0005	0.0010	0.0014	0.0015	0.0016	0.0020	0.0025	0.0030	0.0035	0.0040	0.0045	0.0050	0.0055	0.0060	0.0065	0.0070	0.0075	0.080

FIGURE 6.1

Typical Tensile Test Characteristic for Mild Steel [1]. (For color version of this figure, the reader is referred to the online version of this chapter.)

- toughness (energy of mechanical deformation prior to rupture)
- ultimate tensile strength (UTS)

A typical graphical plot of stress against strain (force against extension will give the same results) is shown in Figure 6.1.

6.3.1 Elastic region

Any material possesses elasticity when stretched and behaves according to Hook's Law. When the material is stretched within the elastic region, upon release, it will spring back to its original position without being permanently deformed.

6.3.2 Yield point

A material that is stretched up to its yield point will always spring back to its original position when released. If the yield point is passed, then the material is said to be in a "plastic" state and will remain deformed.

6.3.3 Ultimate Tensile Strength (UTS)

This is the maximum strength of the material and is often used as a comparison to theoretically calculated stresses.

6.3.4 Fracture point

As the tensile specimen is stretched, it will break at the fracture point.

6.3.5 Young's Modulus *E*

This is sometimes called the "Elastic Modulus." It can be calculated as follows:

$$E = \frac{\text{Stress}}{\text{Strain}}$$

The elastic modulus when calculated thus has units of GN/m^2. Depending on the toughness of the material, the slope of the characteristic within the elastic region will change. If the characteristic is very steep, the material can be considered to be tough. This is the case with steel, where a high stress (or force per unit cross-section) applied to a specimen will give only a small strain (extension). On the other hand, if the angle is very low such as that recorded by stretching an elastic band, a small stress would lead to a very large strain. Clearly, the elastic band is not as tough as a steel specimen.

These and many other valuable parameters that describe the specification of the material allow a comparison of calculated stress against the strength of the material.

Example: Actual Stress Compared with Material Strength

A ϕ 10 mm mild steel tie rod has to be placed in tension in a roof structure and has been shown to be under a load of 10 kN. Using basic stress analysis, the calculated stress σ in the material can be seen as follows:

$$\sigma = \frac{\text{Load}}{\text{Area}} = \frac{\text{Load} \times 4}{\pi \times d^2} = \frac{10 \times 10^3 \times 4}{\pi \times 0.01^2} = 127.3 \times 10^6 \,\text{N/m}^2 \; (\text{MN/m}^2)$$

where d is the diameter of the mild steel bar.

Basic tensile tests on mild steel specimens have shown that the ultimate tensile strength (UTS) is 320 MN/m^2, the value of which can be compared against the theoretical stress in the tie rod. In this case, the theoretical stress is much less than the actual strength of 320 MN/m^2, rendering the tensile rod safe in this application.

Confidence in analytical "predicted" stresses and, further, confidence in the strength of the materials data derived from laboratory tests provides engineers with the tools to show that a component or product is safe under an assumed loading regime.

There are many forms of data that may be provided for a particular material. The tensile test mentioned earlier is one of the most useful of them, but there are others such as torsion tests, fatigue tests, impact tests and many, many more.

For instance, the resistance to a shock load depends on how the energy is absorbed by the material. A material such as steel can be in its natural, malleable state or in a hardened state, so shock loading will have different effects. If, for instance, a mild steel bar is placed in a vice and hit with a hammer, the bar will absorb much of the energy and merely bend. If the same treatment is applied to a bench file, which is hardened and very brittle, the file will shatter. A small amount of energy will have been absorbed but the material cannot absorb much energy because of its brittleness.

The test developed for this application is the "Izod" or "Charpy" test and measures the energy absorption by the material under a prescribed shock load. A diagram of a typical test operation can be seen in Figure 6.2.

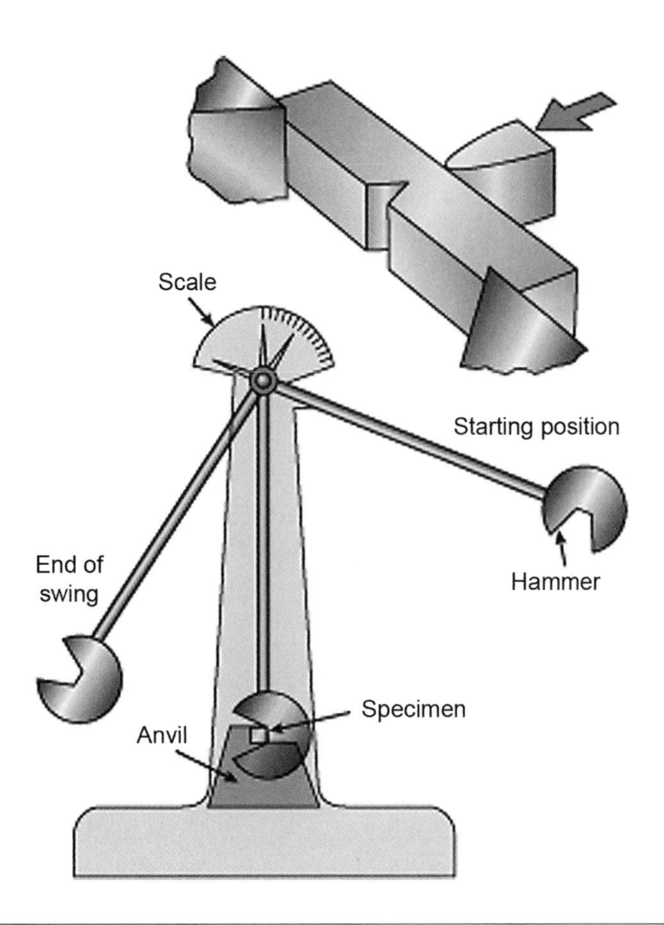

FIGURE 6.2

Typical Izod/Charpy Test [2].

Standard information for particular materials is available from many diverse sources such as from textbooks, data files, and, of course, the Internet. It is incumbent on the designer to build a data file as he/she progresses through his/her career.

6.4 FACTOR OF SAFETY

Once we have the means to calculate theoretical stresses and the ability to compare those results against test data for the material, the inevitable question that arises is "how many times safer a structure is when placed under anticipated loads." Before finding the answer to this question, another question needs to be answered: "Why would a device need to be several times safer than suggested by the anticipated loading?"

Though stress analysis can give the engineer anticipated working stresses within the material and the structure, inaccuracies are inevitable. Generally, it can be assumed that any calculated stresses are only 80% accurate. Furthermore, there are uncertainties in the structural regime which include the following:

- Material integrity (inclusions, blowholes, inaccurate welds, etc.)
- Inaccurate calculations
- Overloading (abuse by the user)
- Aggressive environment (weaknesses caused by early corrosion)
- Consequences of failure (failure of the structure could lead to fatalities)
- Cycling loading leading to a weakened structure
- Change in material microscopic structure as a result of fastening techniques such as welding

A system of factors of safety can be applied to a structure, which will allow for the uncertainties listed here. These general factors of safety are shown in Figure 6.3, but it should be noted that more specialized and detailed safety factors may be employed by professionals working in individual industries.

6.5 CONSOLIDATION OF SAFETY IN STRUCTURES AND DEVICES

The ability to accurately analyze and define the stresses within a structure combined with the ability to derive accurate material strengths from experimental data has inevitably led to the quantifying of structural safety. The use of factors of safety now allows engineers to create a specific safety margin. If, for instance, a structure possesses a factor of safety of 4, it can be said with confidence that the structure is four times safer than the envisaged loading regime.

The three factors, therefore, that allow engineers to predict structural safety are as follows:

1. Accurate predictive stress analysis
2. Accurate structural data derived experimentally
3. Application of factors of safety

   Example: Factor of Safety Applied to a Bridge Structure

A footbridge was to be designed with an overall safety factor of 4, which was intended to be applied to any component within the bridge structure. This was to cover for overloading, errors in manufacture,

Factor of Safety	Explanation of Parameter
1	A very rare application where the theoretical working stress matches the strength of the material. May only be used where there is absolute confidence in stress analysis, material integrity precise use of the components or structures, loading regime.
2	A minimum safety factor applied to compensate for shock loads. Used for ductile materials where there is confidence in the loading regime, material integrity, stress analysis.
3	A more general safety factor often used for brittle materials where there is confidence in loading regime, material integrity, stress analysis and for ductile materials where there is doubt about material properties, loading regime, manufacturing accuracy.
4	Used where there is uncertainty of loading conditions (perhaps abuse or overloading), uncertain material properties or rudimentary analysis
5	Often used where customer reassurance is required or consequences of failure may prove to be expensive. The structure may be used in a corrosive atmosphere, e.g., chemical plant or on board a ship and subjected to a brine environment.
10	Used where there is a combination of uncertainties but in particular where consequences of failure may lead to fatalities such as in the case of critical components of elevators.

FIGURE 6.3

General Factors of Safety [1].

use in the natural environment (wind forces, corrosion, etc.) and gentle cycling loading. The material's ultimate tensile strength was found to be 400 MN/m^2 from lookup tables for the material. The safety factor was applied as follows:

$$\text{FoS} = \frac{\text{UTS}}{\text{Working stress}}$$

$$\text{Working stress} = \frac{\text{UTS}}{\text{FoS}} = \frac{400 \times 10^6}{4} = 100\,\text{MN/m}^2$$

When calculating working stresses, the designers must therefore apply materials to the bridge structure whose working stresses do not exceed 100 MN/m^2.

Example: Multiple Safety Factor

A cargo crane was to be designed to be fitted to an oil rig supply ship for duty in the North Sea. The designers realized that it would be necessary to apply multiple safety factors to cope for various conditions such as those listed here:

- uncertainties with regard to material integrity
- low-temperature brittleness (reducing the ductility of the material)
- corrosive environment

- shock loads
- abusive use (overloading)
- cycling loading

It was decided to apply factors of safety as follows:

5 To cover shock loads
abuse
uncertainties regarding material integrity
abusive use

2 To cover a corrosive environment
2 To cover cycling loading
2 To cover low-temperature brittleness

When these safety factors are combined, the overall safety factor becomes 11.

Example: Pivot Points on the "Car Stacker"

It is sometimes necessary to overdesign components so that they "look right" and the user is reassured that the components possess appropriate strengths. This involves applying a much greater factor of safety than would normally be necessary.

The Car Stacker 3D parking system shown in Plate 6.1 accepts a car on the ramp, which then slides forward to allow a second car to be parked underneath. The pivot, cantilever shafts were calculated at 20 mm diameter, which included an adequate safety factor. However, it was considered that they might not appear to be very strong and that users might not feel confident about parking their expensive cars underneath the ramp when it was loaded with another vehicle. In this case, the consensus was that it would be better to increase the diameter of the pivot pins to 50 mm, thus overdesigning the pivot pin

PLATE 6.1

Car Stacker 3D Parking System. (For color version of this figure, the reader is referred to the online version of this chapter.)

by giving it a safety factor of 15. The ultimate aim was to ensure that the pin was safe, but primarily it was to make it "look" safe and give confidence to the user.

6.6 COMPUTING POWER

The radical improvement in computing power over the last 30 years has powered the development of software that can not only compute stresses but through the finite element method develop a 3D picture of stresses throughout a structure or component. Finite element methods have also been applied to branches of engineering other than stress analysis. Areas such as thermal analysis, fluid dynamics analysis, and vibrations analysis have all benefited from these methods.

The use of specifically designed software has also reformed the way in which engineers consider a problem. If an engineer was designing a beam, for instance, he would consider the loading regime, calculate the point of highest stress, and design the beam to cope with that stress. The engineer would apply a reasonably high factor of safety, which would enlarge the beam. Use of more sophisticated calculation and modeling software could allow the beam to be honed and tuned to fit the loading regime.

When designing a beam, the strength is dependent on the shape of the beam section; a larger beam section generally means more mass. If the greatest stress is at the center of a simply supported beam, then the section has to be largest at that point. This also means that the section can be reduced toward the ends of the beam because there are lower bending stresses experienced at these points. Although standard beam calculations can be used to calculate the section along the length of the beam, computing analysis can greatly speed up the process.

A typical structure that has been designed to these parameters is shown in Plate 6.2.

It can be seen in Plate 6.2 that the beam supporting the roof of the stadium is simply supported, that is, the support points are at each end of the beam. The greatest stress is in the center of the beam at which position is the deepest beam section. Stresses are applied to the beam toward the support points

PLATE 6.2

Stadium Roof [3]. (For color version of this figure, the reader is referred to the online version of this chapter.)

but they are much lower than at the center. It means that the beam section toward the support points does not have to be as large as at the center. This is typical of the tuning that computing power allows engineers to achieve.

6.6.1 Design iteration supported by computer-aided analysis

The use of computer-aided analysis has become an almost indispensable tool. There is often a compromise between the strength and weight of a component or structure. The use of computer-aided analysis is an important factor in reducing the weight while retaining the strength.

Example: The Design of a Foot Pedal Arrangement for a Lightweight Sports Car

A foot pedal arrangement for a lightweight car has to not only possess strength for arduous use, but it also needs to be as light as possible. The design iteration of such a foot pedal box is shown later.

The final concept design for a foot pedal arrangement for a high-performance lightweight vehicle can be seen in Figure 6.4.

During the deign process, it emerged that the need for weight reduction while maintaining performance had become key to the design.

The following figures give a step-by-step illustration of the design process using computational stress analysis techniques to drive the iterative process of design optimization within two apparently conflicting parameters (Figure 6.5).

Computational stress analysis revealed that this shape was low in weight but did not possess the strength required. Figure 6.6 shows where the accelerator pedal would fit in the pedal box design. The next iteration was to start from a solid plate and gradually remove material. The first solid plate iteration can be seen in Figure 6.7.

FIGURE 6.4

Final Concept Design for a High-Performance Vehicle [4]. (For color version of this figure, the reader is referred to the online version of this chapter.)

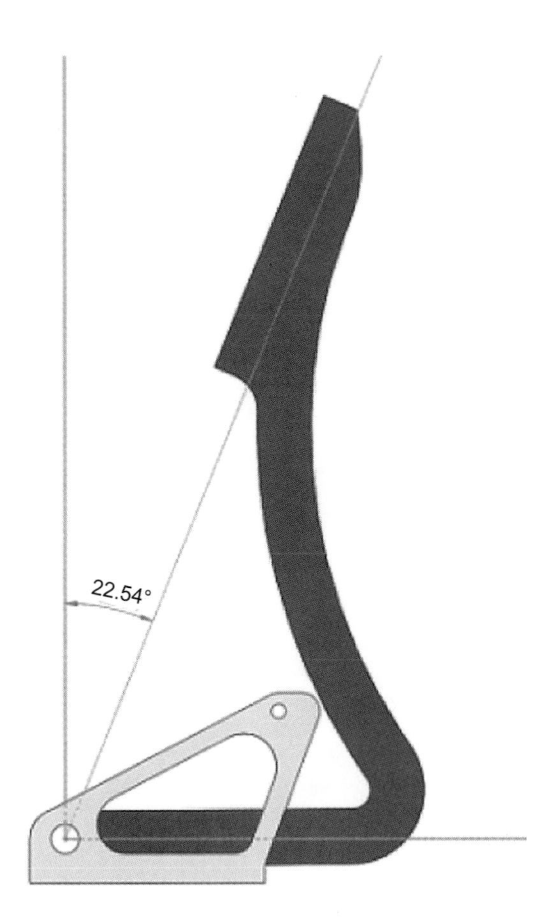

FIGURE 6.5

Side View of the Accelerator Pedal Initial Concept Design [4]. (For color version of this figure, the reader is referred to the online version of this chapter.)

Removing material from the solid plate model using 3D computer modeling, and applying finite element analysis allowed weight reduction while retaining strength in critical areas. After several further iterations, the analytically stressed model that can be seen in Figure 6.8 was built.

6.7 FATIGUE STRENGTH PREDICTION

One of the most critical performance predictions is predicting fatigue strength. As a rule of thumb, it can be considered that if something moves, it will be subjected to cycling loading. An applied load that cycles backward and forward may not initially overcome the strength of the material but constant cycling loading may eventually form a fatigue crack leading to complete failure.

FIGURE 6.6

Initial Concept Design of the Accelerator Pedal Shown *In Situ* in the Pedal Bracket [4]. (For color version of this figure, the reader is referred to the online version of this chapter.)

Take a disused credit card and bend it backward and forward. A crease line will appear and the plastic will tend to turn white. On close inspection, the plastic looks crystalline along the crack. Eventually the card will break along the bend line. This is essentially a fatigue break. The phenomenon is well known to engineers who ignore the occurrence at their peril. There are many examples.

Example: The Comet: The First Jet Airliner

The *de Havilland DH 106 Comet* was the first commercial jet airliner to reach production and ushered in the jet age of air passenger travel. It featured an aerodynamically clean design with four turbojet engines buried in the wings, a pressurized fuselage, and large square windows. Just a year after entering commercial service, the Comets began suffering problems, with three of them breaking up in mid-flight.

The pressurized cabin was a relatively new development. During a normal flight, when the aircraft climbed to its normal cruising height, the cabin was pressurized to around 11 psi, the ambient pressure

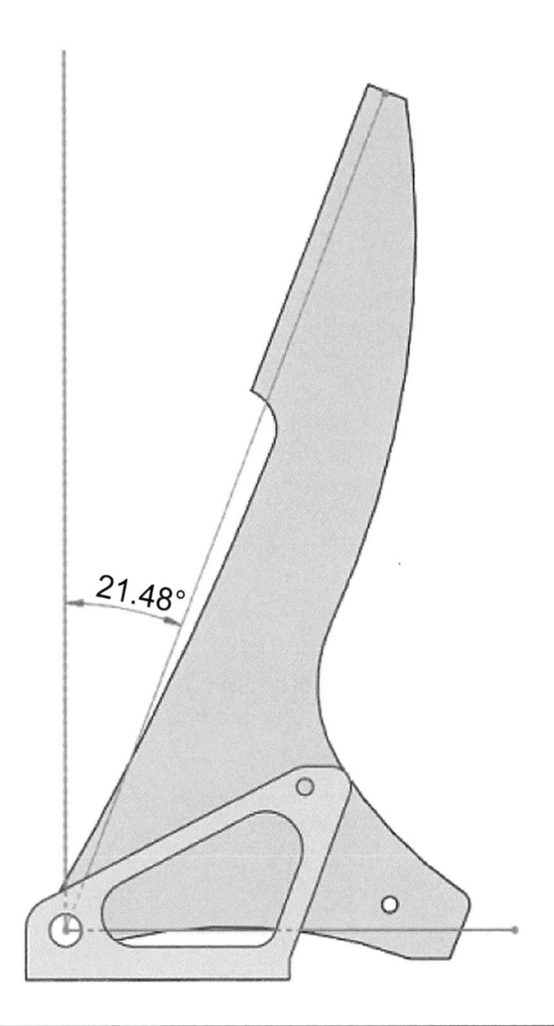

FIGURE 6.7

First Solid Plate Iteration Without Weight Saving [4]. (For color version of this figure, the reader is referred to the online version of this chapter.)

outside the cabin being much less than this. The difference in pressures meant that the fuselage was like a balloon with pressure on the inside stretching the aircraft skin. When the aircraft landed and the cabin depressurized, the stresses in the fuselage were relaxed. This was a classic case of cycling loading.

The Comet was withdrawn from service and extensively tested to discover the cause. Investigations found that the problem was catastrophic metal fatigue, which was not well understood at the time. Design flaws were ultimately identified, including square windows, which created "stress raisers" in the corners of the windows. These allowed fatigue cracks to propagate. The Comet was extensively redesigned with oval windows and structural reinforcement.

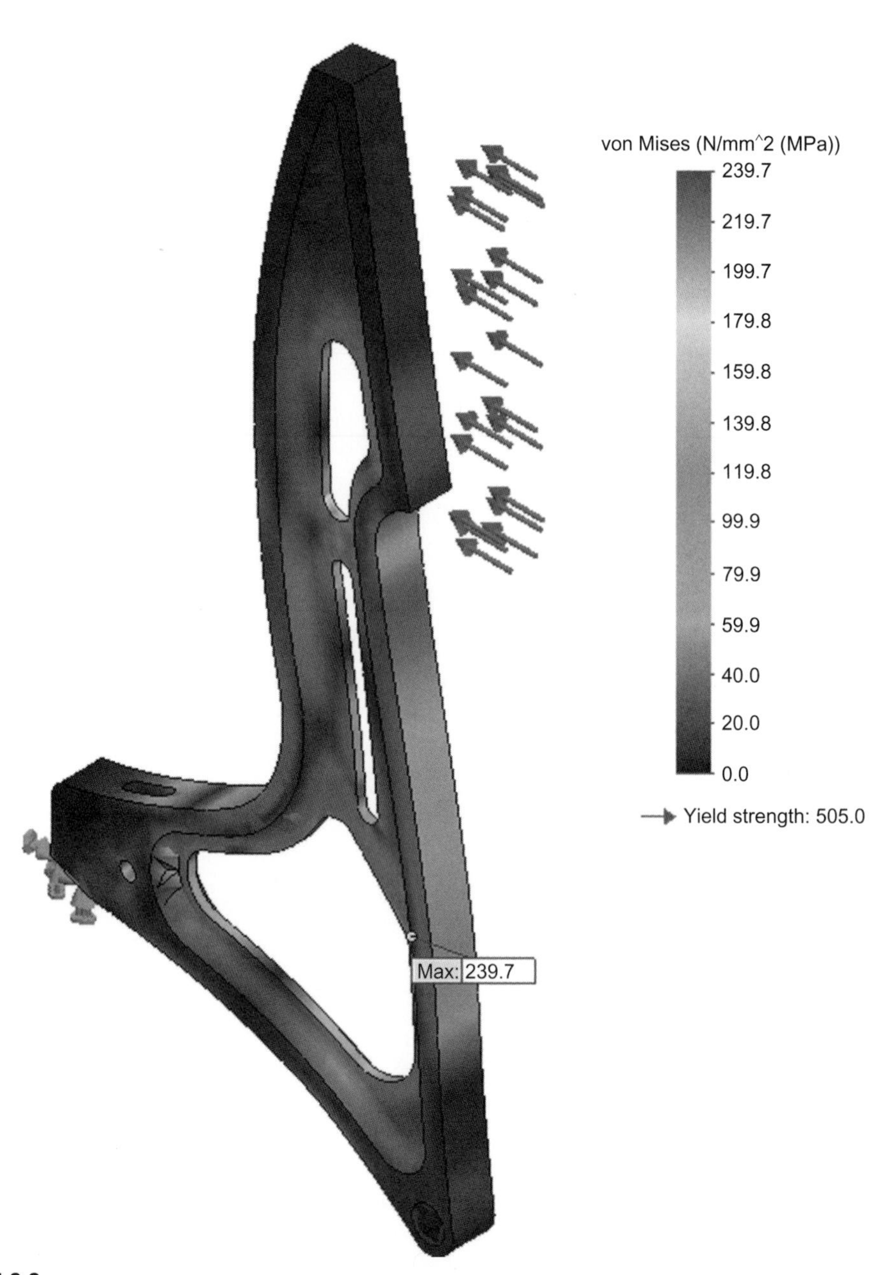

FIGURE 6.8

Finite Element Stress Analysis Model of Concept 2 [4]. (For color version of this figure, the reader is referred to the online version of this chapter.)

6.7.1 Welded joints

If a welder is asked to define the strength of his weld, he will say that it is as strong as the parent metal. The fact is that under cyclic loading, the weld will break because of the changing metallic structure around the welded joint. This was the problem with the Liberty ships in World War II, which sank not due to enemy action, but due to welded joint failure under the cyclical load imposed by the waves. Fatigue cracks propagated within the brittle weld boundaries leading to a catastrophic failure of the ships' hulls. The problem was solved by transforming the metallic structure by annealing so that the welded joint metallic structure became malleable as was the parent metal.

If a welded specimen is placed in a vice and struck with a hammer, the specimen will bend. If it is repeatedly bent backward and forward, it will tend to break along the welded joint. The metallic structure can be transformed back to the original malleable metallic structure by annealing. This is a process in which the material is heated to "a cherry red color" and allowed to cool naturally.

Some structures are too large to anneal unless specialist tooling is used. Often, designers create a welded structure allowing for weld brittleness by increasing the size of the weld. In critical cases where cyclic loading may be a problem, welded joints should be avoided unless there is a secondary retaining system such as bolts.

6.7.2 Quantifying fatigue life

Fatigue prediction is notoriously difficult. Two components manufactured in the same batch and subjected to the same cyclic loading conditions are unlikely to break from fatigue at exactly the same time. Some of the minutest variations may change the way the component behaves. Typically, variations might include loading regime, difference in size (perhaps the difference in a high or low tolerance), material uniformity, and differences in surface finish. Perhaps the two most significant variations between components are in the surface finish and the material integrity.

Welds are usually applied manually, and so can suffer slight variations in heat input, blowholes, uniformity, etc.

Materials, though appearing to be the same, may have dense spots, or metallic inclusions, or blowholes when cast. These differences will serve to change the performance at a microscopic level and will inevitably have an effect on the fatigue life of the component.

Perhaps the most significant effect on fatigue life is that of surface finish. When viewed at a microscopic level, surface finish appears as a series of peaks and valleys, where the bottoms of the valleys are often very sharp. Figure 6.9 shows a typical 3D surface finish profile. The reading was measured on a Talysurf PGI-contacting stylus profilometer (Ametek Taylor Hobson) with an average roughness of 15 μm. The same roughness profile can be seen as a 2D profile in Figure 6.10.

The reader should note that the sample area is 6 mm × 4 mm, but the height of the peaks has been exaggerated to enhance visibility. The rough nature of the surface is fairly typical of surfaces when viewed at a microscopic level. It is clear to see that the acutely sharp nature of the base of the valleys may easily make this area the start of a fatigue crack should cycling loads be applied to that area.

It is often useful, for instance, for measurement purposes, to take a slice through the 3D profile to reveal a 2D profile, as shown in Figure 6.10.

If stresses are applied to, say, a shaft, a crack may propagate from one of these valleys. In fact, any sharp corner, scratch, or interruption (stress raiser) in the smooth surface finish may become the beginning of the propagation of a crack. Fatigue cracks always start at stress raisers, so removing such defects increases the fatigue life.

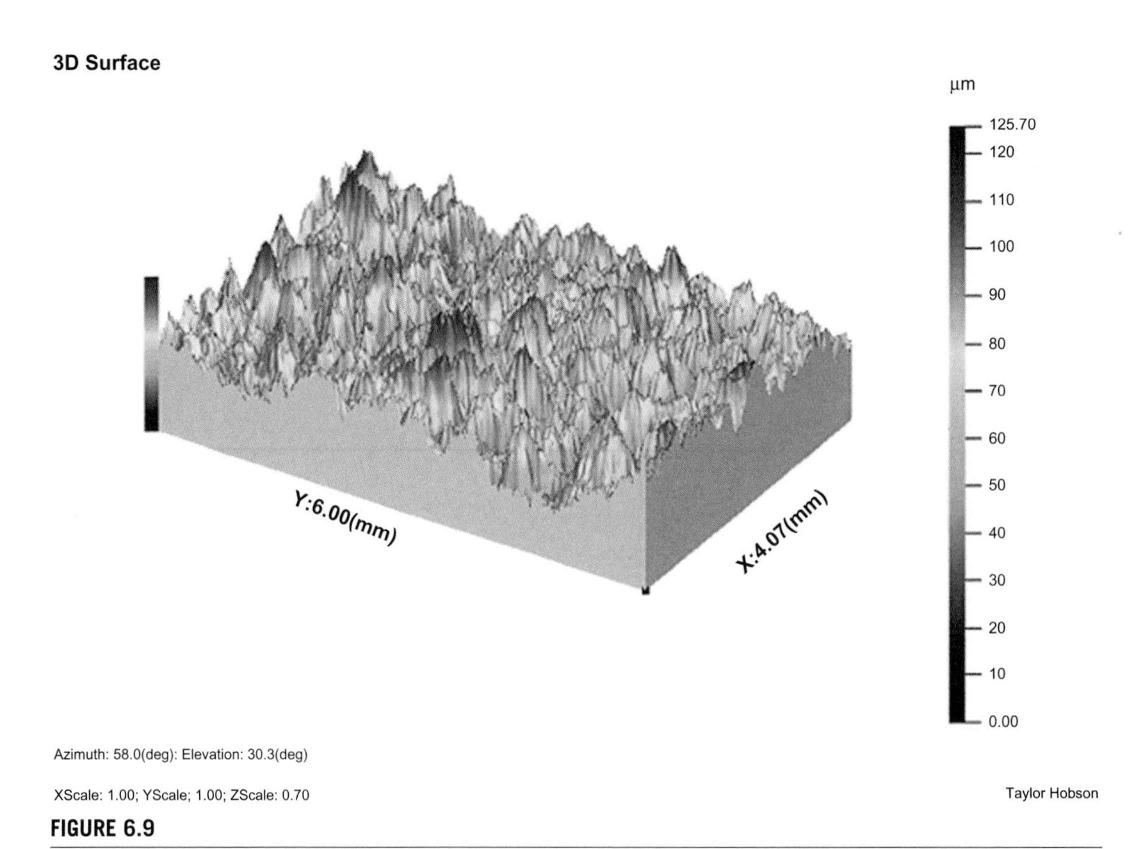

FIGURE 6.9

A Typical 3D Surface Finish Profile with an Average Roughness of 15 µm [5]. (For color version of this figure, the reader is referred to the online version of this chapter.)

A smooth finish is, therefore, very important if it is known that a component will be cyclically loaded. In gear manufacture, when each gear tooth rotates and meshes with the opposite gear tooth, it is cyclically loaded for just a few milliseconds before being released from contact. During the next cycle, the tooth makes contact once again. Gear teeth, therefore, can suffer severe cycling loading. Understanding this phenomenon, precision gear manufacturers manufacture the gear teeth to a very smooth surface finish, thus extending the life of the gear teeth (Plate 6.3).

6.7.3 S-N Curves (Goodman Curves)

It is well known that the bending of a piece of material backward and forward eventually leads to the breaking of the material. The breaking of the credit card is just such an example. A great deal of work has been done in this field to determine the number of cycles that materials may suffer before breaking. The diagram in Figure 6.11 shows generic S-N curves (stressed/number of cyclic reversals); sometimes these graphs are called Goodman curves after the engineer who accomplished much work in the area of fatigue.

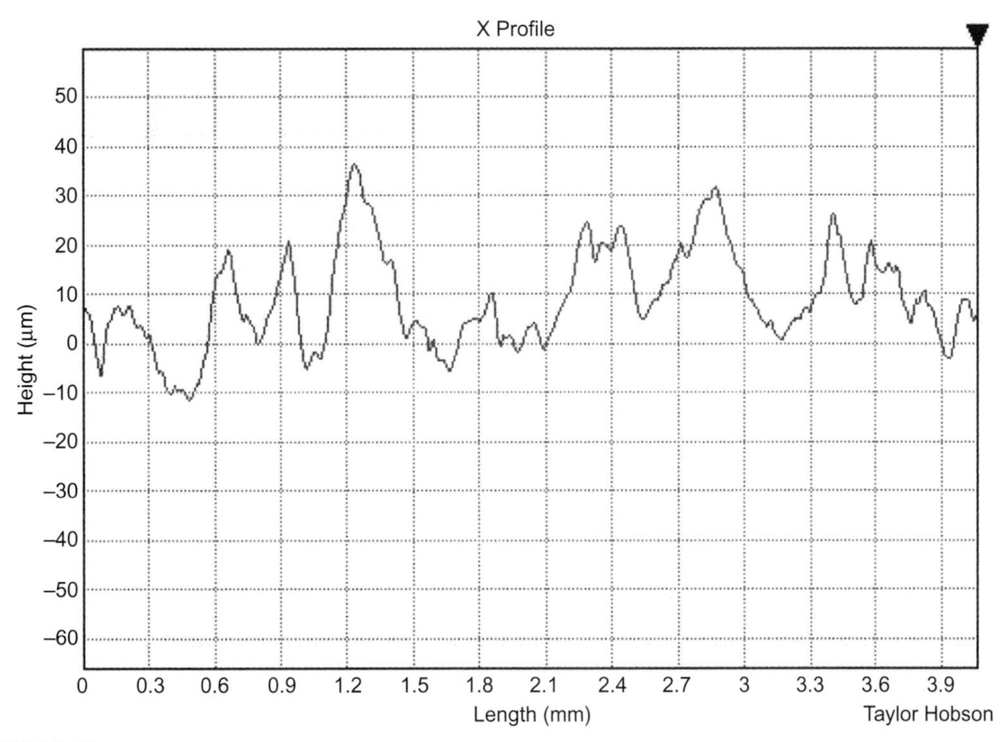

FIGURE 6.10

The Surface Profile of Figure 6.9 Shown as a 2D Surface Finish Profile with Average Roughness of 15 μm [5]. (For color version of this figure, the reader is referred to the online version of this chapter.)

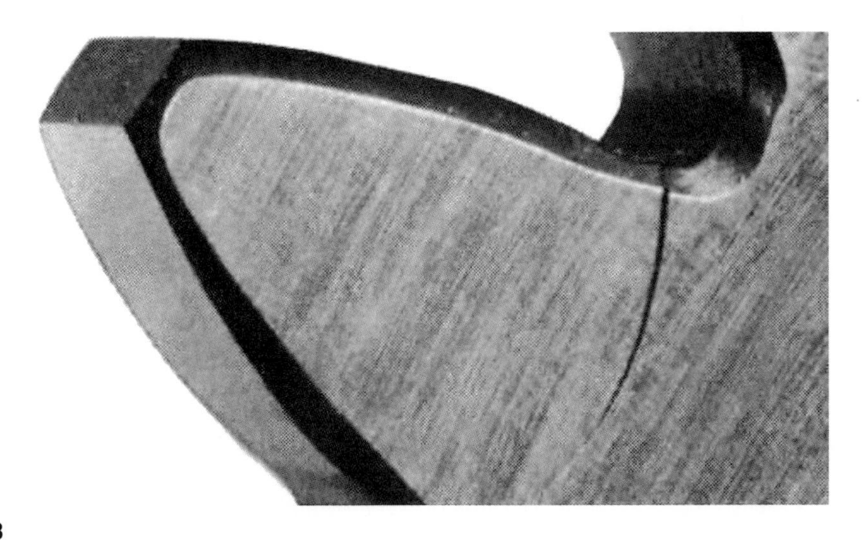

PLATE 6.3

Gear Tooth Crack Propagation [6].

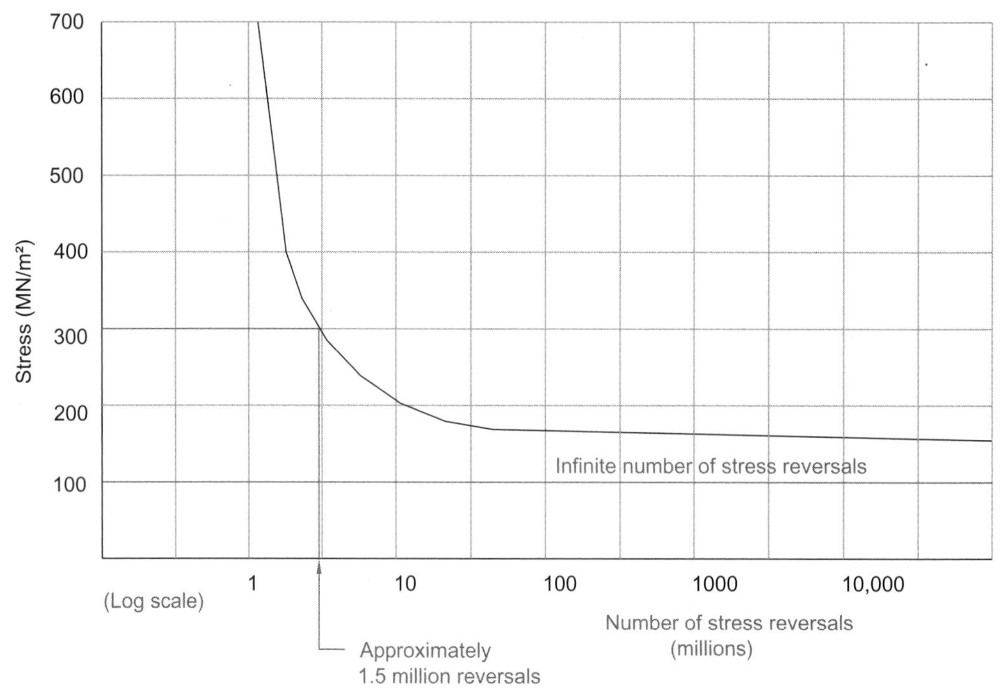

FIGURE 6.11

Idealized S-N Curve (Goodman Curve) [1]. (For color version of this figure, the reader is referred to the online version of this chapter.)

The graph line in Figure 6.11 plots applied stress on the vertical axis and the number of stress reversals on the horizontal axis. If the applied stress on a component is, say, 300 MN/m^2, the number of stress reversals before failure would be in the region of 1.5 million.

If now the material section is modified so that the applied stress becomes 100 MN/m^2, it can be seen that the material can cope with an infinite number of stress reversals because the S-N curve indicates that stresses at that level will never become dangerous.

The graph shown in Figure 6.11 is an idealized curve. Normally these curves are produced from empirical data for specific materials in different states. For instance, EN 8 (080M46) will have different S-N curves for its natural state compared to its hardened state. These curves can normally be obtained from materials suppliers.

Example: Shaft with Cycling Loading

A shaft protrudes from Bearings 150 mm to the center of a chain drive sprocket.

Consider a notional stressed element of the shaft shown in Figure 6.12. The load applies a tensile stress to the outer layer of the shaft at the element, but as the shaft rotates through 180°, the same part of the shaft under the element sees a compressive stress. As the shaft rotates back to its original position, the element sees a tensile stress once again. Through every revolution of the shaft, this "stress reversal" continues to happen. It is the same condition that the credit card experiences when it is bent backward and forward until it breaks (Figure 6.13).

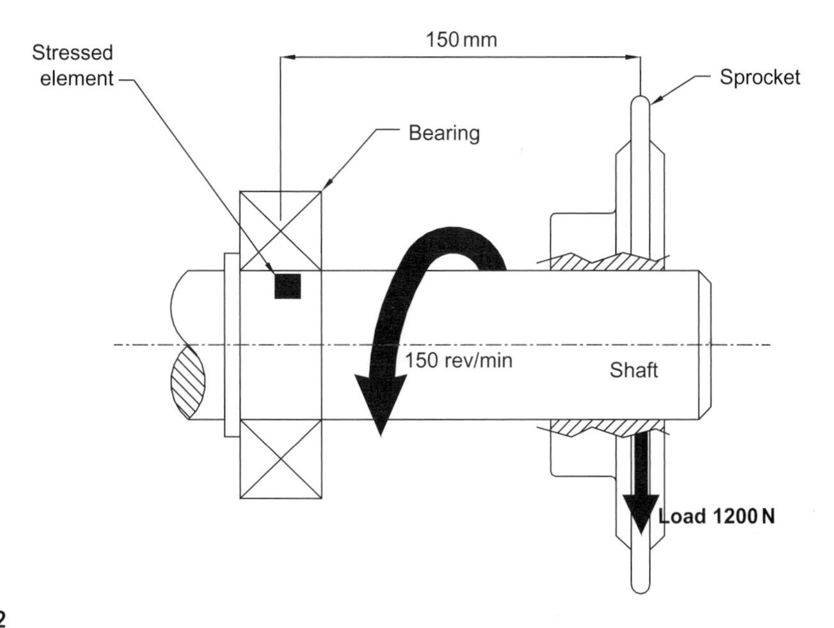

FIGURE 6.12

Cantilever at Shaft Showing Stressed Element [1].

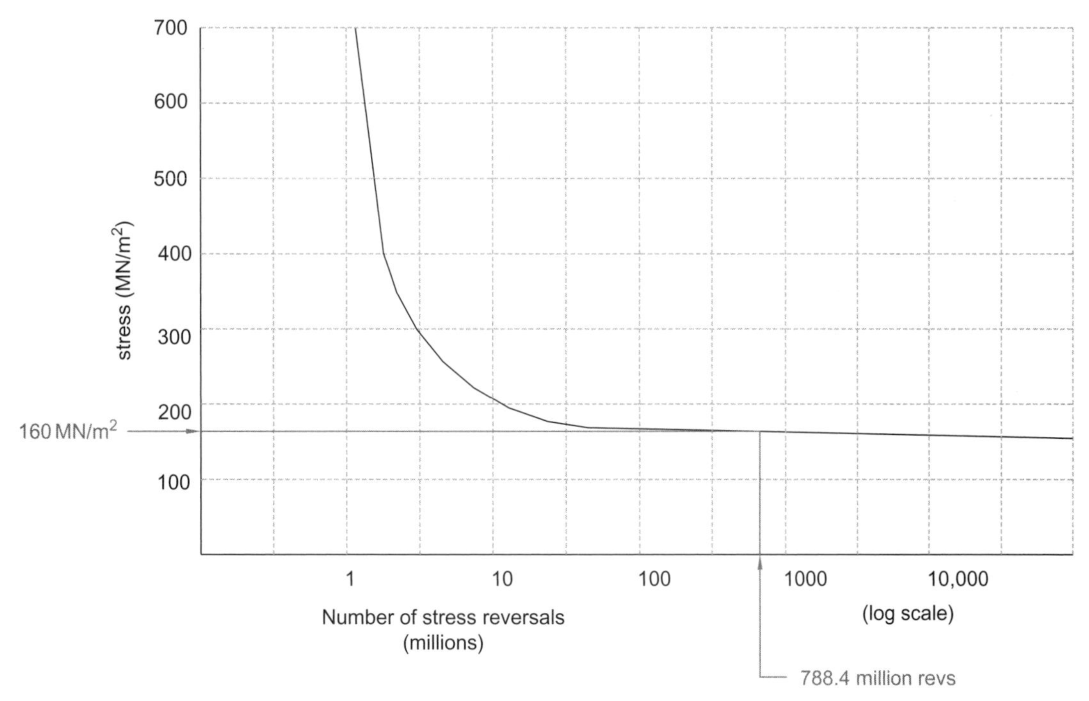

FIGURE 6.13

Idealized S-N Curve for a Medium Carbon Steel. (For color version of this figure, the reader is referred to the online version of this chapter.)

Shaft parameters
150 rev/min
Factor of safety (FoS) of 3
Shaft material: medium carbon steel $\sigma_{UTS} = 600\ MN/m^2$

Calculate the shaft diameter when a static load is applied
Find the working stress σ_w

$$FoS = \frac{\sigma_{UTS}}{\sigma_w}$$

or

$$\sigma_w = \frac{\sigma_{UTS}}{FoS} = \frac{600 \times 10^6}{3} = 200 \times 10^6$$

Find the maximum bending moment
$Mo = \text{Load} \times \text{moment arm} = 1200 \times 0.15 = 180\ Nm$

Find the shaft diameter
From

$$\frac{\sigma}{y} = \frac{m}{I} = \frac{E}{R}$$

$$\sigma_w = \frac{my}{I} \quad \text{or} \quad \frac{m \times \cancel{d} \times \cancel{64}^{32}}{\cancel{2}\pi \cancel{d}^4 d^3}$$

$$d = \text{Cube root} \frac{(32 \times m)}{(\sigma_w \times \pi)}$$

$$d = \text{Cube root} \frac{(32 \times 180)}{(200 \times 10^6 \times \pi)}$$

$d = 0.0209$ m or 21 mm

The shaft diameter has been calculated at ϕ 21 mm assuming a static load with a factor of safety of 3. This would normally be assumed to be safe, but in this case cycling loading is applied with the incumbent stress reversals.

Determine the shaft diameter taking into consideration the requirement for an almost infinite number of stress reversals.
Application parameters
Continuous operation for 10 years
150 rev/min
Shaft material: medium carbon steel $\sigma_{UTS} = 600\ MN/m^2$

Determine a number of stress reversals (*N*) within the expected life
$N = 150$ rev/min $\times$ 60 min $\times$ 24 h $\times$ 365 days $\times$ 10 years $= 788.4$ million rev

Apply the expected number of stress reversals, 788.4 million, to the diagram in Figure 6.12. It can be seen that the maximum stress to which the shaft can be submitted is 160 MN/m^2. To determine some safety margin and to extend the life of the shaft, it is proposed to make the maximum working stress 100 MN/m^2.

Determine the shaft diameter using a working stress σ_w of 100 MN/m^2
Previously calculated bending moment = 180 Nm

From

$$\frac{\sigma}{y} = \frac{m}{I} = \frac{E}{R}$$

$$\sigma_w = \frac{my}{I} \quad \text{or} \quad \frac{m \times \cancel{d} \times \cancel{64}^{32}}{\cancel{2}\pi \cancel{d}^4 d^3}$$

$$d = \text{Cube root}\frac{(32 \times m)}{(\sigma_w \times \pi)}$$

$$d = \text{Cube root}\frac{(32 \times 180)}{(100 \times 10^6 \times \pi)}$$

$d = 0.0264$ m or 27 mm

The result of the consideration of stress reversals is to increase the diameter of the shaft, which reduces the applied stress and thus increases the number of stress reversals to which the shaft can be subjected.

6.7.4 Stress Concentrations

The "S-N" curve described earlier is an excellent method of predicting the life of a component in which there are stress reversals; however, it is important to realize that fatigue cracks usually start from an abrupt change in the surface condition of the material such as small holes, screw threads, scratches, machine tool striations, and many other surface defects. It is incumbent on the design engineer to understand the stresses applied at the surface of a component and to eliminate stress concentrations at the most highly stressed positions on the surface.

There are, however, discontinuities on the surface of the component, which are necessary, perhaps to house other components such as bearings or gears. These elements could be shoulders or undercuts that effectively leave sharp corners from the machining process. An example could be the 90° corner left at the bottom of a keyway or perhaps a shoulder turned onto a shaft. These "designed-in" surface discontinuities can be built in to the stress calculations for the shaft.

Example: A spur gear is held in an axial position along the shaft by a spacer on one side of the gear and on the other side by a shoulder machined onto the shaft. In any machining operation, the tool tip radius would leave a very small radius at the base of the shoulder, shown in Figure 6.14, preventing the spur gear from properly contacting the shoulder. A relief groove may therefore be applied such as that shown in Figure 6.15. Since this is effectively a stress concentration, which is designed into the shaft, it may be included in the stress calculations for the shaft using data from *Peterson* [7]. Please also see Figure A6.3, also in Appendix.

Stress analysis. In order to compare the shaft with and without the relief groove, there are two sets of calculations to be performed. These calculations can be seen in Appendix A6.1 with a synopsis indicated below:

(1) Maximum allowable stress for the shaft shown in Figure 6.14 = 101.9 MN/m^2
(2) Maximum allowable stress for the shaft shown in Figure 6.15 = 11.5 MN/m^2

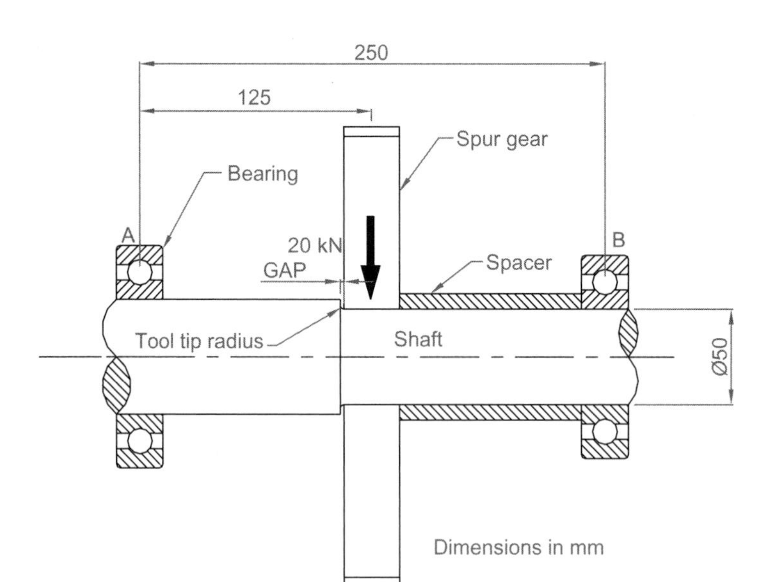

FIGURE 6.14

Spur Gear Mounted on a Shaft Between Two Bearings [1]. (For color version of this figure, the reader is referred to the online version of this chapter.)

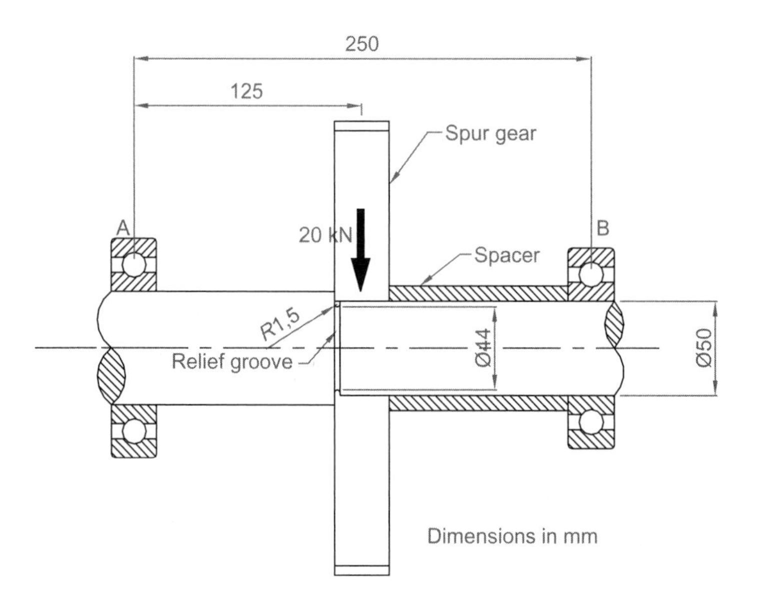

FIGURE 6.15

Spur Gear Mounted on a Shaft Showing Relief Groove [1]. (For color version of this figure, the reader is referred to the online version of this chapter.)

The value of $\sigma_{max} = 11.5$ MN/m^2 can be considered to be the maximum stress the shaft can undergo when such a relief groove is applied to the shaft. Consider the maximum stress the shaft can undergo when there is no relief groove. The value of 101.9 MN/m^2 is a factor of 10 greater than the maximum stress in the shaft with a relief groove. The reduction in strength is a consequence of designing into the shaft a legitimate stress raiser, but it should be considered that the analysis has allowed a precise design to be created that caters to a reduction in strength.

6.8 PERFORMANCE PREDICTION METHODOLOGY AND APPLICATION

In the early pages of this chapter, it was stated that there are generally two reasons for performance prediction

- Performance prediction to ensure the product operates within safe limits
- Performance prediction to obtain a precise working specification

Examples have thus far centered on stress analysis, which is normally used to ensure that structures or devices possess the strength to cope with the applied loads. A great deal of analysis, however, is carried out to predict how a product might behave and to enable the correct selection of components that may be purchased from an outside source.

Behavioral analysis might be used, for instance, to calculate the parameters of a bought-out product such as a vibration isolator. An example can be seen in Appendix A6.6. This example outlines a manual method of selecting an appropriate spring stiffness, which can then be used to select an isolator from a catalog. Vibration isolation vendors often offer an online selection process where the purchaser can enter appropriate parameters into the software. This will then calculate and automatically select a suitable isolator.

Many component vendors such as bearings and seals suppliers offer similar on-line services. Since the design information is the key to efficient design creation, the ease of selection of components not only assists the designer but also ensures that the designer will continue to go back to the particular supplier who offers the most easy-to-access information.

Behavioral analysis may also be used to predict the product's performance when performance targets have been set perhaps by the marketing team or perhaps within the Product Design Specification (PDS). The example outlined in Appendix A6.3 describes the analysis of a torsion meter attached to a ship's propeller shaft. The eventual outcome is predicting the total angular deflection, 9.83° along the whole 13 m length of the propeller shaft.

Appendix A6.4 describes the dynamics of powering a bicycle and, under certain circumstances, predicts the forces at the brake blocks, the driving forces at the road-tire interface, the work done by the rider, and the power requirement.

Appendix A6.5 describes the power requirement for lifting a mass via a winding drum. This kind of analysis can be used to select the power plant such as the diesel engine.

Performance prediction analysis can also be used to assist the build process and help to prescribe the method of use of the product. Predictive analysis was successfully used in the case of the Denby Dale pie challenge. Denby Dale is a village in West Yorkshire renowned as the village that creates the largest meat and potato pies in the world. The last attempt was the Millennium Pie challenge in 2000, which was to cook 12 tons of pie filling and distribute it to 40,000 people on

PLATE 6.4

The Denby Dale Pie Transported by the Low Loader Emerging from the Cooking Tent [1]. (For color version of this figure, the reader is referred to the online version of this chapter.)

"pie day." See Plate 6.4. The author, along with his close colleagues, was set the task of designing the "pie cooker." The cooker dimensions were 12 m × 2.3 m × 1 m and it housed 24 vats, each holding 500 kg of pie filling and each being heated by individual elements.

One of the criteria was that the whole pie should be paraded around the village before it was returned to the serving field where the contents would be distributed to the audience. Initial estimates put the contents at 12 tons and the cooker structure at 6 tons resulting in a total mass of 18 tons. This clearly required a low loader truck systems for the transportation through the village and back to the "pie field." In fact, the cooker dimensions were determined by the size of the low loader transporter.

The cooker was designed as a beam with the normal standard beam analysis, which can be seen in Appendix A6.7. This is clearly an example of analysis that has been performed on a structure so that it can be used safely.

It was essential that ingredients were added to the pie filling in a precise fashion so that temperatures did not fall below 65 °C, which would trigger the growth of the *Clostridium perfringens* bacterium. This is the bacterium that causes food poisoning and typically causes victims to develop diarrhea and sickness. The cooking process involved heating the ingredients of each VAT to 100 °C and then adding further ingredients such as chopped potatoes, onions, seasoning, etc. It was the task of the designer to calculate, using the principles of heat transfer, the mass of ingredients which could be added before the pie filling temperature was reduced to 65 °C. This analysis can be seen in Appendix A6.8 and was performed primarily for safety reasons and also to define the methods by which ingredients were added to the mix of the pie filling.

6.8.1 Computational analysis overview

Before the advent of computer aids, analysis was achieved largely using manual calculations. These calculations were not only tedious but also very limiting in that they could not allow analytical sophistication beyond the use of a good calculator. Furthermore, assumptions made during the calculation process often reduced the accuracy of the analytical model. Safety factors were, therefore, increased to compensate for these inaccuracies.

Modern analytical software often eliminates the need for these rule-of-thumb safety factors, allowing more efficient use of materials and better performance prediction.

6.8.2 Basic electronic calculation tools

The advent of scientific calculators provided engineers with a very powerful calculation tool, which was more accurate and more convenient than the previous and very common slide rule. Calculation tools became more and more powerful with the advent of personal computers and the complementary software. Spread sheet packages allowed engineers to create lengthy and repetitive analysis, thus avoiding writing specific programs in languages such as Fortran or C++. Though these languages are useful and may still be relevant to the engineer's armory, it is often much more convenient to use spreadsheet packages in which engineers can input equations and link cells to other cells.

6.8.3 Advanced computational analytical techniques

Advanced computational analysis relies on the theory of the finite element and originated from the need to solve complex elasticity and structural analysis problems. The origin of the finite element can be traced back to work originally put forward by Hrennikoff and Courant, who approached the problem from different directions but whose complimentary work resulted in "mesh generation." Typical mesh generation can be seen in Figure 6.16.

As each of the elements is affected by forces, displacements, temperatures, etc., these effects are passed to the next element. The transmutation can then be solved by second-order partial differential equations, which were originally devised and introduced by early analytical engineers such as Rayleigh, Ritz, and Galerkin.

A typical application involves dividing the structure into a collection of subdomains. Each subdomain is then represented by a set of elements and appropriate equations. The subdomains and elements are then recombined into a global system to enable the final calculation.

Meshes can be generated with different degrees of coarseness. Critical areas of the component may have a finer mesh generated leading to a more accurate analysis. Non-critical parts of the component still require mesh generation even if only to link critical areas to each other. It should be noted that a finer mesh will lead to the requirement for greater computing power in order to cope with the larger volume of calculations.

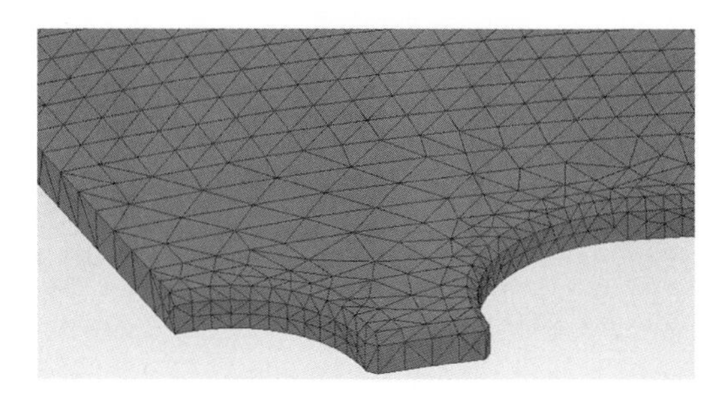

FIGURE 6.16

Typical Mesh Generation for Part of a Vehicle Brake Disc [8].

6.8.4 Computational analysis potentials

The first step in any computational design/analysis process is to create the shape of the component. This may start as a two dimensional shape but if full advantage is to be taken of computational analysis, then a 3D shape needs to be created. It is normal, therefore, for designers to use a 3D CAD package, which usually has built-in basic analytical finite element (FE) options as standard. The basic packages would normally include stress analysis, deflections analysis, buckling analysis, heat transfer, natural frequency, and product durability. Extendable packages may include fluid flow simulation, dynamic analysis, mechanisms analysis, and many more.

6.8.5 Examples of practical applications

There follow several examples of the analysis that can be achieved with a 3D package equipped with the appropriate analytical software.

The example of a brake pedal bracket can be seen in Figure 6.17. The model shows the 3D component in which arrows indicate where the driver's foot pressure is applied. The different colors indicate the stresses within the bracket according to the key to the right of the figure. Red-colored areas are the most highly stressed portions of the bracket, while the deep blue areas are the lowest stressed elements. Even though the red-colored areas may look seriously overstressed, the actual value depends on the setup procedure and may well be within appropriate safety limits.

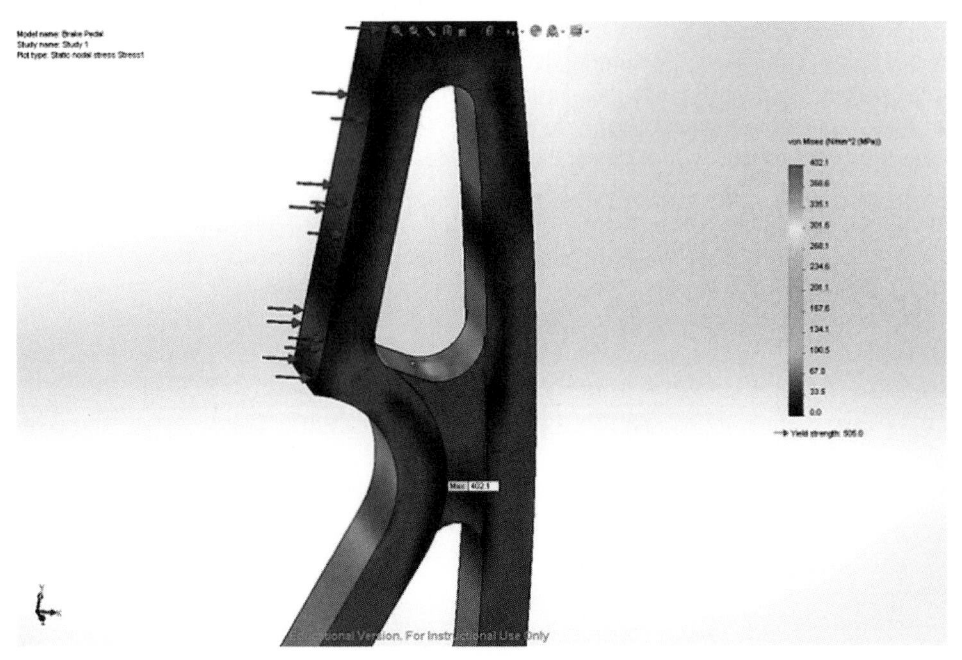

FIGURE 6.17

Example of Brake Pedal Bracket: Stressed Component Using Finite Element Stress Analysis (Formula Student Vehicle 2013 Entry). (For color version of this figure, the reader is referred to the online version of this chapter.)

PLATE 6.5

Pedal Box *In Situ* Showing the Brake Pedal Bracket (Center) (Formula Student Vehicle 2013 Entry) [1]. (For color version of this figure, the reader is referred to the online version of this chapter.)

PLATE 6.6

Brake Disc Shown *In Situ* (Formula Student Vehicle 2013 Entry) [1]. (For color version of this figure, the reader is referred to the online version of this chapter.)

Plate 6.5 shows the whole pedal bracket built into the vehicle. The brake pedal is the at the center of the three pedals.

Careful consideration was also given to the brakes on the vehicle. Plate 6.6 shows the brake disk *in situ* on the vehicle prior to being fitted with wheels. A great deal of work was applied to the brake disc in order to reduce weight and retain strength. The initial stage for the analysis of the brake disc was, first of

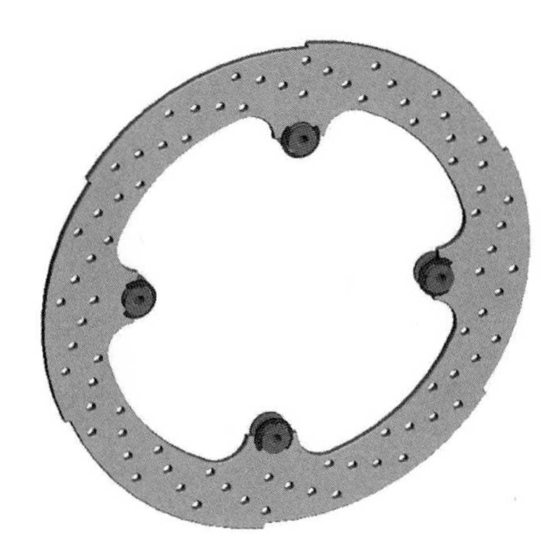

FIGURE 6.18

3D Rendition of a Brake Disc (Formula Student Vehicle 2013 Entry) [9]. (For color version of this figure, the reader is referred to the online version of this chapter.)

all, to create a 3D model shown in Figure 6.18, from which the finite element mesh, Figure 6.16, could be generated.

Having generated the finite element mesh, several analyses can be performed on the brake disc. For instance, the thermal analysis can be seen in Figure 6.19. This shows the distribution of temperatures under certain loading conditions. Red colors indicate high temperatures and blue colors, the lowest temperatures. The temperatures as indicated may, or may not, be critical as this depends on the calibration. Analysis such as this allows the designer to determine whether excessive temperatures may warp the disc or whether the heating of the disc may be an acceptable risk.

Computational fluid dynamics (CFD) is an extremely powerful tool that allows designers to apply fluid flow to a component within the computer environment. Since the term "fluid" refers to liquids and gases, CFD can be applied to any application where there is mass flow. Applications such as liquid flow through pipes or perhaps a flow around turbine blades are quite common. The example in Figure 6.20 is that of CFD analysis performed on the whole vehicle assembly for the FSV 2013 entry. This was executed in "ANSYS Fluent." Here the parameter was static pressure applied to the vehicle as if it was moving through still air, but other parameters may be examined such as fluid velocities, accelerations, temperature changes, and many more.

Often impact behavior is the most difficult to predict, but again, using finite element methods, impact can be simulated. Figure 6.21a-e shows the progressive impact for an impact attenuator, which was intended to be fitted to the nose cone of the FSV 2013 entry. It can be seen that the attenuator is a pyramid shape where the gray disc is the external impact element. As the impact progresses, the cone gradually crushes, thus distorting the finite element mesh until in Figure 6.21e, the attenuator is completely crushed. Analysis such as this allows the designer to design an attenuator that absorbs energy progressively. Not only can the designer see the progressive shape change of the attenuator, but he can also quantify loadings and energy absorption using the scale built into each diagram.

Information generated within the 3D CAD system can be used for many things. As indicated in previous chapters, one of the main purposes of constructing components within the CAD system is

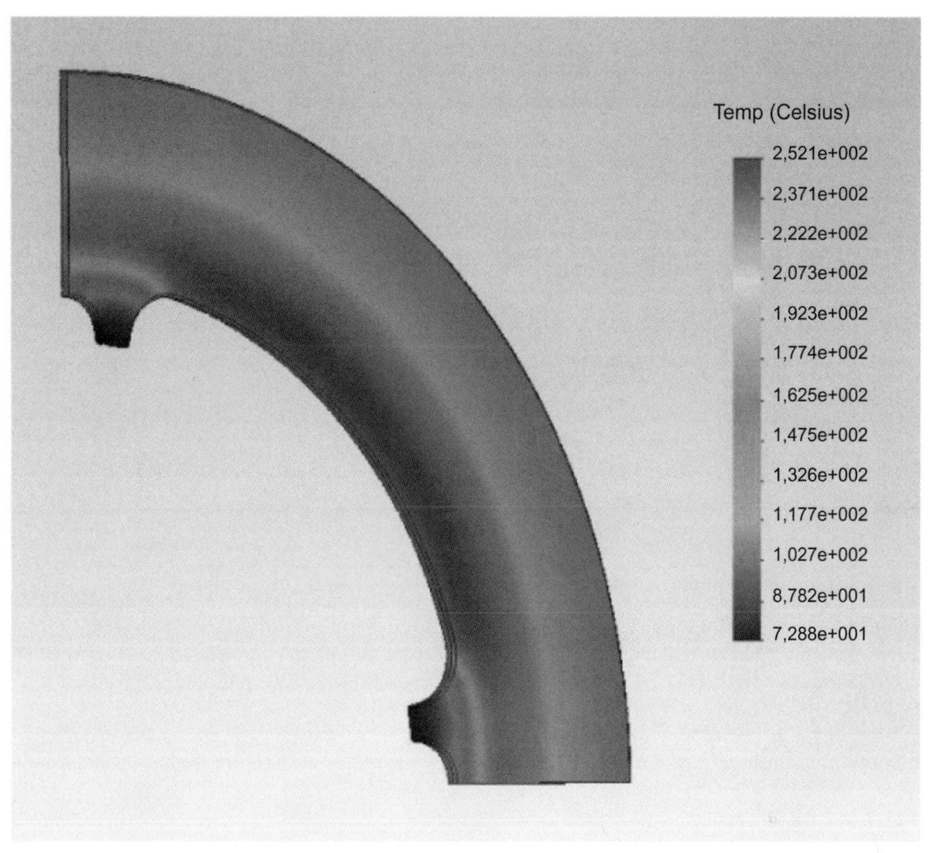

FIGURE 6.19

Examples of Thermal Analysis of the Brake Disc (Formula Student Vehicle 2013 Entry) [9]. (For color version of this figure, the reader is referred to the online version of this chapter.)

to provide engineering drawings. Once the component is digitized within a computer's memory, the file can be used for many other applications, analysis being just one of them.

One of the major functions of a 3D CAD system is that of assembly. Each component that is developed within the system can be assembled as if it were a manufactured component. The software creates an assembly where parts can be fitted together so that the whole assembly can be created within the computer environment and then displayed on-screen as a complete assembly. Figure 6.22 is just such an assembly where all the components have been designed within the computer environment and have then been compiled into a single assembly. Some features are built into the software, which assists in the assembly process. One such item is that of "collision detection" sometimes called "dynamic interference detection." When two parts are fitted together, say, a shaft is larger than the hole into which it is inserted, the designer needs to know if there is a clash between the material of the shaft and the material of the hole. Dynamic interference detection will alert the designer when this happens. This ensures that when the assembly is completed as a physical prototype, the parts will fit together as originally designed.

FIGURE 6.20

Fluid Flow Analysis for the Whole Vehicle Assembly for Fluid Flow (Formula Student Vehicle 2013 Entry) [10]. (For color version of this figure, the reader is referred to the online version of this chapter.)

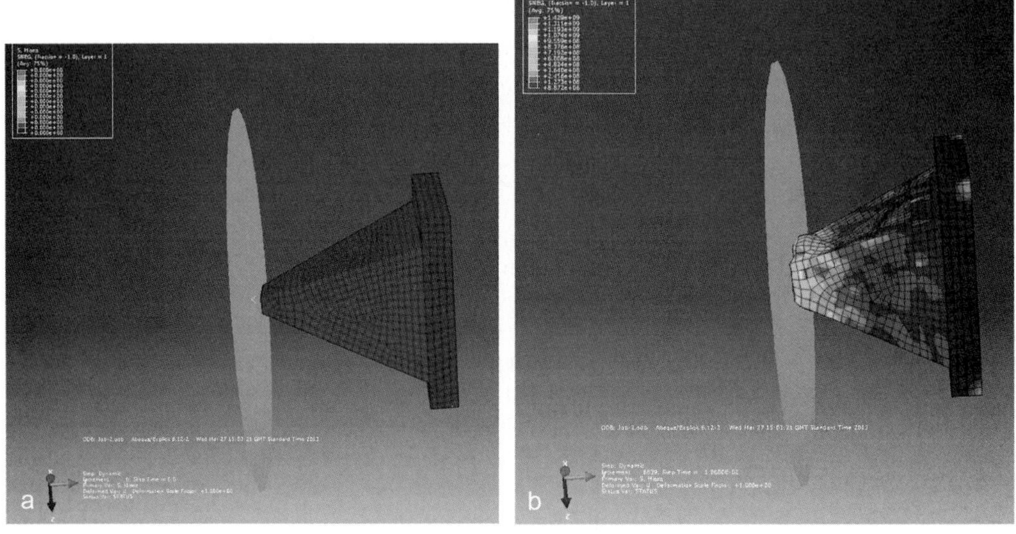

FIGURE 6.21

(a) and (b) FEA Modelling of Progressive Impact for a Carbon Fiber Impact Attenuator [11].

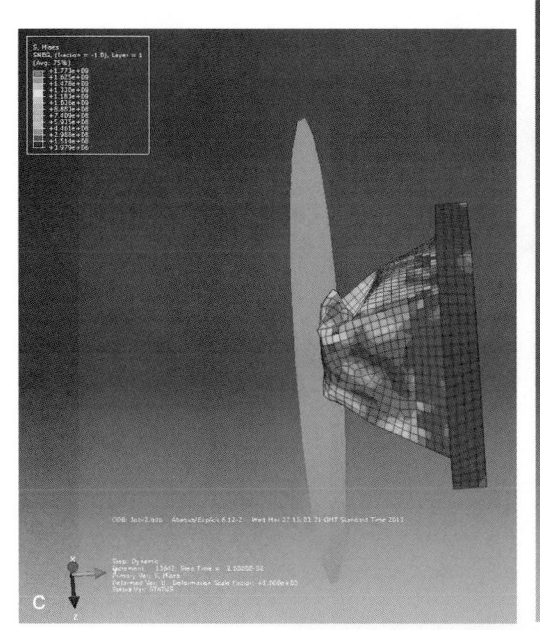

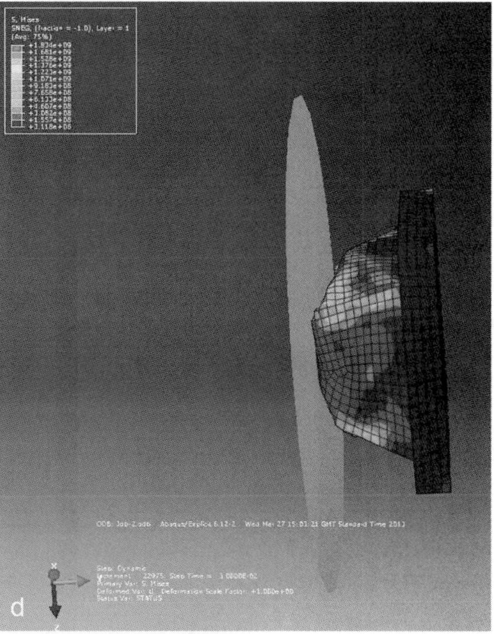

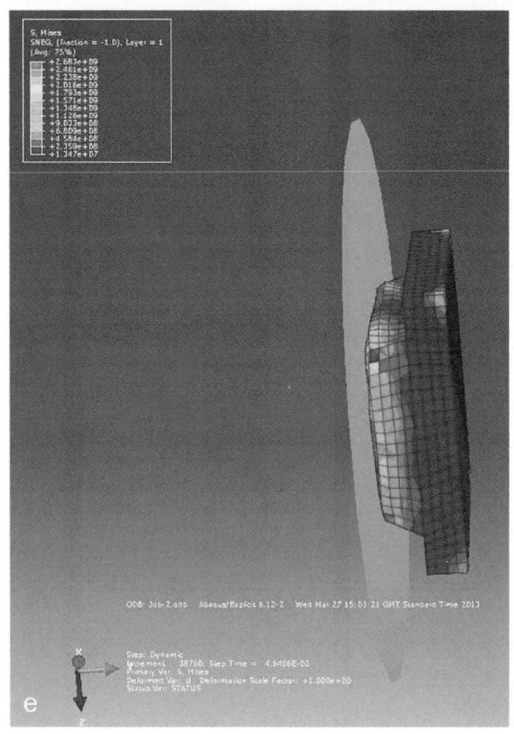

FIGURE 6.21, **con't** (c) through (e) FEA Modelling of Progressive Impact for a Carbon Fiber Impact Atenuator [11]. (For color version of this figure, the reader is referred to the online version of this chapter.)

FIGURE 6.22

Photo Rendered Image of the University of Huddersfield 2013 Entry for the Formula Student Vehicle Contest Run by the Institution of Mechanical Engineers [12]. (For color version of this figure, the reader is referred to the online version of this chapter.)

Most clients are very pleased to receive 3D images of the product they have instigated.

To this end, 3D CAD packages have several built-in functions. Most CAD packages offer a free download of a viewer so that the customer can easily view a 3D, read-only image of the component. The marketing function may require pictures to be used with brochures, posters, etc., and to this end, rendered images can be generated. Figure 6.22 is a photographically rendered image of the FSV 2013 entry, which was used on promotional data before the vehicle was built.

6.9 CHECKS AND BALANCES

The use of computer-aided analysis is relatively easy, but it is also dangerous when used by a novice, who may not understand the design challenge and, therefore, miss subtle but important elements. Checks and balances and even engineering intuition must be built into any analysis so that the designer can declare with confidence that the information generated by the computer-aided analysis is correct and accurate.

6.9.1 Computer-Aided Failure (CAF)

There is a belief among some novice engineers that if the computer can create a model, then it is correct. This assumption is very dangerous since the computer will only provide data based on the input. Furthermore, the novice may not totally understand the problem in which case his/her assumptions will be flawed from the beginning.

There have been several well-documented cases where failures have been traced back to the designer who has merely accepted information developed by computer-aided analysis. This approach could have fatal consequences if, for instance, a supermarket (Mall) roof were to collapse, which is exactly what happened several years ago. The fault was traced back to the designer who had merely accepted the results of his computer simulation.

Failure to understand and appreciate the engineering behind analysis can lead to computer-aided failure (CAF). To avoid CAF, there must be checks and balances put in place even with the most complex calculations. The checks may involve hand calculations to confirm the output from the analysis or the use of results from experimentation and a prototype's work.

6.9.2 Stainless steel pipeline junction (by Caroline Sumner, University of Huddersfield)

Whenever a junction is placed into a pressurized pipe, the joint creates a discontinuity in the integrity of the major pipe. This is effectively a weak point and is ideal for the application of finite element stress analysis. The situation, however, is difficult to assess using hand calculations, so in this particular case, a small portion of pipe was built with an appropriate tee-junction. This can be seen in Plate 6.7. The test rig was then instrumented with strain gauges around the predicted weak area, the neck of the tee-junction as shown in Plate 6.8.

The test rig was then pressurized to an appropriate test pressure whereupon the strain readings were taken and converted into values of stress (N/m^2). In the meantime, the researcher was creating a 3D model and applying finite element stress analysis. The result of the analysis can be seen in Figure 6.23.

PLATE 6.7

Stainless Steel Pipe with Integral Tee-Joint Test Rig [1]. (For color version of this figure, the reader is referred to the online version of this chapter.)

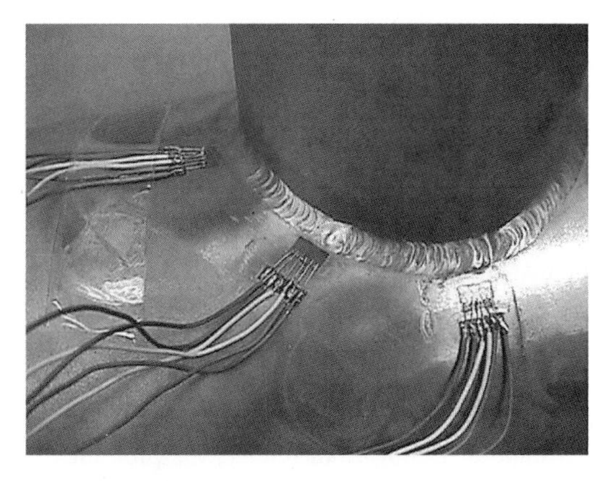

PLATE 6.8

Application of Strain Gauges to the Neck of the Tee-Junction [1]. (For color version of this figure, the reader is referred to the online version of this chapter.)

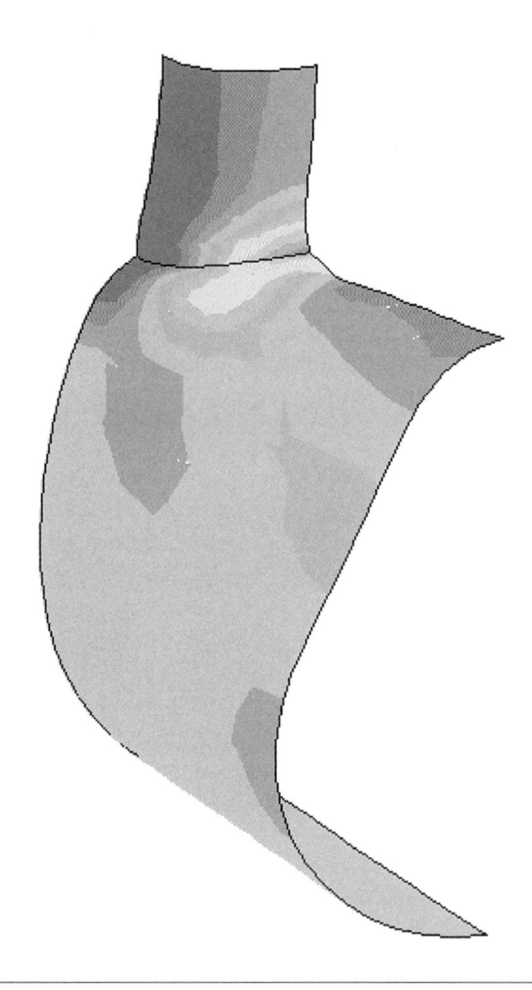

FIGURE 6.23

Results of Finite Element Analysis on the Pressurized Tee-Junction [13]. (For color version of this figure, the reader is referred to the online version of this chapter.)

The model shows that there is a high stress area at the neck of the tee junction. The calibration within the software determined whether this high-stress area was unsafe. It would have been dangerous, however, to accept the results of the finite element analysis without comparing those results to the results generated by the pressurizing of the test rig.

The test rig results were able to confirm the accuracy of the finite element stress analysis, effectively proving the model, which could then be used for similar applications. The reader's attention is also drawn to the distortion in the main pipe. The entry of the tee junction creates a discontinuity in the surface of the main pipe, and the pipe distorts to compensate the spreading of the load and stresses by attempting to change the shape of the pipe. In reality, the pipe may not visibly change shape at all because of the inherent strength of the material, but the model, having zero strength may show this anomaly.

6.9.3 CFD analysis verified by a wind tunnel model (by Jay Medina, University of Huddersfield)

CFD flow analysis was applied to a fairly crude 3D model of a near-standard passenger vehicle shape so that the energy of the displaced air could be measured as the vehicle traveled at motorway speeds.

The three-dimensional model was originally created within a 3D CAD package, shown in Figure 6.24.

The digitized version of the 3D model was then imported into the CFD package attached to the CAD software. The firsts analysis performed was to consider the location and value of the air pressure profile as it was applied to the vehicle. This analysis can be seen in Figure 6.25 and indicates the high-pressure area on the vehicle bonnet and windscreen.

It was realized that as the vehicle pushed through still air, there would be differing values of air pressure and velocity at varying heights from the ground upward. Analysis therefore created a velocity profile at 100 mm intervals upward from the road surface. Figure 6.26 shows a series of analysis "snapshots" at 100 mm intervals from the ground. It was discovered that the greater deflected air velocity and, therefore, greater energy dissipation, came from the lower levels corresponding to the least aerodynamic points on the vehicle. The output from this analysis indicated the position of the energy extraction device intended to absorb some of the energy from a passing vehicle.

The analysis appeared to be very conclusive as the red-colored areas indicated the level of maximum air disruption. Though the analysis generated a great deal of confidence, it was necessary to confirm that the analysis was correct. The consequences of the analysis being incorrect could be costly development effort spent on an energy generation device that did not work. It was, therefore, necessary to confirm the analysis by conducting wind tunnel tests.

The original 3D model shown in Figure 6.24 was fed into a rapid prototyping machine that printed 1/10 scale models. This was then used in wind tunnel tests, which indicated that the simulation model possessed some inaccuracies. After knowing the inaccuracies in the model, compensation could be made so that the model could still be used to complete the project.

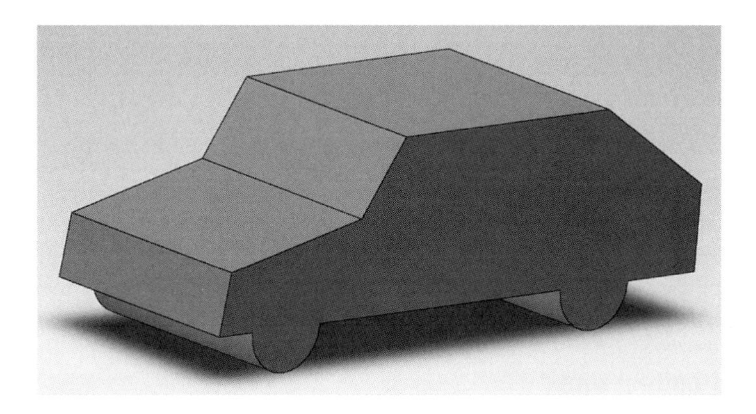

FIGURE 6.24

Three-Dimensional Model of a Standard Passenger Vehicle [1]. (For color version of this figure, the reader is referred to the online version of this chapter.)

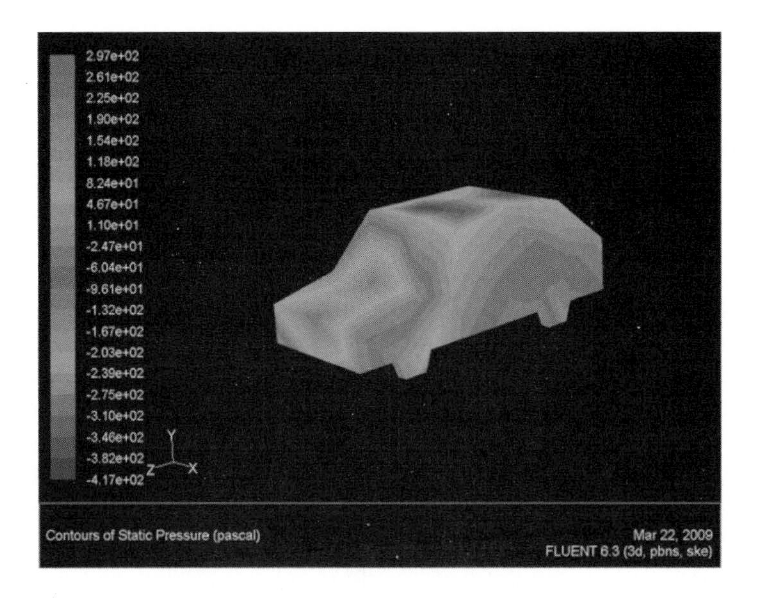

FIGURE 6.25

Air Pressure Profile Applied to the Vehicle at a Velocity of 70 Mph [14]. (For color version of this figure, the reader is referred to the online version of this chapter.)

6.10 CONCLUSION

It is essential that any computer-aided analysis be verified by whatever means. It is dangerous to take the computer-generated data as correct without this corroboration.

The designer must always be conscious that his/her design decisions and analytical results have consequences that can be related to the triple bottom line introduced in Chapter 1. To reiterate the triple bottom line, it is as follows:

First bottom line	Profit
Second bottom line	People and society
Third bottom line	Environmental impact and sustainability

6.10.1 First bottom line

The aim of any product creation is to create value and, therefore, profit. Design decisions are aimed at increasing the value and reducing the risk. The designer should always approach the design challenge with enthusiasm and positivity, but consequences of poor design and poor analysis may lead to loss of profit and, in extreme circumstances, the demise of the company.

6.10.2 Second bottom line

Most design challenges have the intention of improving the lot of people and society in general. The consequences of poor design decisions and poor analysis could mean injury to the user and, in worst cases, even death. The responsibilities of design can be quite heavy, especially where the safety of the user is concerned.

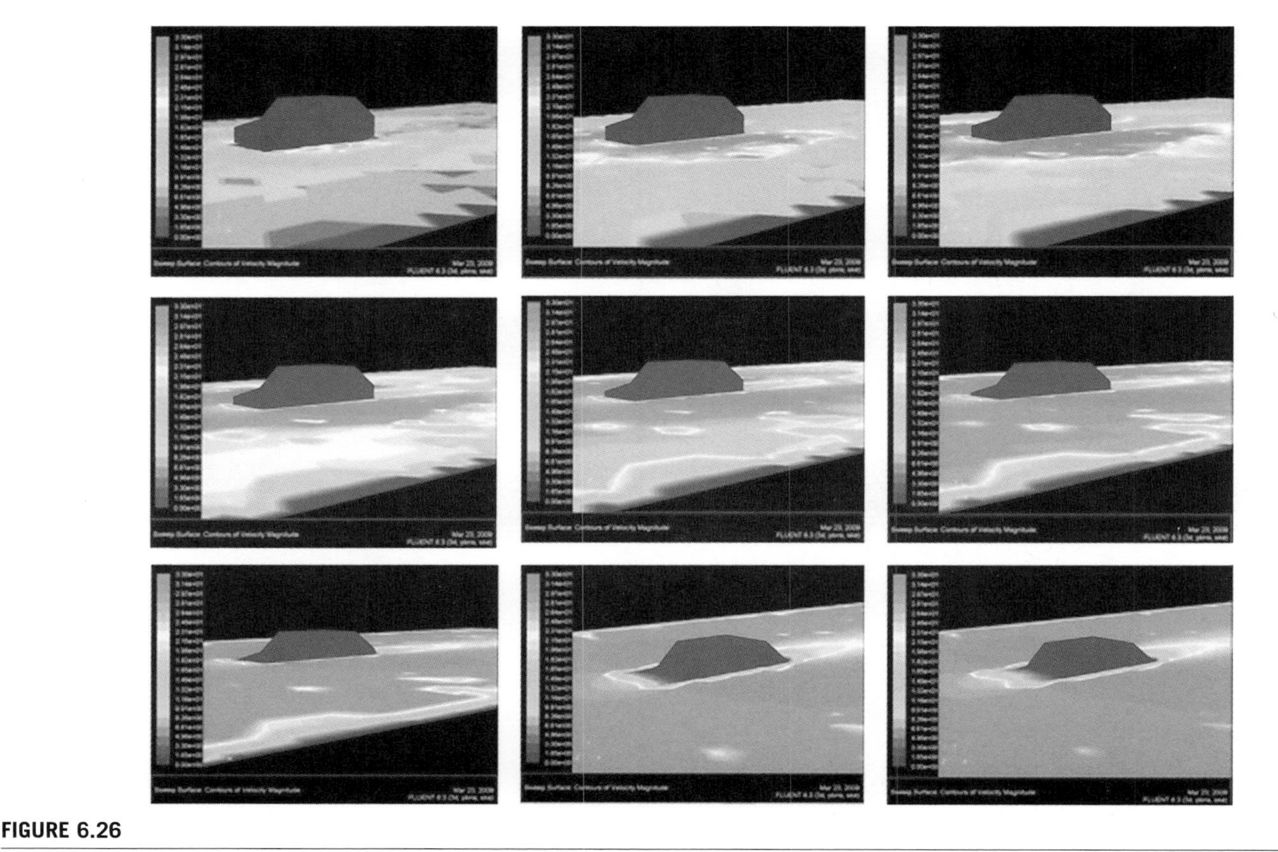

FIGURE 6.26

A Series of Analytical Snapshots from CFD Analysis Relating to 100 mm-Height Intervals from the Road Surface [14]. (For color version of this figure, the reader is referred to the online version of this chapter.)

6.10.3 Third bottom line

Awareness and consideration of the environment is an increasingly important aspect of design. The design function in the creation of new products can play an important part in reducing environmental impact. When approaching a new design challenge, the designer must develop an overview of consideration whereby the new product can be created with a reduced impact on the environment.

It can be argued that design is an art form, but Mechanical Engineering Design is an art form that is constrained by scientific principles and prediction of performance. If either or both of these two fundamentals are ignored, the consequences could be, at best, costly and, at worst, lethal.

APPENDIX

A6.1 STRESS IN A SHAFT CONSIDERING STRESS CONCENTRATION FACTORS

Determination of the stress in a gear-mounted shaft

Gear-mounted shaft with no relief groove

Find reactions

20 kN is applied in the center of the shaft. By observation, the reactions R_A and R_B are

$$R_A = R_B = \frac{20 \times 10^3}{2} = 10 \times 10^3 \text{ or } 10\text{kN}$$

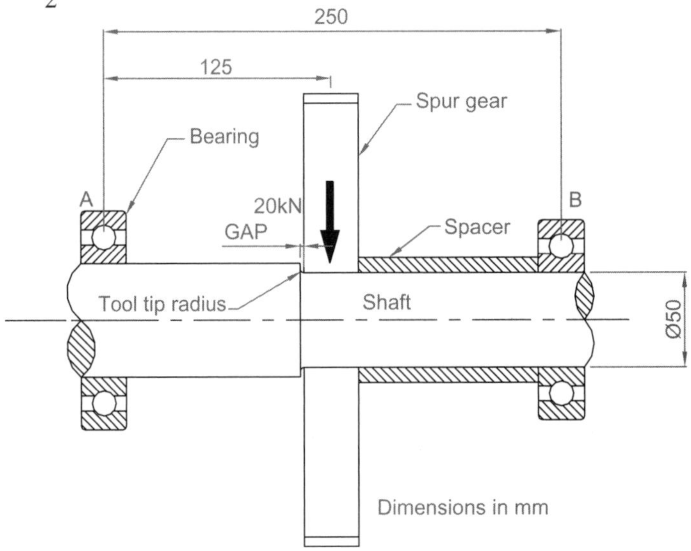

FIGURE A6.1

Shaft Showing Tool Tip Radius [1]

Find the bending moment
The maximum bending moment occurs underneath the gear.

Assumption: the maximum bending moment occurs at the shoulder.
$\text{Bm} = 0.125 \times 10 \times 10^3 = 1.25\,\text{kNm}$

Find the second moment of area
where D = shaft diameter of 50 mm

$$I = \frac{\pi \times D^4}{64} = \frac{\pi \times 0.05^4}{64} = 3.068 \times 10^{-7}\,\text{m}^4$$

Find the bending stress
$y = 0.025$ m = distance from neutral axis to extreme fibers (radius)

$$\sigma = \frac{\text{Bm} \times y}{I} = \frac{1.25 \times 10^3 \times 0.025}{3.068 \times 10^{-7}} = 101.9\,\text{MN/m}^2$$

Determine stress with relief groove

Bm = 1.25 kNm, previously calculated.

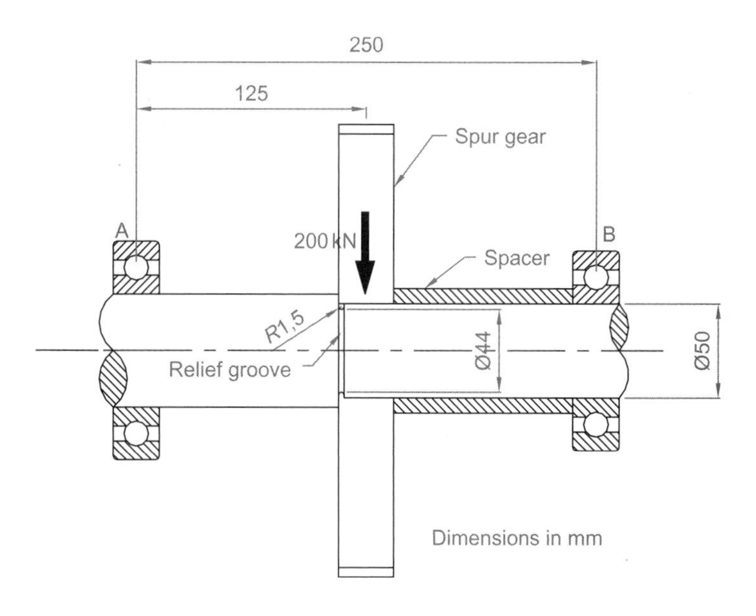

FIGURE A6.2

Shaft Showing Relief Groove [1]. (For color version of this figure, the reader is referred to the online version of this chapter.)

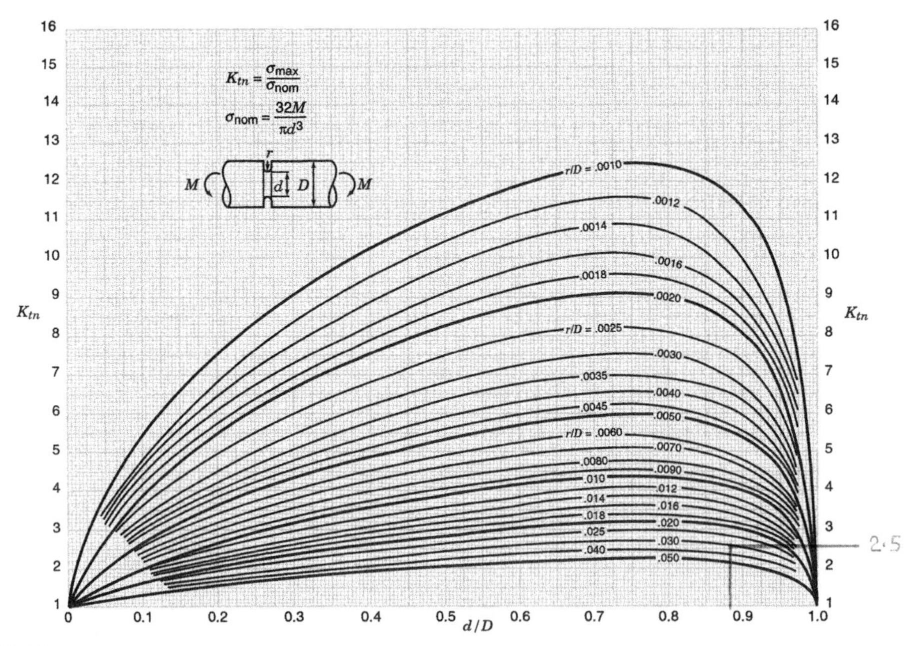

FIGURE A6.3

Stress Concentration Factors for Shaft with Relief Groove [7].

To compensate for the relief groove, the chart below can be used. See Figure A6.3. This shows stress concentration factors K_{tn} for circular shafts in bending with a U-shaped groove [7]. The stress concentration factors K_{tn} when combined within an appropriate analysis act in a manner similar to a factor of safety. The chart in Figure A6.3 requires certain minor calculations to be performed before the chart can be used. They are calculated as follows:

Key
D = major diameter of shaft = 50 mm (0.05 m)
d = diameter at the base of the groove = 44 mm (0.044 m)
r = radius at the base of the groove = 1.5 mm

Find d/D ratio

$$\frac{d}{D} = \frac{44}{50} = 0.88$$

Find r/D ratio

$$\frac{r}{D} = \frac{1.5}{50} = 0.03$$

The parameters have been marked on the chart in Figure A6.3. The K_{tn} can now be read as $K_{tn} = 2.5$. This parameter can now be used to determine the nominal stress σ_{nom}.

Determine nominal stress σ_{nom}

$$\sigma_{nom} = \frac{32 \times Bm}{\pi \times d^3} = \frac{32 \times 1.25 \times 10^3}{\pi \times 0.044^3} = 4.6 \times 10^6 N/m^2$$

Determine σ_{max}

$K_{tn} = 2.5$

$$K_{tn} = \frac{\sigma_{max}}{\sigma_{nom}}$$

or

$\sigma_{max} = \sigma_{nom} \times K_{tn} = 4.6 \times 10^6 \times 2.5 = 11.5 \times 10^6 N/m^2$

Conclusion

The value of $\sigma_{max} = 11.5$ MN/m^2 can be considered to be the maximum stress that the shaft can undergo when such a relief groove is applied to it. Consider the maximum stress the shaft can undergo when there is no relief groove. The value of 101.9 MN/m^2 is a factor of 10 greater than the maximum stress in the shaft with a relief groove. The reduction in strength is a consequence of designing into the shaft a legitimate stress raiser, but it should be considered that the analysis has allowed a precise design to be created that caters to the reduction in strength.

A6.2 STRESS IN BEAMS

In a factory refurbishment, new machinery is to be installed. In particular, large loads are to be suspended from a particular roof beam. The beam, complete with the loading regime, is shown in Figure A6.4. The beam is a standard "I" section and can be taken as being pin-jointed at each end. The beam section can be seen in Figure A6.5.

As one of the engineers working on the refurbishment, your task is to analyze the beam and ensure that a minimum safety factor of 20 is applied to it when it is fully loaded.

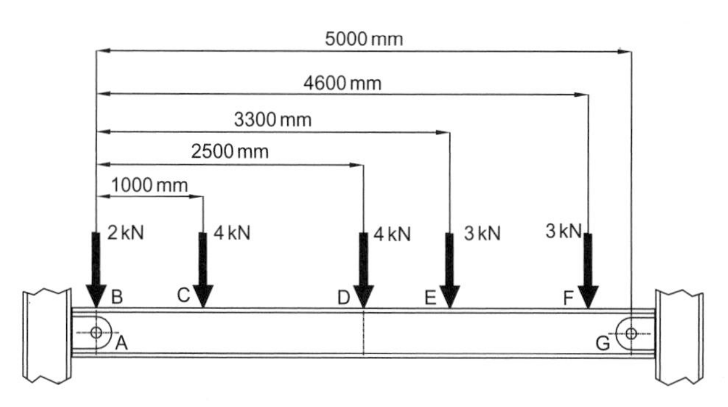

FIGURE A6.4

Roof Beam Loading Regime [1].

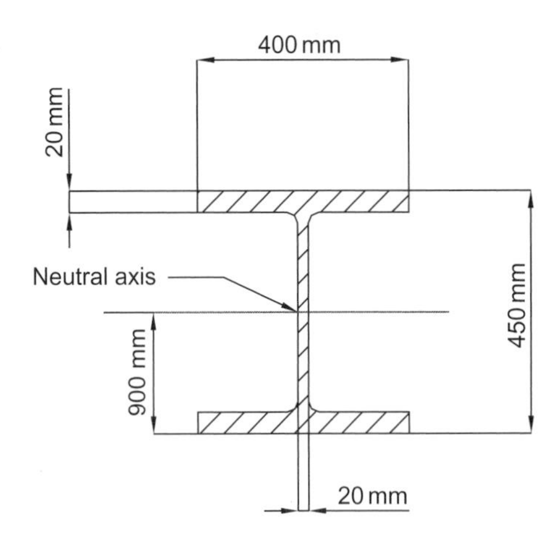

FIGURE A6.5

Roof Beam Section [1].

In order to meet the criteria set out in the original brief, the analyst/designer would need to follow the following procedure:

- **a.** Find the reactions at A and G
- **b.** Draw the shear force diagram
- **c.** Indicate the position of the maximum bending moment
- **d.** Calculate the value of the maximum bending moment
- **e.** Indicate the position of the neutral axis on a sketch of the beam section
- **f.** Determine the second moment of area for the section
- **g.** Determine the maximum bending stress in the beam
- **h.** If the ultimate tensile strength of the cast iron material is 150 MN/m^2, determine the factor of safety. State whether the actual factor of safety complies with the design factor of safety.
  - **a.** Find the reactions at A and G

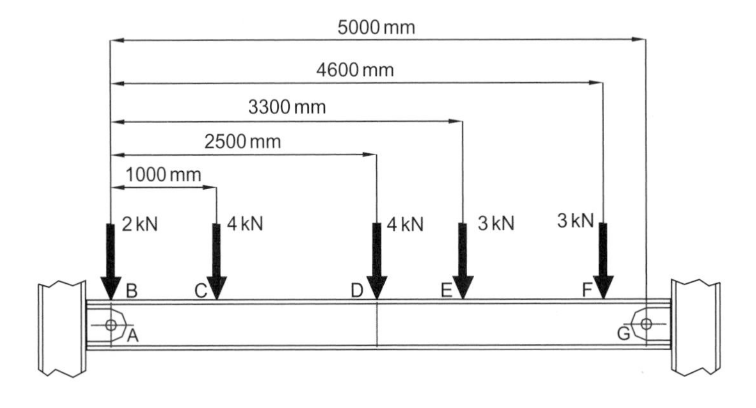

FIGURE A6.6

Roof Beam Loading Regime [1].

Clockwise moments = anticlockwise moments

To find R_A

Moments about R_G

$$R_A \times 5 = (2 \times 5) + (4 \times 4) + (4 \times 2.5) + (3 \times 1.7) + (3 \times 0.4)$$

$$R_A = \frac{10 + 16 + 10 + 5.1 + 1.2}{5} = 8.46\,\text{kN}$$

To find R_G

Upward forces = downward forces

$$R_A + R_G = 2 + 4 + 4 + 3 + 3$$

$$R_G = 16 - 8.46 = 7.54\,\text{kN}$$

b. Draw the shear force diagram

c. The maximum bending moment is indicated in Figure A6.7

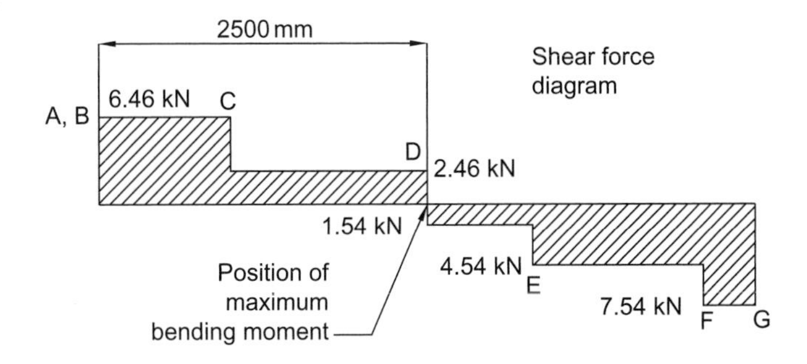

FIGURE A6.7

Shear Force Diagram [1].

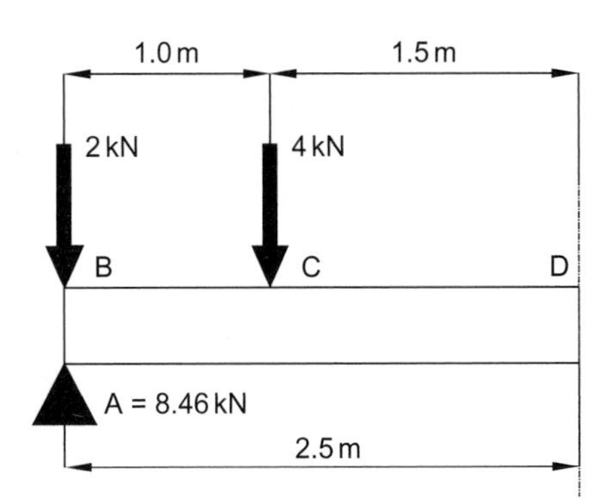

FIGURE A6.8

Bending Moment Space Diagram [1].

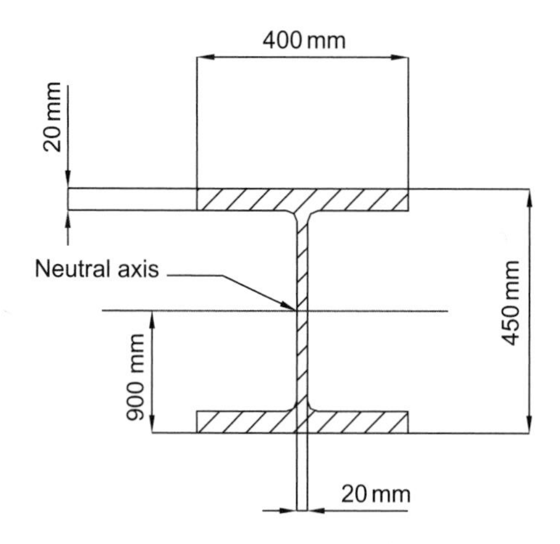

FIGURE A6.9

Sectional View of Intended Beam [1].

d. Calculate the value of the maximum bending moment

Clockwise moments = anticlockwise moments
Clockwise moments are +ve
Moments about D
$M_D = (8.46 \times 2.5) - (2 \times 2.5) - (4 \times 1.5)$
$M_D = 21.15 - 5 - 6 = 10.15\,\text{kNm}$

e. Indicate the position of the neutral axis on a sketch of the beam section

f. Determine the second moment of area for the section

$$I = \frac{BD^3}{12} + Ah^2$$

$$I_{\text{Upper flange}} = \frac{0.4 \times 0.02^3}{12} + (0.4 \times 0.02) \times 0.215^2 = 3.7 \times 10^{-4}\,\text{m}^4$$

$$I_{\text{Upperflange}} = I_{\text{Lowerflange}} = 3.7 \times 10^{-4}\,\text{m}^4$$

$$I_{\text{Vertical web}} = \frac{0.02 \times 0.39^3}{12} = 9.887 \times 10^{-5}\,\text{m}^4$$

$$I_{\text{Total}} = I_{\text{Upperflange}} + I_{\text{Lowerflange}} + I_{\text{Verticalweb}} = 8.39 \times 10^{-4}\,\text{m}^4$$

g. Determine the maximum bending stress in the beam

$$\frac{\sigma}{y} = \frac{M}{\text{Ixx}} = \frac{E}{R}$$

$$\sigma = \frac{M \times y}{\text{Ixx}} = \frac{10.15 \times 10^3 \times 0.225}{8.39 \times 10^{-4}} = 2.72\,\text{MN/m}^2$$

h. Find the factor of safety

$$\text{FoS} = \frac{\sigma_{\text{UTS}}}{\sigma_{\text{working}}} = \frac{150 \times 10^6}{2.72 \times 10^6} = 55$$

Calculated FoS is above a design FoS of 20 and is therefore adequate

A6.3 TORSION IN A SHAFT

A torsion meter whose schematic is shown in Figure A6.10 is attached to a ship's propeller shaft. The torsion meter operates by comparing the position of two flanges set a certain distance apart on the shaft. The torsion meter is set so that the flanges and, therefore, the reading surfaces are set 3 m apart. The readings therefore allow the engineers to judge whether the shaft is being overstressed in torsion. The diesel engine is able to apply a maximum torque of 25 kNm to the shaft. As one of the engineers developing the torsion meter for this particular ship, you are required to calculate the stresses and torsional displacement to confirm the accuracy of the torsion meter.

The parameters of the shaft are as follows:

outside diameter: 200 mm
inside diameter: 192 mm
distance between measurement points: 3 m
material: medium carbon steel, modulus of rigidity $= G = 80$ GN/m^2
maximum torque $= 25$ kNm.

The design engineer is required to follow the procedure set out here in order to fulfill the project requirements.

a. Find the polar second moment of area for the shaft.
b. Find the angle of twist between the measurement points of the torsion meter in radians.
c. The angle of twist in the shaft should not exceed 3° in any 3 m length. Compare the calculated value with the design value and indicate if the system is within designed parameters.
d. Determine the shear stress in the shaft caused by the applied torque of 25 kNm.
e. The ultimate shear strength for the propeller shaft material is 270 MN/m^2. Find the factor of safety.

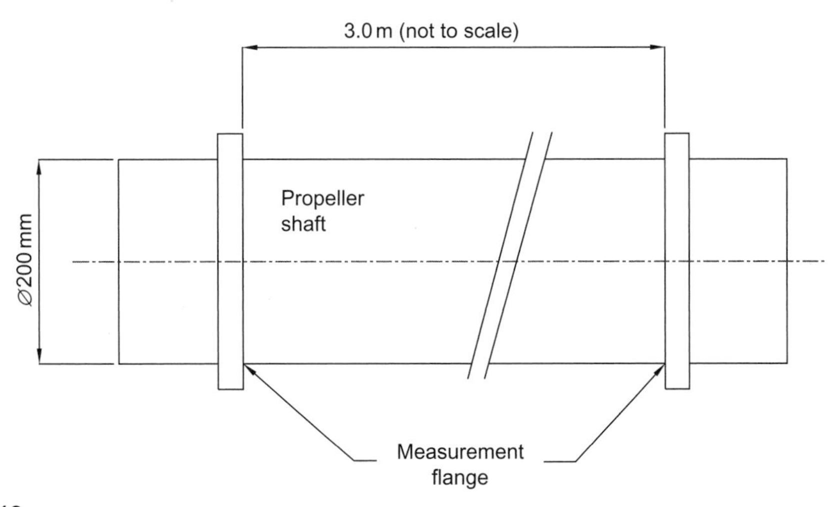

FIGURE A6.10

Measurement Setup for the Torsion Meter [1].

f. The factor of safety set by the designers is 3. State whether the stresses in the shaft meet this criterion.

g. Find the total angle of twist between the engine and the propeller if the distance is 13 m.

a. Determine the polar second moment of area for the shaft

Outside diameter = 200 mm
Inside diameter = 192 mm
$G = 80\ \text{GN/m}^2$
Torque = 25 kNm

$$J = \frac{\pi \times (D^4 - d^4)}{32} = \frac{\pi \times (0.2^4 - 0.192^4)}{32} = 2.366 \times 10^{-5}\,\text{m}^4$$

b. Determine the angle of twist between the measurement points of the torsion meter in radians.
Length = 3.0 m, $G = 80\ \text{GN/m}^2$
From

$$\frac{\tau}{r} = \frac{T}{J} = \frac{G\theta}{l}$$

$$\theta = \frac{Tl}{GJ} = \frac{25 \times 10^3 \times 3.0}{80 \times 10^9 \times 2.366 \times 10^{-5}} = 0.0396\,\text{rad}$$

c. The angle of twist in the shaft should not exceed 3° in any 3 m length. Compare the calculated value with the design value and indicate if the system is within designed parameters.

$$\theta_{\text{degrees}} = \frac{0.0396 \times 360}{2 \times \pi} = 2.27^\circ$$

The shaft will twist 2.27°, which is less than the maximum allowable twist of 3°. The shaft is within design parameters.

d. Determine the shear stress in the shaft caused by the applied torque of 25 kNm
Torque = 25 kNm, $r = 0.1$ m
From

$$\frac{\tau}{r} = \frac{T}{J} = \frac{G\theta}{l}$$

$$\tau = \frac{T \times r}{J} = \frac{25 \times 10^3 \times 0.1}{2.366 \times 10^{-5}} = \underline{105.6 \times 10^6\,\text{N/m}^2}$$

e. Determine the factor of safety if the ultimate shear strength is 270 MN/m^2

$$\text{FoS} = \frac{\tau_{\text{UTS}}}{\tau_{\text{working}}} = \frac{270 \times 10^6}{105.6 \times 10^6} = \underline{2.84}$$

f. The factor of safety set by the designers is 3. State whether the stresses in the shaft meet this criterion.
The new safety factor is 2.84, which is adequate to meet the design factor of safety of 3

g. Determine the total angle of twist between the engine and the propeller if the distance is 13 m.

Angle of twist over a 3 m length = 2.27°

$$\text{Angle of twist over 13m} = \frac{13 \times 2.27}{3} = \underline{9.83^\circ}$$

A6.4 DYNAMICS OF SOLID BODIES/MECHANICS: BICYCLE PARAMETERS

You are a designer of bicycles and have been tasked with performing analysis relating to the worst conditions that a particular size of bicycle will see in service. They relate to the deceleration involving brake efficiency and acceleration involving the rider's ability to inject energy. The analysis thus performed will be used to improve the new bicycle design. Please see Figure A6.11 for masses. Figure A6.12 shows the wheel dimensions.

The following parameters prevail:

General

Rider mass	95 kg
Bicycle mass	10 kg
Wheel radius road/tire interface	315 mm
Radius of brake shoes	280 mm

Deceleration analysis

Initial velocity u	40 km/h
Final velocity v	0
Stopping distance	20 m

Acceleration analysis

Pedal radius	200 mm

Pedal sprocket diameter = drive wheel sprocket diameter

The design engineer is required to follow the procedure set out here in order to fulfill the project requirements.

a. Find the deceleration in m/s^2 when bringing the bicycle to a halt. Refer to the parameters specified earlier.

b. Determine the value of the deceleration force at the road/tire interface assuming that the entire braking force is delivered to one wheel only.

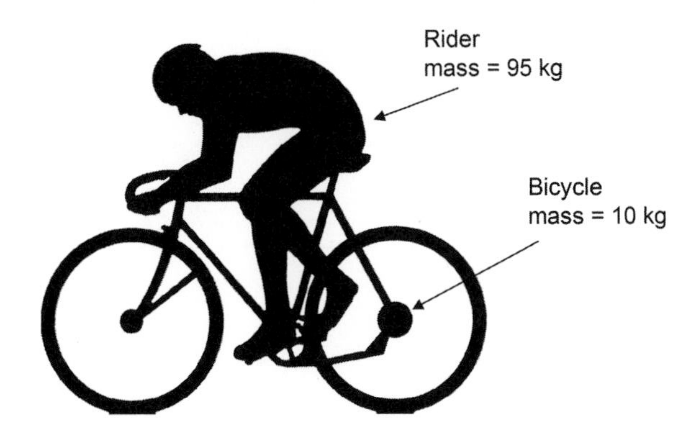

FIGURE A6.11

Bicycle Mass Elements [1]. (For color version of this figure, the reader is referred to the online version of this chapter.)

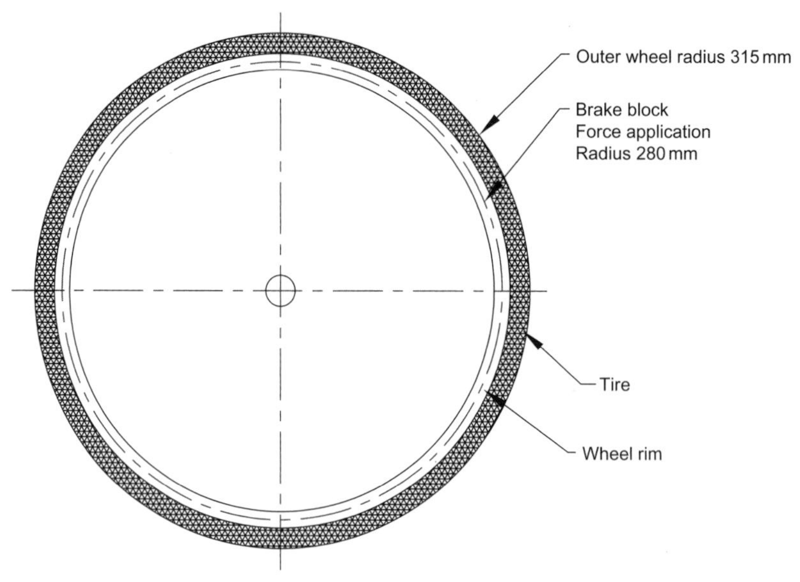

FIGURE A6.12

Wheel Dimensions [1].

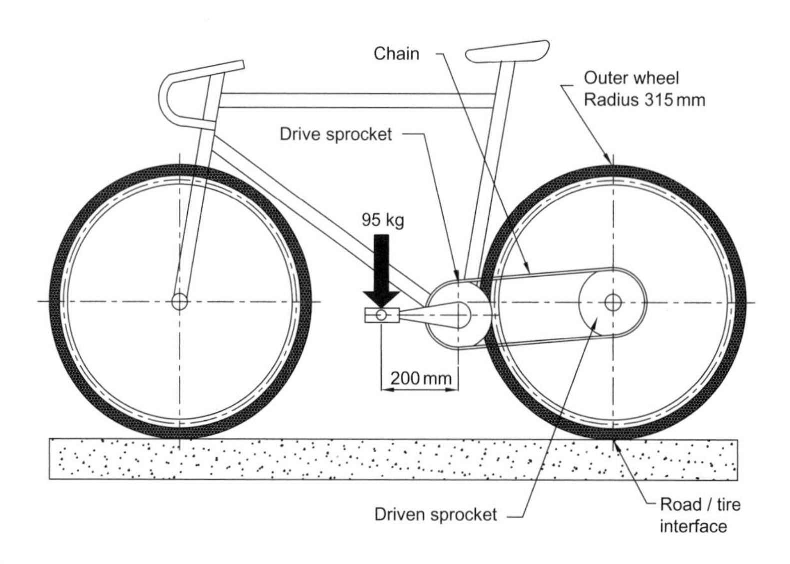

FIGURE A6.13

Pedal-Wheel Relationship [1].

c. Determine the force required at the brake block noting the difference between the road/tire interface radius and the brake block radius (in N).

d. The rider now accelerates by standing on a pedal with all his mass as indicated in Figure A6.13. Determine the driving force at the rear wheel at the road/tire interface (*N*). The pedal radius is

200 mm and the road/tire interface radius is 315 mm. Assume the drive sprocket and the driven sprocket are the same diameter.

e. Determine the acceleration of the bicycle and rider in m/s^2.

f. Determine the final velocity of the bicycle if the initial velocity is zero and the acceleration is maintained for 1.6 s.

g. The resistance to motion is a combination of the rolling resistance of tires, bearing friction, and aerodynamic resistance. The resistance to motion of the bicycle and rider can be taken as 2.5 N/kg. Find the total resistance to motion.

h. Find the work done in 20 min if the maximum velocity of 9 m/s is maintained.

i. Find the power applied by the rider during the maximum velocity (9 m/s) period.

a. Find the deceleration in m/s^2

$$\text{Initial velocity } u = 40\,\text{km/h} = \frac{40 \times 10^3}{60 \times 60} = 11.1\,\text{m/s}$$

Final velocity $v = 0$

Stopping distance $S = 20$ m

From $v^2 = u^2 + 2aS$

Transpose for a

$$a = \frac{v^2 - u^2}{2 \times S} = \frac{0 - 11.1^2}{2 \times 20} = \underline{-3.086\,\text{m/s}^2}$$

Note that -ve indicates deceleration

$v = 0$

b. Determine the value of the deceleration force at the road/tire interface (N)

Mass = rider + bicycle = 95 + 10 = 105 kg

From $F = ma$

$F = 105 \times 3.086 = \underline{324\,\text{N}}$

c. Determine the force required at the brake block

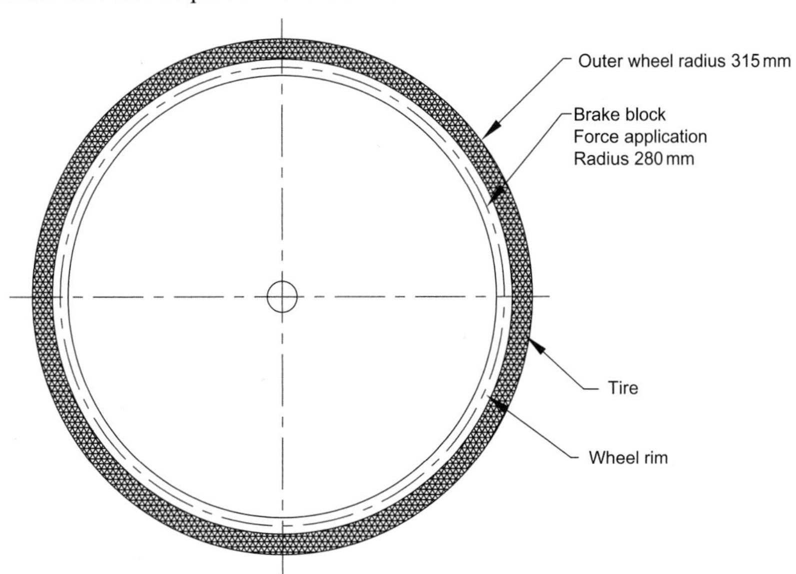

FIGURE A6.14

Wheel Dimensions [1].

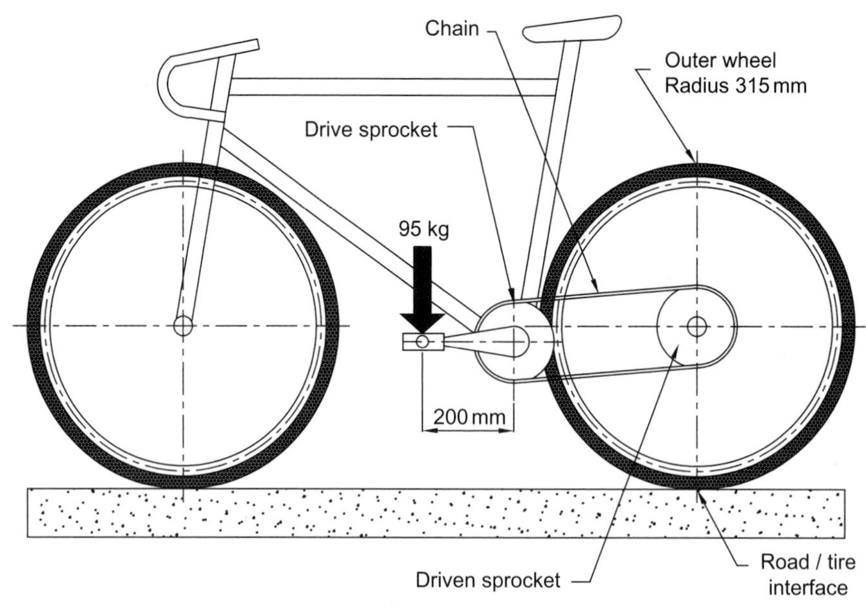

FIGURE A6.15

Bicycle Relative Components [1].

Wheel torque = brake torque

Force at wheel × wheel radius = force at brake × brake radius

$F_w \times R_w = F_b \times R_b$

$$F_b = \frac{F_w \times R_w}{R_b} = \frac{324 \times 0.315}{0.28} = \underline{364.5\,\text{N}}$$

d. Determine the driving force at the rear wheel at the road/tire interface (N)

For equilibrium

Torque at pedal = Torque at wheel

Note: 95 kg × 9.81 m/s^2 = 932 N

$F_p \times R_p = F_w \times R_w$

$$F_w = \frac{F_p \times R_p}{R_w} = \frac{932 \times 0.2}{0.315} = \underline{591.7\,\text{N}}$$

e. Determine the acceleration of the bicycle and rider in m/s^2

Mass = Rider mass + bicycle mass = 95 + 10 = 105 kg

From $F = ma$,

$$a = \frac{F}{m} = \frac{591.7}{105} = \underline{5.63\,\text{m/s}^2}$$

$v = 0 + 5.63 \times 1.6 = \underline{9\,\text{m/s}}$

f. Determine the final velocity of the bicycle

$u = 0$

$a = 5.63\,\text{m/s}^2$

$v = ?$

$t = 1.6\,\text{s}$

From $v = u + a \times t$

g. Find the total resistance to motion

Resistance $= 2.5\,\text{N/kg}$

Mass $= 105\,\text{N}$

Total resistance $= 105 \times 2.5 = \underline{262.5\,\text{N}}$

h. Find the work done in 20 min

constant velocity $= 9\,\text{m/s}$

Time $= 20 \times 60 = 1200\,\text{s}$

Distance covered $= 1200 \times 9 = 10{,}800\,\text{m}$

WD $=$ resisting force $\times$ distance

WD $= 262.5 \times 10{,}800 = \underline{2835\,\text{kJ}}$

i. Find the power applied by the rider during the maximum velocity (9 m/s) period

force $= 262.5\,\text{N}$

velocity $= 9\,\text{m/s}$

Power $=$ force $\times$ velocity

$P = 262.5 \times 9 = \underline{2362.5\,\text{W} \quad \text{or} \quad 2.3625\,\text{kW}}$

Alternative

$$\text{Power} = \text{WD} = \frac{2835 \times 10^3}{1200} = \underline{2.36\,\text{kW}}$$

A6.5 INERTIAS AND POWER PREDICTION

A diesel engine is configured to drive a rotating drum as shown in Figure A6.16. In operation, the engine is instantaneously coupled to the drum via a friction clutch.

The moment of inertia of the engine is equivalent to a mass of 25 kg acting at a radius of gyration of 393 mm. The engine rotates at 200 rev/min before engagement with a drum winch through the clutch.

The drum has a mass of 60 kg acting at a radius of gyration of 406 mm and is initially at rest.

The drum is the winding end of a winch having a rope wrapped around its diameter of 900 mm. It is designed to lift 1000 kg pallets through the floors of a warehouse.

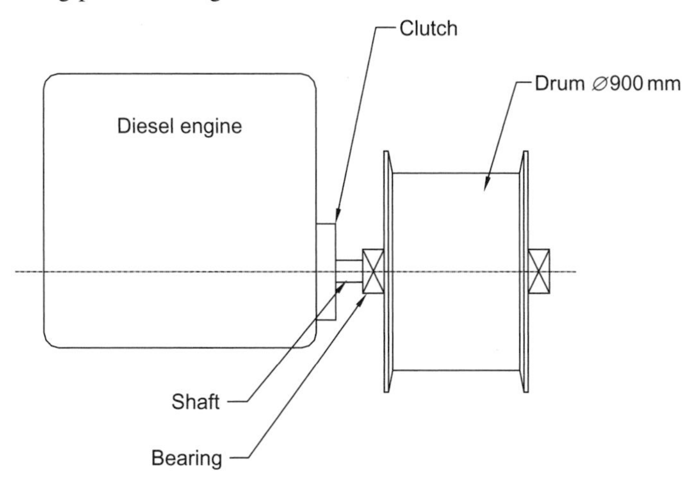

FIGURE A6.16

Diesel Engine, Clutch, and Winding Drum Configuration [1].

a. Determine the moment of inertia of the engine
b. Determine the moment of inertia of the drum
c. Determine the combined angular velocity in rev/min immediately after connection. Assume the drum and the engine do not separate after connection
d. Determine the power required at the engine to lift a 1000 kg pallet using the combined angular velocity calculated in part (c). Assume no losses in the system.

(a) Determine the moment of inertia of the engine

$k_E = 0.393\,\text{m}$

$m_E = 25\,\text{kg}$

$I_E = mk^2 = 25 \times 0.393^2 = \underline{3.86\,\text{kg m}^2}$

(b) Determine the moment of inertia of the drum

$m_D = 60\,\text{kg}$

$r_D = 0.406\,\text{m}$

$I_{\text{drum}} = m_D r_D{}^2$

$I_{\text{drum}} = 60 \times 0.406^2 = \underline{9.89\,\text{kg m}^2}$

(c) Determine the combined angular velocity in rev/min immediately after connection.

$$\omega_E = \frac{200 \times 2\pi}{60} = \underline{20.94\,\text{rad/s}^2}$$

$\omega_D = $ zero (initial angular velocity of drum)

Using the principle of conservation of momentum

Momentum before connection = momentum after connection

$$(I_E \times \omega_E) + \cancel{(I_{\text{drum}} \times \omega_{\text{drum}})} = (I_E \times \omega_{\text{Combined}}) + (I_{\text{drum}} \times \omega_{\text{combined}})$$

$(I_E \times \omega_E) = (I_E + I_{\text{drum}})\omega_{\text{combined}}$

$\omega_{\text{combined}} = $ combined angular velocity

$$\omega_{\text{combined}} = \frac{I_E \times \omega_E}{I_E + I_{\text{drum}}} = \frac{3.86 \times 20.94}{3.86 + 9.89} = \underline{5.88\,\text{rad/s}}$$

$$\omega_{\text{rev/min}} = \frac{5.88 \times 60}{2\pi} = \underline{56.1\,\text{rev/min}}$$

(d) Determine the power required at the engine to lift a 1000 kg pallet

$\omega = 5.88\,\text{rad/s}$

Mass $= 1000\,\text{kg}$

Drum radius $= 0.45\,\text{m}$

Find torque at the drum

Torque = Force × drum radius

Torque $= 1000 \times 9.81 \times 0.45 = \underline{4414.5\,\text{Nm}}$

Find power

$$P = \text{Torque} \times \omega = 4414.5 \times 5.88 = \underline{25.96\,\text{kW}}$$

A6.6 DYNAMICS: VIBRATION ISOLATION

Diesel engine motive power is usually used for large items of plant such as that shown in Figure A6.17. Such large engines need to be placed on engine mounting systems so that the induced vibrations are isolated from the chassis, thus preventing unwanted vibrations reaching sensitive equipment and personnel. Diesel engines are normally mounted on four vibration isolators, also shown in the schematic in Figure A6.17. In this case, assume the engine mass is 2600 kg, which is distributed evenly between the four isolators. The speed of rotation of the engine is 1800 rev/min.

Considering only oscillation in the vertical direction and using the isolation efficiency graph in Figure A6.18, calculate the individual stiffness of the isolators if the isolation efficiency is to be 90%. Assume that the isolators are identical and that *damping can be neglected* for this selection process.

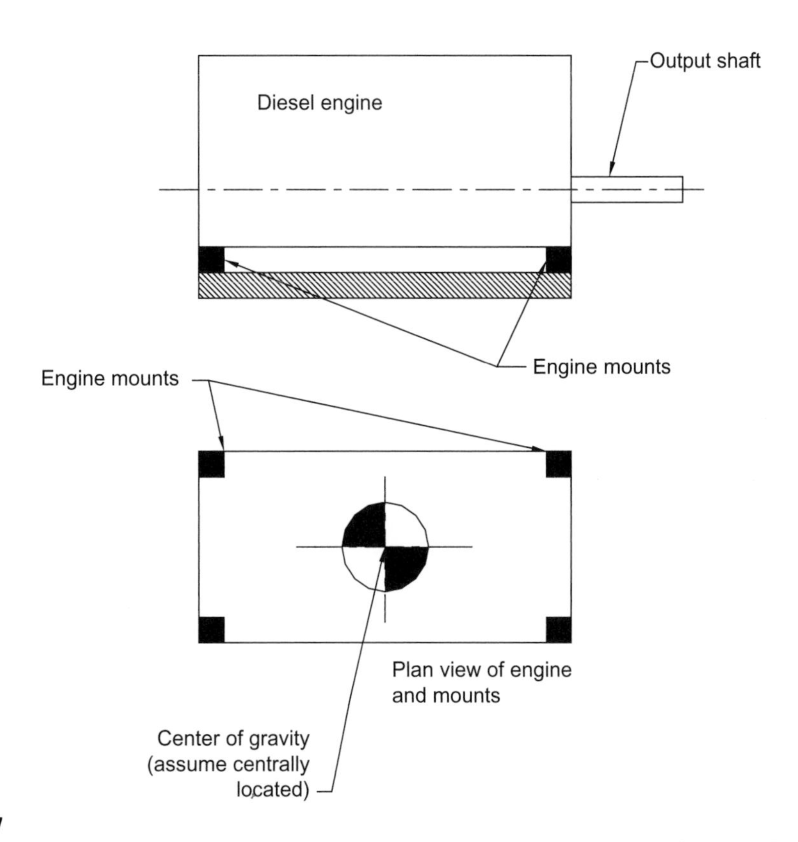

FIGURE A6.17

Engine Mount Arrangement for Large Diesel Engine [1].

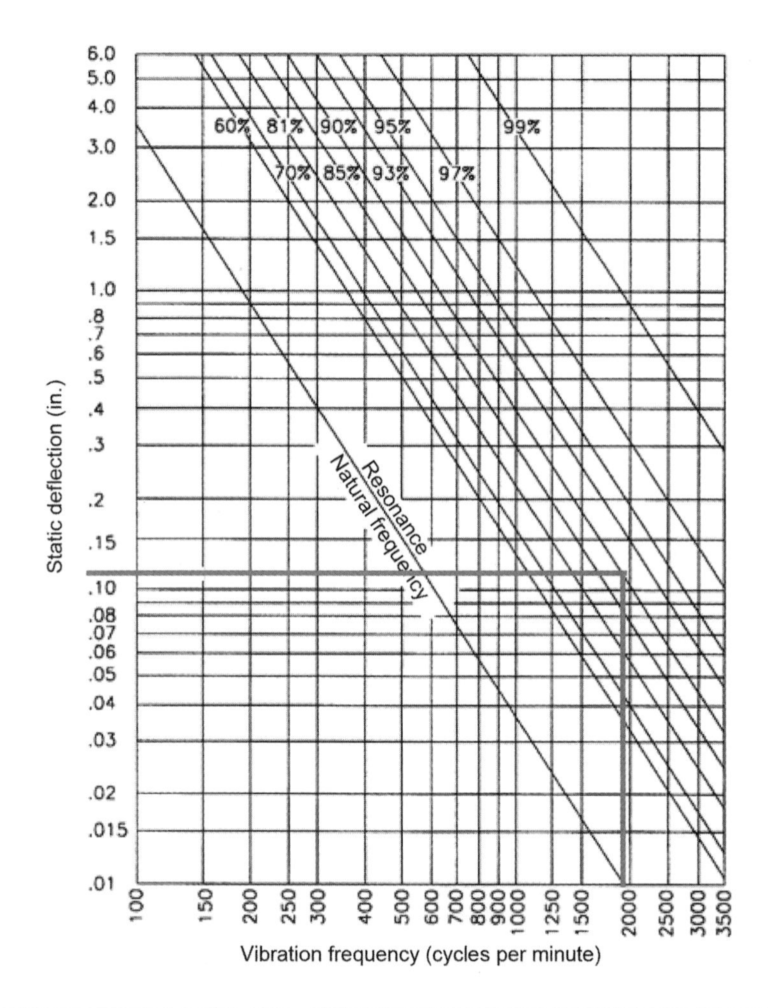

FIGURE A6.18

Isolation Efficiency Graph [8] (Note 1 in. = 25.4 mm).

Find load on each isolator

$$\text{Load/isolator} = \frac{\text{engine mass}}{4} = \frac{2600}{4} = 650\text{kg}$$

Load/isolator = 650 × 9.81 = $\underline{6380\text{ N}}$

Find the deflection from the Isolation Efficiency graph (Fig A6.18)

rev/min (cycles/minute) = 1800 rev/min

Isolation 90%

Insert vibration frequency and read deflection from the Graph

Deflection 0.12 in.

Deflection (metric) 0.12 x 25.4 = *3.05 mm (0.00305 m)*

Find spring stiffness

$$K = \frac{\text{Load}}{\text{Deflection}} = \frac{6380}{0.00305} = \underline{2.09\,\text{MN/m}}$$

The spring stiffness can now be used to select an appropriate mount from a supplier of anti-vibration mounts

A6.7 STRUCTURAL DESIGN OF THE DENBY DALE PIE COOKER

In the year 2000, the Denby Dale Pie cooker took the form of a welded steel structure with 24 compartments containing the stew. Figure A6.19 shows the layout of part of the structure excluding the compartments.

The 2004 attempt at the "Biggest Pie in the World" was to have a mass of 12 tons and cooked in a specially made "pie dish."

The whole structure was considered as a simply supported beam, 12 m long, whose total mass was 6000 kg and which was treated as a uniformly distributed load (UDL) of 4900 N/m.

The designers were required to ensure that the structure was safe by performing stress analysis and determining the factor of safety. The beam can be taken as simply supported with a loading regime as shown in Figure A6.20.

The designers followed the route given here by calculating the following:

a. The position of the neutral axis of the structure.
b. The reactions at the supports.
c. The shear force diagram indicating the position of the maximum bending moment.
d. The maximum bending moment
e. The total second moment of area for the beam
f. The maximum stress at the position of the maximum bending moment
g. The factor of safety given that the material used has an ultimate tensile strength (UTS) of 480 MN/m^2

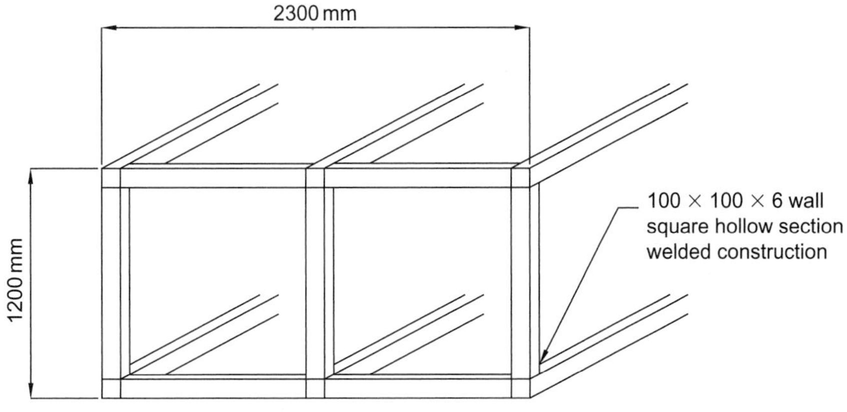

FIGURE A6.19

Basic Structure of the Denby Dale Pie Cooker [1].

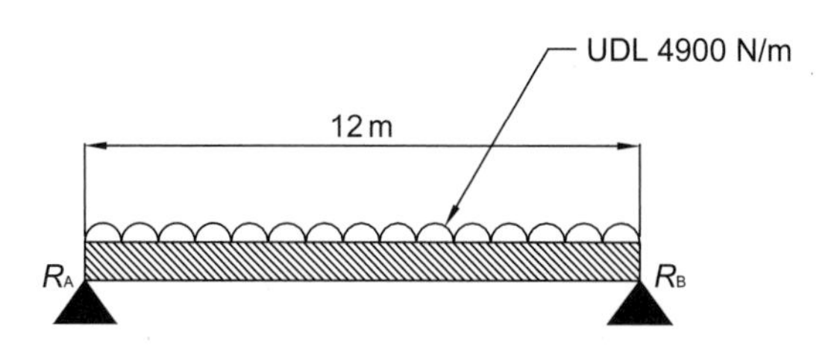

FIGURE A6.20

Loading Regime for the Denby Dale Pie Cooker [1].

a. The position of the neutral axis of the structure.

The structure is 1.2 m high and is uniform

By observation, the neutral axis *N/A acts at 0.6 m* from the base at the center of the section

b. The reactions at the supports.

Clockwise moments = anticlockwise moments

Moments about R_B

$R_A \times 12 = [(4900 \times 12) \times 6]$

$$R_A = \frac{352,800}{12} = \underline{29,400\ \text{N}}$$

$$R_A = R_B = \underline{29,400\ \text{N}}$$

c. The shear force diagram indicating the position of the maximum bending moment.

d. The maximum bending moment

Clockwise moments = anticlockwise moments

Moments about the center of the beam

$$\text{Bm} = (29,400 \times 6) - (6 \times 4900 \times 3) = \underline{88,200\ \text{Nm}}$$

e. The total second moment of area for the beam

From

$$\text{Ixx} = \frac{BD^3}{12} + Ah^2$$

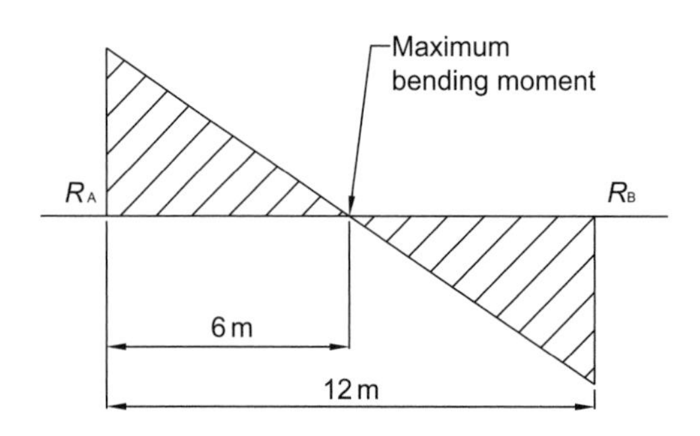

FIGURE A6.21

Shear Force Diagram Showing Position of Maximum Bending Moment [1].

Find Ixx for a single beam element as shown in Figure A6.22

$$\text{Ixx} = \frac{B(D^3 - d^3)}{12} = \frac{0.1 \times (0.1^3 - 0.088^3)}{12} = \underline{2.65 \times 10^{-6}\,\text{m}^4}$$

Find cross-sectional area

Area $= 0.1^2 - 0.88^2 = \underline{2.265 \times 10^{-3}\,\text{m}^2}$

Find total second moment of area IxxT

Refer to Figure A6.23

From

$$\text{Ixx} = \frac{BD^3}{12} + Ah^2$$

$$\text{IxxA} = 2.65 \times 10^{-6} + (2.265 \times 10^{-3} \times 0.55^2) = \underline{6.88 \times 10^{-4}\,\text{m}^4}$$

$$\text{IxxA} = \text{IxxB} = \underline{6.88 \times 10^{-4}\,\text{m}^4}$$

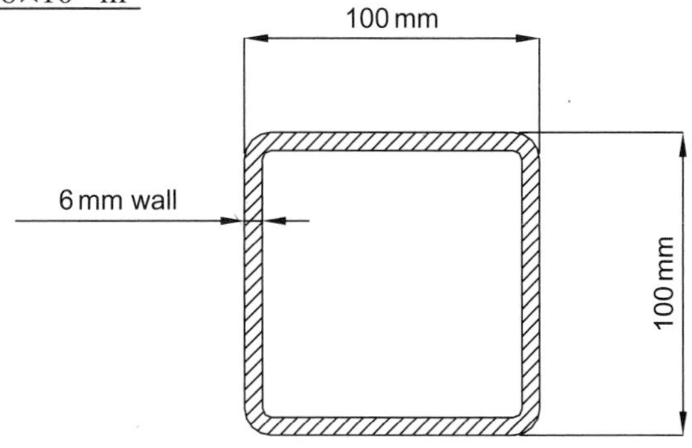

FIGURE A6.22

Cross-Section of a Single Beam Element [1].

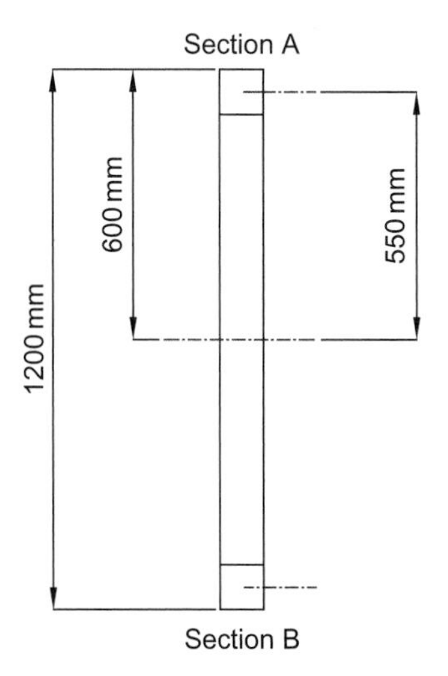

FIGURE A6.23

Structure of One Single Column Within the Fabrication [1].

Second moment of area for a single column IxxSC

$IxxSC = IxxA + IxxB = 2 \times 6.88 \times 10^{-4} = \underline{1.376 \times 10^{-3}\ m^4}$

But there are three columns

Total $IxxT = 3 \times 1.376 \times 10^{-3} = \underline{4.128 \times 10^{-3}\ m^4}$

f. The maximum working stress σ_w at the position of the maximum bending moment, where y = N/A to extreme fibers = 0.6 m

From

$$\sigma_w = \frac{m \times y}{IxxT} = \frac{88,200 \times 0.6}{4.128 \times 10^{-3}} = \underline{12.82\ MN/m^2}$$

g. The factor of safety given that the material used has an ultimate tensile strength (UTS) of 480 MN/m^2

$$FoS = \frac{\sigma_{UTS}}{\sigma_w} = \frac{480 \times 10^6}{12.82 \times 10^6} = \underline{37.44}$$ and is considered to be very safe

A6.8 HEAT TRANSFER

Denby Dale Pie: maximum rate of ingredients addition

During the cooking of the pie, it was essential that the ingredients were added to the stew in a precise fashion so that temperatures did not fall below 65 °C. This would trigger the growth of the *C. perfringens* bacterium. This is the bacterium that causes food poisoning and typically causes victims to develop diarrhea and sickness. It was, therefore, imperative that the stew was cooked in such a way as to avoid the growth of the bacterium.

The cooking process involved heating the ingredients of each of the twenty four vats to 100 °C, then adding further ingredients such as chopped potatoes, onions, and seasoning. It was the task of the designer to calculate, using the principles of heat transfer, the mass of ingredients that could be added before the stew temperature was reduced to 65 °C. The parameters were as follows:

Temperature of frozen ingredients	−4 °C
Specific heat capacity of stew (derived through experimentation), C	4200 J/kg K
Temperature of main body of stew	100 °C
Mass of main body of stew	500 kg

Determine the Energy Capacity of 500 kg of Boiling Stew

$C_p = C \times \text{mass} \times (\text{temperature change})$

where C_p = energy capacity (J)

$C_p = 4200 \times 500 \times (100 - 0) = \underline{210 \times 10^6\ J \text{ or } 210\ MJ}.$

Determine the mass that can be added

Determine the mass of ingredients that can be added to bring the general temperature down to 65 °C.

Conversely, determine the mass that will increase the temperature of the ingredients to 65 °C using the available energy of 210 MJ.

$C_p = 210$ MJ

$t_1 = -4$ °C

$t_2 = 65$ °C

$C = 4200$ J/kg K

From

$C_p = C \times \text{mass} \times (t_2 - t_1)$

isolate mass

$$\text{Mass} = \frac{C_p}{C \times (t_2 - t_1)} = \frac{210 \times 10^6}{4200 \times (65 - (-4))} = \underline{724\,\text{kg}}$$

The mass that could be added to each vat was 724 kg. This was more than was needed in total for each individual vat. In practice, no more than 50 kg of ingredients were added at any one time, thus keeping the stew temperature well above the minimum 65 °C.

References

[1] Author supplied.
[2] www.welding web.com 2013.
[3] The Stadium, Architecture for the New Global Culture, Periplus Editions, Hong Kong, 2005.
[4] D. Warburton, Final Year undergraduate design exercise, Huddersfield University, UK, 2012.
[5] Dr. Leigh Fleming, University of Huddersfield.
[6] Michele Bandini. E-mail: info@peenservice.it: www.peenservice.it.
[7] W.D. Pilkey, D.F. Pilkey, Peterson's Stress Concentration Factors, third ed., John Wiley & Sons, Hoboken NJ, 2008. ISBN:9780470 048245.
[8] W.T. Thompson, Theory of Vibrations, Nelson Thornes Ltd., Chelrenham UK, 2012.
[9] Barron Luke, University of Huddersfield, UK, 2013.
[10] Hague S, University of Huddersfield, UK, 2013.
[11] King Oliver, University of Huddersfield, UK, 2013.
[12] FSV 2013 Team.
[13] Abbott Caroline, University of Huddersfield, UK, 2013.
[14] J. Medina, University of Huddersfield, UK, 2013.

CHAPTER

Design for Total Control

7

The role of the engineering designer grows more and more complex with every passing year. There are more duties and constraints built on this role than any other role involved in the development of a product. The designer or design function must be creative, must produce a conceptual product that has to fulfill the market needs, must design it in such detail that it can be manufactured, and, while all this is taking place, the designer has to consider how to keep the cost of the product low. Recent years has seen the designer take on more roles, especially where the environment and sustainability are concerned.

Traditionally designers have designed equipment and manufacturers have built it. This is no longer possible since there are so many demands placed on the design function that the designer now has to take *total* control of the whole life of a product, from sourcing materials to disposing of the product at the end of its life.

7.1 TRADITIONAL APPROACHES

Design for manufacture has been a theme of designers for many years, but the demands and expectations placed on designers of new products mean that the scope of the design function has to be expanded. There has always been a drive to reduce the creation costs of products. Indeed, cost reduction of new products is a primary objective for designers and manufacturers alike. A quote from an anonymous industrialist defines the problem: "Everything costs money."

Recent years have seen a growing emphasis on providing products that are environmentally friendly (that is, sustainable). It is a fact that many businesspeople, designers, and manufacturers consider this effort an expensive enterprise, but in reality the design and manufacture to sustainable values and requirements often lead to lower-cost production.

7.2 THE SUSTAINABILITY UMBRELLA MODEL

The traditional design and manufacture goal of designing to cost has now been joined by the need to design and manufacture for sustainability. The quote from the anonymous industrialist can be enlarged as follows: "Everything costs money, and everything has an environmental impact." Perhaps this idea can be redefined as:

- All new products cost money.
- All new products "take" from the environment.
- All new products therefore need to be developed for *low cost* and *high sustainability*.

Sustainability and low cost often go hand in hand. It can be argued that products designed with sustainability as one of the primary objectives can also be designed under an umbrella of sustainability

Sustainability in Engineering Design. http://dx.doi.org/10.1016/B978-0-08-099369-0.00007-8

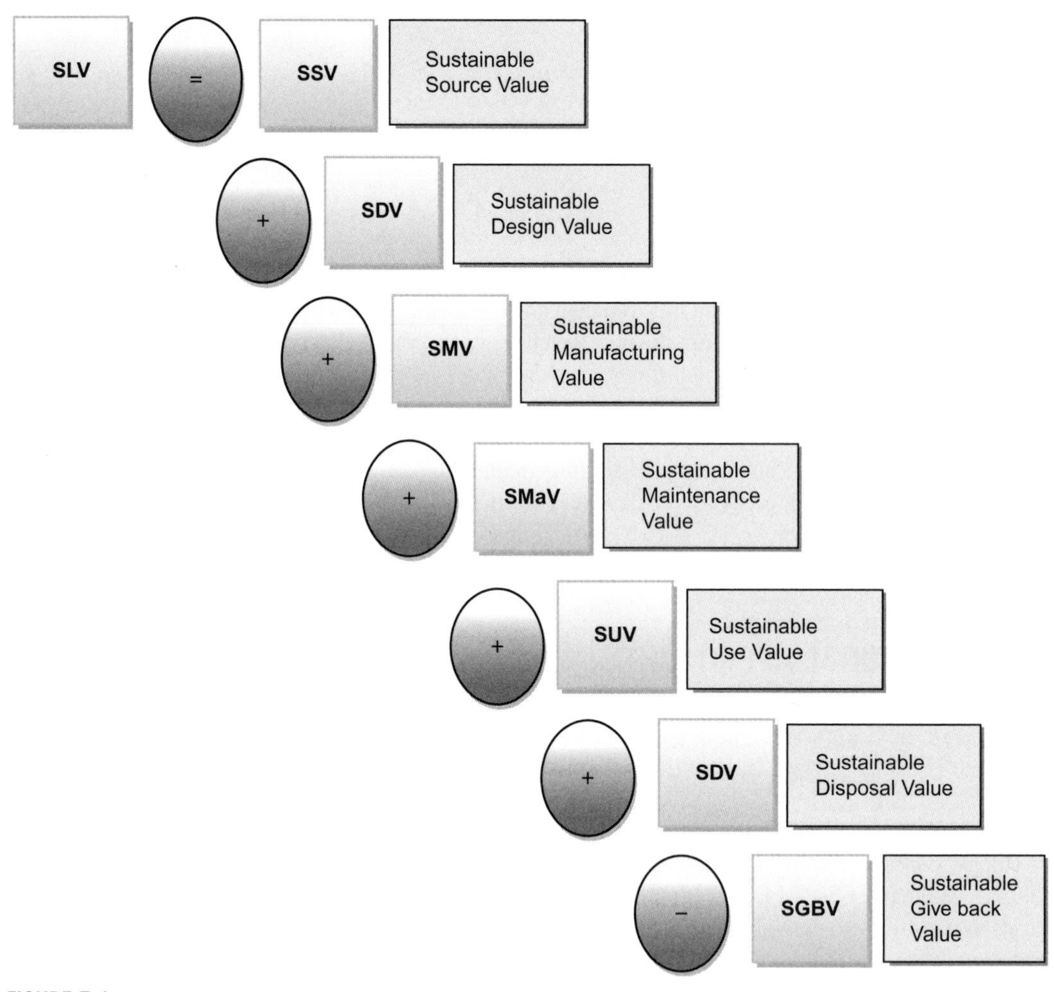

FIGURE 7.1

Sustainable Whole-Life Model [5]

covering *all* other facets of the design process. Consider the sustainable whole-life model previously put forward and now shown in Figure 7.1.

Design for manufacture has been a constant element within the design process for many years, but the demands and expectations placed on new products mean that the design function has to be expanded further.

7.3 TOTAL DESIGN CONTROL

Any manufactured product will have used a certain amount of energy in its manufacture; this energy could be derived from several sources. For instance, much of the energy could be derived from fossil fuels, which can be considered synthetic energy and will possess a carbon footprint. Increasingly, the

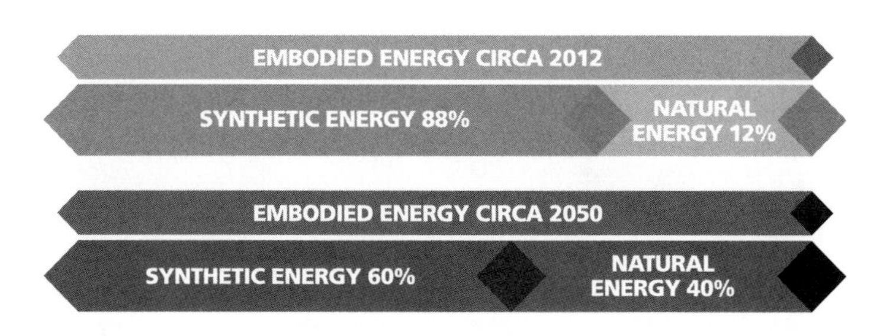

FIGURE 7.2

Embodied Energy Proportions [1, 2]

energy used in the creation of a product will be derived from a renewable source such as hydroelectric or wind or solar energy. These sources can be considered natural energy.

Energy is required whenever a process is applied to a material. A finished product has had expended on it a certain amount of energy, which is normally considered to be *embodied energy* [14]. This value of energy is a combination of synthetic energy and natural energy. The embodied energy diagram in Figure 7.2 indicates the likely proportions of synthetic versus natural energy within the embodied energy of a product [1, 2].

It is important that the embodied energy value is quantifiable. Since every aspect of the design and manufacture of a product demands that energy is applied, it seems that a value of energy per process is an appropriate measurement value. This complicated process has been much simplified by Granta Design Ltd. of Cambridge, U.K., which has created a very sophisticated software tool that calculates the embodied energy at various stages of a product's development [3].

We can see from the sustainable whole-life model in Figure 7.1 that the overall design of a product requires a whole-life approach and can be achieved in its entirety by the designer or the design team. Furthermore, the designer must not design in isolation. The designer or design team has to be in control of all aspects of the design, from instigation through manufacture and in some cases marketing. This, then, is *total design control.*

7.4 A NEW DESIGN APPROACH (THE UMBRELLA OF SUSTAINABLE DESIGN)

The traditional design and manufacture goal has always been to design to a cost, but recent years have seen this model being joined by the need to design and manufacture to sustainability values. The general perception of design with sustainability as a criterion is that it is an expensive process. The truth is that any product will be expensive if created with inefficient or ill-considered processes.

Any product that is brought to market has had energy applied to it in the form of manufacturing and other processing activities. This then gives the product a value of *embedded energy.* If the value of embedded energy could be reduced, so would the cost of processing. A reduction in embedded energy is also a major goal of design for sustainability and is therefore symbiotic with a desire to create products at a low cost.

The design and manufacture process should now involve both:

- Design for low-cost product creation
- Design for sustainable product creation

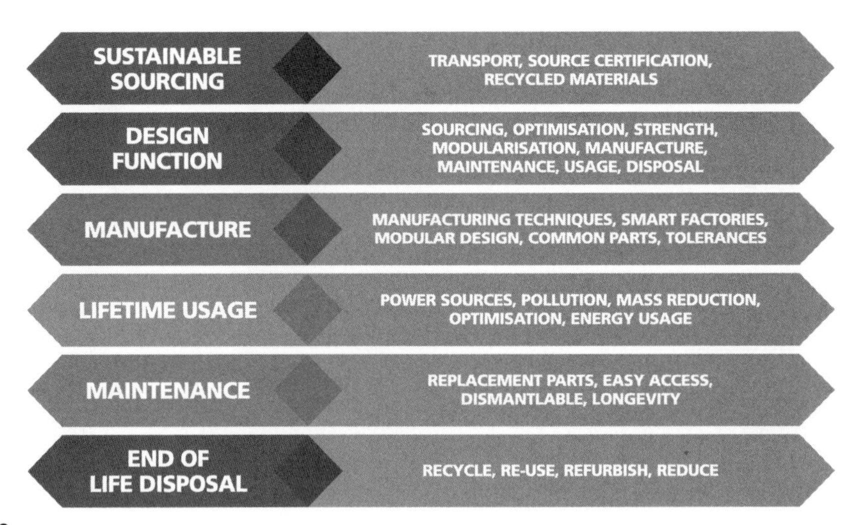

FIGURE 7.3

Sustainable Design Objectives Model [5]

Products designed and created with sustainability as the primary objective can be generated under the umbrella model of sustainability, which encompasses all the facets of design and manufacture.

Consider the sustainable whole-life model shown in Figure 7.1. This can be used as a guide to show the elements that the designer needs to consider and that are set out in the sustainable design objectives model shown in Figure 7.3.

Applying sustainability techniques at each of the stages in the sustainable design objectives model ensures that all sustainability and cost requirements are present in the final product. These requirements include already standard design requirements such as efficient manufacturing and low manufacturing cost.

The application of the sustainable whole-life model in conjunction with the sustainable design objectives model ensures that the designer controls the whole design process and in doing so includes all the design objectives, old and new, that are required of a new product. The great advantage of adopting this model is that the designer can oversee the whole process, integrating appropriate procedures and techniques throughout the life of the product. This integration will, in turn, create appropriate efficiencies.

7.4.1 Sustainable sourcing

It is important that the designer can specify and control the raw material he intends to use in the new product. Raw materials need to have their sources identified, and for the designer to quantify the value of sustainability (embodied energy), this source identifier requires the embodied energy value applied to the raw material. This may seem a tall order, but the process is already in place in several industries.

The aircraft manufacturing industry, for example, requires a certificated origin of materials that are sourced for consistency in strength, leading to safer components. Materials from a noncertified source may at first be suitable but may prove unreliable, since they might not comply with purity requirements, chemical composition requirements, or strength requirements.

One example of this scenario is where wing root pins (pins that hold the wing to a plane fuselage) were found to be cracked. After following the trail back to the material source through the certification process, it was found that the material origin was an unreliable source and documents had been forged. This was clearly a breach of safety procedure and resulted in legal prosecutions. This case may be safety oriented but shows that sourcing procedures and certification are already in place out of necessity.

Another example is that of the sourcing of timber used in street furniture. This timber is used on external wooden walkways, wooden bollards, and seating on street benches. This timber may also be used as decorative elements on other street furniture such as litter bins. The materials are used in an outdoor environment where the temperatures may vary from –20 °C to +50 °C. Rain may lash the timber, or perhaps it may suffer an extended dry period. Timbers such as oak, ash, pine, and sepele would disintegrate within a very short time if placed in such environments. The best material for this purpose is iroko, a very sturdy and durable mahogany grown along the west coast of Africa.

When refurbishing urban environments, U.K. local councils specify iroko only when there is a certificate declaring that the timber is from a sustainable source. This policy clearly warrants that retailers, importers, and growers ensure that the product is sustainable or they would simply not make any sales.

Once the system has been established for other materials such as steel, rubber, and plastics, designers could specify materials, confident that they would be from sustainable sources and would have a minimal impact on the planet's resources. This would then be part of designers' control over selecting those raw materials certified with a certain sustainable source value (SSV).

7.4.2 Recycled materials

Recycled materials are gleaned from products that have come to the end of their life. Energy is required to convert the recycled material into a usable raw product, but this energy use is likely to be a fraction of the energy used in obtaining the commodity from an original source.

The remaining rubber in used vehicle tires is removed from the tire using a grinding process, resulting in a recyclable granulated rubber. This raw material can be applied to many different products, from road speed humps to soft children's playground floors. Though it requires energy to convert the recycled material into new products, there is notably less energy expended compared to obtaining the original material and transporting it across the globe to a factory.

Recycling for many materials is already the norm. Some of the more common recycled materials are:

- Steel
- Rubber
- Glass
- Plastics
- Building materials
- Wood

Many products combine differing materials, creating difficulties in later separating those materials for recycling. Such products include passenger vehicles and computers. However, emerging techniques enable useful pure materials to be extracted from such products. It would be helpful to the end-of-life disposal engineer if material separation was a feature built in at the design stage. This is already being achieved in the passenger vehicle industry, where designers aim for a recycled value of 95% of the vehicle.

For designers to take advantage of recycled materials, those materials must be made available in a pure form, with a certificate of authenticity showing the embedded energy value. This certification would allow the designer to control the sourcing of recycled raw materials so that he could quantify the SSV.

7.4.3 Reduction of haulage dependence

It is inevitable that the goods will have to be transported. Raw materials require delivery to manufacturing centers, and new products need to be shipped to their point of sale. Haulage of goods can never be eliminated, but certain measures can be taken to reduce the embedded energy required in that transportation.

To understand why materials have to be transported, it is useful to roughly categorize particular materials and the reason for the transport. The breakdown of categories of transported materials is as follows:

1. Materials that are created near their extraction point, such as aluminum, timber, or certain foods such as coffee, tea, and wine.
2. Materials and products that are manufactured overseas and imported and that could include such items as passenger vehicles, golf bags, motorcycles, lawnmowers, and barbecues.
3. Those items manufactured in a particular country, then exported.

These activities will always be present; indeed, the prosperity of some countries is based on export and import of goods and other raw materials. The energy used in transporting these goods will always be required, whether generated from an artificial source or a natural source. In many cases the appropriate application of sustainable methods will reduce the dependence on artificial energy in favor of naturally generated energy.

Consider those materials in paragraph (1) where there is no alternative but to transport the materials, often globally. The appropriate application of sustainable methods would apply natural power such as that used in the past by sailing ships. In past centuries these were used to great effect and were completely sustainable, since they relied on natural wind power rather than the artificial power derived from, say, diesel engines.

In the modern era, when the convenience of diesel power overshadows many other options, employing natural energy such as that used by sailing ships may seem implausible. However, some enterprising companies are offering systems that supplement diesel power for ships with sails and solar collectors [3.6, 3.7]. Some short over-water voyages are powered completely using solar collected power. Such is the case with Eco-Marine Power, which operates a ferry service across Hong Kong Harbor [3.6].

These initiatives can greatly reduce the emphasis on carbon-fueled transport but require quantifying and certification so that embodied energy values are available to the designer. This system would put designers in control by offering them the means of selecting materials with the lowest environmental impact.

Now consider imports and exports outlined in points the earlier (2) and (3). Many consumer items are manufactured overseas, often because the cost of doing so is very low. Typically, many consumer items are manufactured in the Far East countries such as China, Japan, and South Korea, then shipped to destinations such as Europe and the United States. Unfortunately this enterprise does not usually consider the environmental cost of transporting the goods.

As the tendencies of recycled commodities become more the norm, materials may be gleaned from local sources, reducing transport costs and environmental impact. Appropriate certification with values of embodied energy can then quantify the sustainability value of a product. This system will give designers much more control, since they will be able to select appropriate materials and thus manage the environmental impact within their particular design projects.

This method not only gives designers control over material selection, it encourages other beneficial aspects. The use of local materials and local labor leads to improved local economies. Although it is inevitable that the goods will have to be transported, certification of these goods and the measurement of embodied energy enable designers to take control of the material selection element of the design, thus improving the SSV while promoting local prosperity.

7.5 THE SUSTAINABLE DESIGN FUNCTION

The design function is the only function in the entire product creation process that has a total overview; as such, it has the power to influence the whole process. Traditional design and manufacturing models tend to isolate the design function from the manufacturing function. Indeed, the manufacturing function has often dominated the whole process, and it is suggested that these isolated functions lead to inefficiencies in cost and sustainability. Corbett and Dooner [4] suggest that 70% to 80% of manufacturing costs are defined at the design stage. It should be noted that this estimate is for manufacturing costs only. As has been shown, manufacturing is only one element of the whole product life process.

The sustainable design objectives model shown in Figure 7.3 outlines the general elements for which the design function is responsible. However, certain aspects of design influence many other elements throughout the whole-life model and can only be considered at the design stage. The design considerations are as follows:

- Sourcing
- Optimization
- Strength
- Modularization
- Manufacture
- Maintenance
- Usage
- Disposal

Some of these elements will be considered later in more detail, but it is incumbent to discuss several of these aspects since they are a pure design function. Such elements as optimization, strength, modularization, and maintenance should be considered at the design stage, since they influence processes, usage, and expedience of disposal in later stages of the product's life.

Whenever a design is created, there has to be an element of optimization. In some cases this could be an exercise in selecting the best compromise. For instance, there is always a compromise between adding mass and adding strength. An aircraft requires strength, but it also requires low mass. Aircraft designers always have to deal with the compromise between the requirements of high strength and low weight.

Optimization is a technique whereby the designer creates a product that is tuned for a particular purpose. A vehicle designer may design a particular vehicle for high speed, in which case it will possess the following qualities:

- High-performance engine
- Streamlined shape (high-speed aerodynamic profile)
- Two-seat capacity
- High-performance suspension and tires
- Low profile
- High-performance gearbox

The same designer could also design a particular vehicle for driving in a congested urban environment, in which case the vehicle will have the following attributes:

- Low-speed performance engine
- Low fuel consumption
- Low-speed aerodynamic profile
- Two-plus-two passenger capacity
- Maneuverability for parking

The two vehicles may both have the usual elements, such as engine, four wheels, passenger compartment, and brakes, but they will be optimized for very different uses, resulting in very different vehicles.

The optimization process can become extremely complicated, often using statistical analysis to effect the best optimized design. In this book we outline a more simplistic approach aimed at understanding the process to achieve more practical aims.

The brief for a new design sets out the requirements and, through investigation, the product design specification is formulated. The designer may not be consciously aware, but she will have exercised *design optimization*. Design optimization is present in every design, to a greater or lesser degree. Its function is to guide the design so that its performance is tuned for a particular purpose.

A cantilever lifting device, for example, was originally intended to lift small pallets within a factory environment. It was manufactured as an all-welded construction with cantilever arms that were 20-mm-thick plasma-cut fabrications. In this industrial situation, the designer did not perform stress analysis to the cantilever and consequently overdesigned the component. This was a case of "If it looks right, it is right," and it fell into the trap of overdesign because humans generally overestimate. The lifting device was optimized for use in a factory environment, however, where there is plenty of space and noise is not a problem.

These optimized parameters are acceptable in a factory where noise, finish, and size would not normally be a problem. Manufacturing the unit for domestic use is much more likely to be a bolted assembly, so that components can be maneuvered through small spaces such as stairways and doorways and into bedrooms.

An entrepreneurial businessman had the inspired idea of adapting the same device for use as a bed-lifting device in studio apartments. The lifting device was therefore optimized so that it would lift a bed to the ceiling and as such could be used in bedrooms to create living space during the day and sleeping space during the evening. Though the lifting mechanism was the same, there were very different parameters to the two constructions. The list of optimizations follows:

- Assembled by two workers in four hours
- Components should be light enough to be lifted by one person
- Components should be small enough to be carried upstairs and fit through doorways
- Assembled on-site in a bedroom environment
- The mechanism had to be low noise
- The finish had to be 100% domestic (as you would expect a finished product to be in your own home)

Though the general principles of the industrial device were used on the domestic bed, there was a great deal of optimization required, since it had to fit the preceding list of parameters. A schematic of the bed-lifting device is shown in Figure 7.4.

Many optimization elements can be applied to a new product or an old product that is being rejuvenated. The particular optimization goals depend on the type of product and its end use. Figure 7.5 gives a very general selection of optimization goals related to typical products.

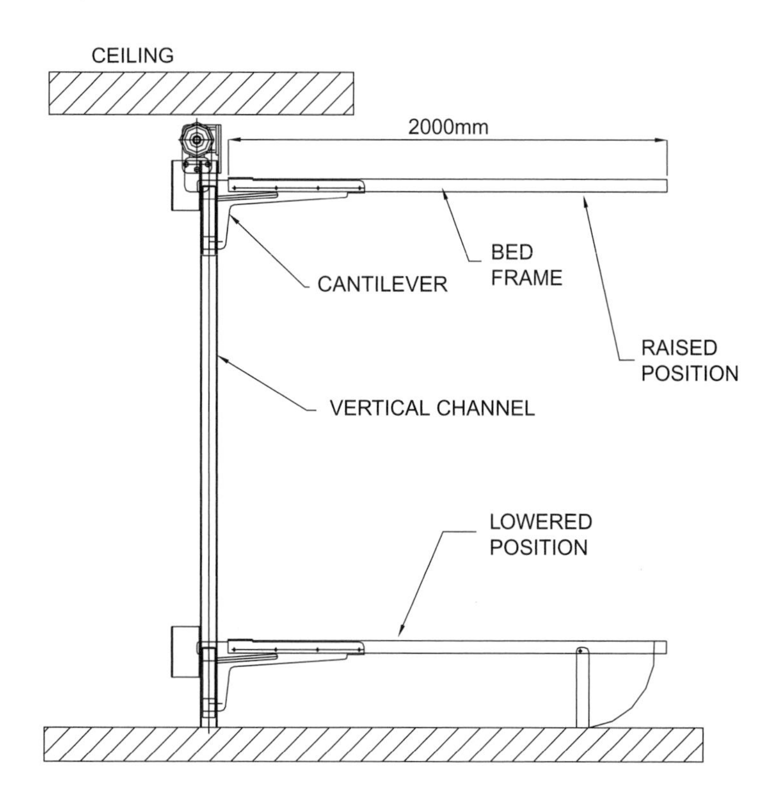

FIGURE 7.4

The Bed-Lifting Device Optimized for Bedrooms [5]

	Low mass	Low Manufacturing cost	Sustainable	Low Fuel Consumption	New Materials	Strength	Recycling	Re-use	Aerodynamic	Ergonomic	Low noise
Motorcycle	#	#	#	#	#	#	#	#	#	#	#
Passenger Vehicle	#	#	#	#	#	#	#	#	#	#	#
Digger Back Hoe	#	#	#		#	#	#	#			
Boat 4 passenger	#	#	#	#	#	#	#	#	#		#
Truck	#	#	#	#	#	#	#	#	#		#
Brick Grab	#	#	#		#	#		#			
Caravan	#	#	#		#	#	#	#	#		
Trailer	#	#	#		#	#	#	#	#		
Motorcycle Crash Helmet	#	#	#		#	#			#		

FIGURE 7.5

Product Types and the Major Optimization Approaches [5]

We can see from the chart that low mass, low cost, and sustainability can be applied to all of the products. Low mass and low costs are standard design parameters, but sustainability has many facets and can be subdivided into the elements shown in the sustainable design objectives list in Figure 7.3.

The chart in Figure 7.5 is merely an example of optimization goals applied to familiar products. In reality, a major optimization goal should be set. Minor optimization goals will automatically contribute to the major goal.

7.5.1 Optimization strategy

Design optimization is almost always a multicriteria design approach in that several aspects of the design are optimized to achieve the main optimization goal. The strategy should be to define the major optimization goal and utilize several minor optimizations to achieve the major goal.

For instance, the major optimization goal for a touring motorcycle would be that of rider comfort. To achieve this goal, there would be minor optimizations to apply to the design. Figure 7.6 indicates how these minor optimizations contribute to the major optimization goal.

It should further be noted that even the minor optimization goals may be further subdivided so that, for instance, the goal of low fuel consumption may involve redesigning the engine or selecting an alternative engine.

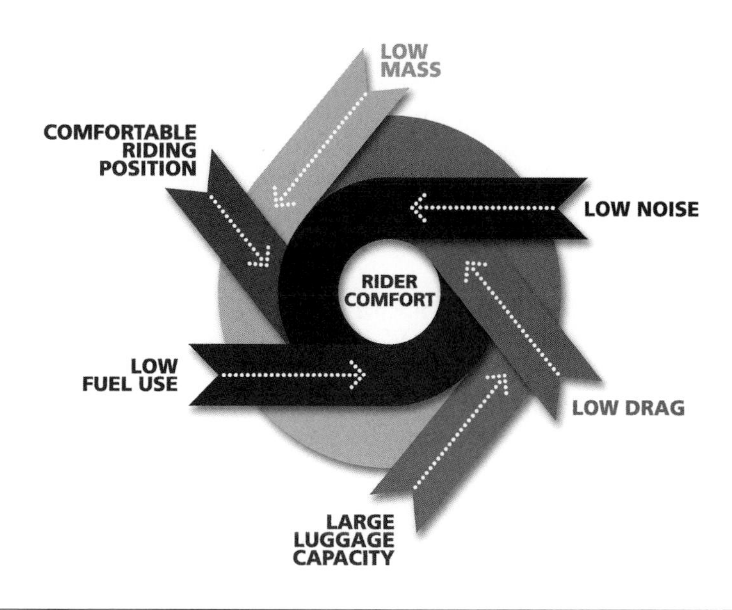

FIGURE 7.6

Minor Optimization Goals Contributing to the Major Optimization Goal of Rider Comfort for a Touring Motorcycle [5]

Optimization tunes a product for a particular use and in so doing filters out many irrelevancies. For instance, a vehicle optimized for urban use would tend to have a smaller engine than one developed for high speed. The smaller engine would therefore suit the environment in which it was meant to function and would have minimal irrelevancies that would consume extra resources.

In tuning a product for a particular function, savings can be made throughout the life of the product and could involve savings in sourcing, manufacture, usage, maintenance, and disposal. Optimization therefore minimizes embodied energy and contributes to the sustainable whole-life model in Figure 7.1.

7.5.2 Strength

Many products require strength to some degree in order to perform their designed duty. For example, a teapot requires strength in the handle and the ability to stay rigid while a person is carrying hot liquid in it.

Some devices. However, require in-built strength in order to perform safely. In these cases the consequences of failure could be fatal. An excellent example is the roof structure shown later in the chapter in Plate 7.2, the strength of which is critically important in ensuring the safety of the audience seated beneath. Another example is that of an aircraft wing, which requires strength no matter what extremes of use. Failure in flight would lead to fatal consequences.

During the design of such devices it is necessary to perform structural analysis to ensure that appropriate strength is built into the structure so that it can perform its design duty. Modern analytical tools allow design engineers to calculate the performance of devices with high accuracy, eventually leading to the reduction of material without compromising strength. Furthermore, the flexibility of such tools gives designers the opportunity to devise more complex and stronger structural shapes, further reducing material content. The analytical approach using such powerful tools now allows designers to promote radically new designs and apply innovative process methods.

The development of the passenger vehicle body is an excellent example of the application of this process. A vehicle manufactured in the 1950s normally comprised a chassis and a separate body. The body material was manufactured from relatively thick steel sheets, which made the vehicles rather heavy compared to their modern equivalents. Though various components of vehicles are made from different thicknesses of sheet metal, typically the sheet metal used in body panels in 1950s vehicles was around 1 mm thick. Reduction in vehicle mass in modern vehicles has been achieved by clever design to give appropriate strength but mainly because of reduction in material thickness, which in current vehicles is typically around 0.75 mm.

The old-style chassis design incorporating a separate body has been replaced in modern vehicles by the structural element becoming the car body itself. Even the windscreen is a structural element in modern vehicles.

Though this section of the chapter predominantly deals with strength and structures, it should be remembered that other areas of engineering are benefiting from digital analytical prediction, including:

- Dynamic analysis
- Mechanisms analysis
- Fluid flow analysis
- Thermal analysis

Previous chapters suggested that humans tend to increase structural size "just to be on the safe side." This approach is commonly known as a "rule of thumb" and increases factors of safety above those that are necessary. This approach classically adds mass but relieves the designer of performing precise prediction analysis. Applied correctly, hand calculations verifying digital analysis allow precise prediction of performance, reducing safety factors, reducing mass, and more important, reducing the embodied energy required to process the material. In terms of vehicles, a low-mass body requires a smaller engine and smaller brakes, resulting in lower fuel consumption—a suitably sustainable approach.

Reduction in mass can also be achieved using a uniform stress technique. Many easily accessible components, such as structural steels, are supplied as a uniform section simply because it is easier to manufacture uniform sections. A typical example of a 100 mm × 200 mm rectangular hollow section is shown in Figure 7.7.

A simple beam, however, has little bending stress applied to the support points, whereas in the middle there are high stresses, perhaps because of applied loads from above. An example of the beam

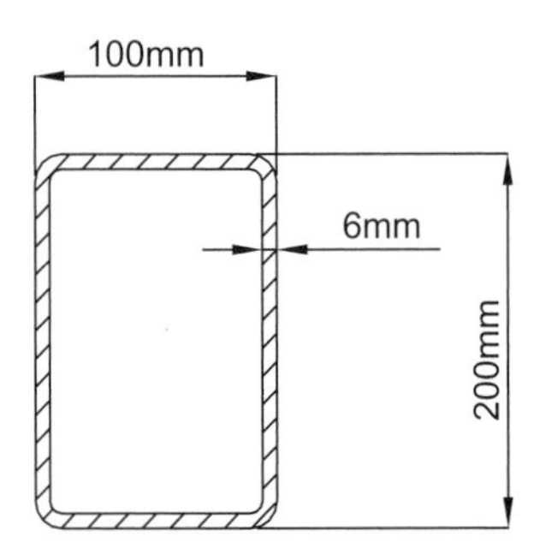

FIGURE 7.7

A Typical Example of a Uniform Section, 100 × 200 mm(Rectangular Hollow Section) [5]

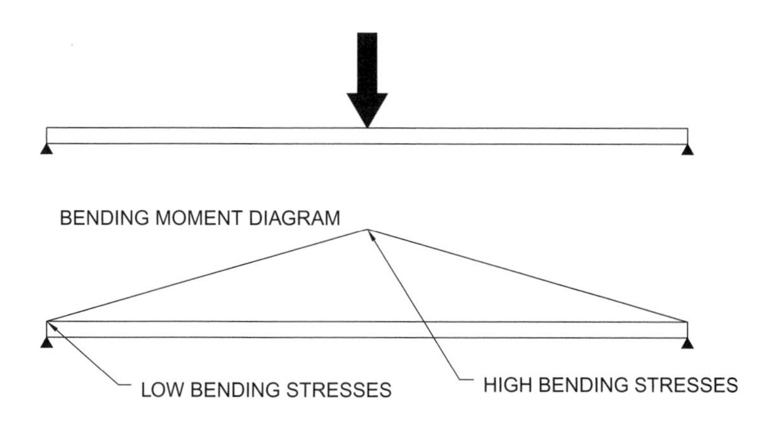

FIGURE 7.8

Typical Simply Supported Beam with Central Load Showing Bending Moment [5]

supported at each end and loaded by a point load in the center is shown in Figure 7.8. The figure also includes a bending moment diagram, which is essentially a graphical representation of the loading across the beam. We can see that the greatest bending moments, and therefore bending stresses, can be found toward the center of the beam. There will obviously be stresses at the support points, but the bending moment and therefore bending stresses are very low.

It seems reasonable, therefore, to apply material where the stresses are highest. In the case of the beam, the greatest strength would be required at the center of the beam. Though it may seem appropriate to increase material where there is high stress, the subtle approach would be to increase the "depth of section." In general beam theory, the greater the distance between the top and bottom surfaces of a beam, the stronger the beam becomes. This can often be achieved with only a small percentage increase in material. If the depth of the section is increased where the loads are highest, such as at the center of the beam, it follows that the depth of section can be much reduced at the support points.

The bridge shown in Plate 7.1 is essentially a beam supported at each end. The weight of the beam is applied as a uniformly distributed load (UDL). The bending moment for this type of loading is shown in

PLATE 7.1

Typical Fabricated Foot Bridge [5]

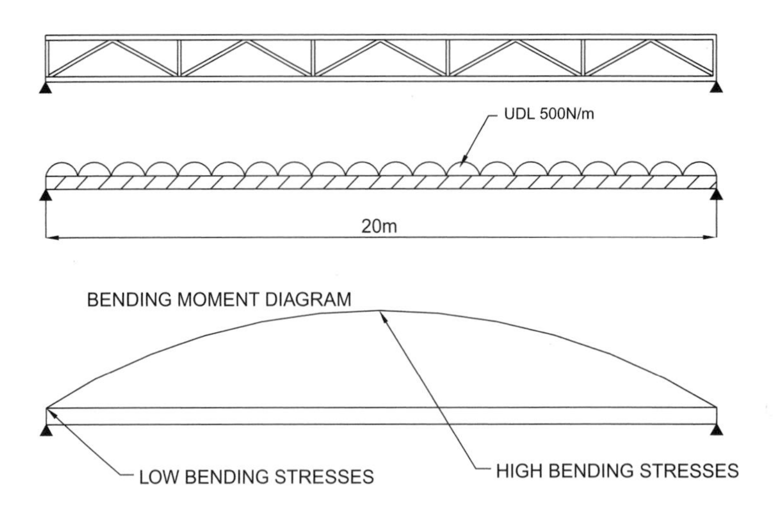

FIGURE 7.9

Typical Bending Moment for a Beam with a Uniformly Distributed Load [5]

Figure 7.9. We can see that the greatest stresses are applied toward the center of the beam, so this is a nonuniformly stressed beam in which stresses are higher at the center.

In a uniformly stressed design, the material and hence the depth of section are applied according to the value of the stresses in the beam.

The theory presented so far is that of a beam designed to encounter nonuniform stresses between the support points. Material on a standard section B is therefore wasted toward the support points, where there are very low bending stresses. If, however, the beam were to be designed so that the stresses between the support points were uniform, the shape of the beam would change so that it would have a greater depth of section in the middle and a lower depth of section toward the support points. This design feature has been used successfully on many occasions. A classic example of this uniformly stressed design approach is that of a particular football stadium roof structure, shown in Plate 7.2.

PLATE 7.2

Football Stadium Roof Structure [2]

The loadings applied to such stadium roofs take the form of UDLs incorporating such loading elements as settling snow, heavy rain, and winds. It is, however, a structure that follows fairly closely the assumed typical bending moment by applying a greater depth of section and therefore strength toward the center of the beam. Though the ends of the beam at the support points also require strength, the stresses are different from those applied to the center of the beam, and therefore a smaller depth of section can be applied.

This kind of technique can reduce material usage, though it may marginally increase the embodied energy required to manufacture.

Uniformly stressed beams can be used to great effect when large quantities of items can be produced. A classic example here is that of a digger backhoe (see Plate 7.3), which can be treated in one mode of operation as a simply supported beam and which possesses a near uniformly stressed sectional design.

The backhoe's secondary arm that supports the bucket is wider at the head, where the hydraulic ram and pivot are located, than at the bucket end. Though there are significant loads applied to the bucket end, the major bending moments and therefore high stresses are at the ram and pivot locations. Rather than use a standard section beam, parallel for its full length, manufacturers have created a deeper section where the highest stresses are located and tapered the section toward the bucket, where there are lower stresses.

Manufacturing costs for a single-unit production could prove very high for this type of fabrication, since components would need to be cut to shape and welded. A much cheaper option would be to use a standard uniform section such as that shown in Figure 7.7. Backhoes are manufactured in quantity, however, which means that the cost of creating a particular shape is outweighed by the saving in materials over hundreds of units.

The great advantage of structural analysis to define the strength of the structure is that the appropriate strength can be applied without overdesign or including excessive material "just in case." The whole process is based on a scientific process rather than a "rule of thumb" approach, which wastes material and often leads to excessive energy use in the building of the product. The analytical approach therefore offers a safe product with a minimal embodied energy value, thus contributing to the sustainable manufacturing value (SMV).

PLATE 7.3

Classic Digger Backhoe [6]

7.5.3 Modularization

Traditional design methods design and manufacture a single product, but in markets where the product has different end uses, it is necessary to effect changes to the product without starting the design from scratch. The key to this conundrum is to design and build different elements of the product separately. This modular design method is an approach that subdivides the product into smaller components (modules) that can be independently created and applied so that the end result can be used in different ways. A modular system can be characterized by the following:

- Partitioning of a product into discrete scalable, reusable modules. Each module consists of isolated, self-contained, independently functioning elements.
- Rigorous use of well-defined modular interfaces, which could include mechanical, electrical, or software interfaces.
- Connection or attachment interfaces designed for ease of assembly, and where possible use of industry standards for key interfaces should be applied. These key interfaces could include standard electrical or computer sockets and plugs, standard flange sizes for pipes, or industry-standard locators such as those applied to vehicle wheels or motor flanges.

Modularization offers flexibilty without the need of customization. Modules are often mass produced, offering low-cost production and a great deal of flexibility in design. It is also possible to easily add new solutions in the form of new modules. A great example of modular design and manufacture is that of passenger vehicles. One of the reasons the cost of passenger vehicles is kept relatively low is because a great deal of modularization is applied. A vehicle body may be produced for a premium vehicle type, but with a different modularized engine, upgraded brake modules, paint, badges, and modular interior, the vehicle can be converted into, say, a sporty version. This flexibility can be achieved without returning to the beginning of the design process.

Modularization can also be applied across vehicle marks. Chassis, brake components, engines, tires, wheels, and many more components have all been used across vehicle types and across vehicle makes.

One of the classic uses of modularization, and one that transformed vehicle manufacture, was that of the design of the Mini, which was first released in 1959. Alec Issigonis designed the modular chassis and body system to improve manufacturing and increase space inside the vehicle. The concept consisted of three major elements: a front subframe, a rear subframe, and the vehicle body. The front subframe housed the engine, drive shafts, brakes, and suspension; the rear subframe supported the rear suspension and brakes. The real design breakthrough was that the vehicle body was used as the chassis structure. This was probably one of the first production instances of the application of a *monocoque* design, which incorporated the vehicle body as the structure. The cleverness of this design approach was in the modularization of major components. Figure 7.10 outlines the configuration of the subframe modules and vehicle body for the Mini.

In a more modern concept car, Thomas Budde Christiensen [7] reported that General Motors developed the Chevy Volt. The main vehicle architecture is taken from the Chevrolet Cruz and the Saab 9-3 in a classic case of modularization across vehicle makes. The innovative modularization concept behind the Chevy Volt is that there is a selection of three different propulsion options:

- A pure battery electric motor option
- A combined electric combustion engine option
- A combined electric fuel cell option

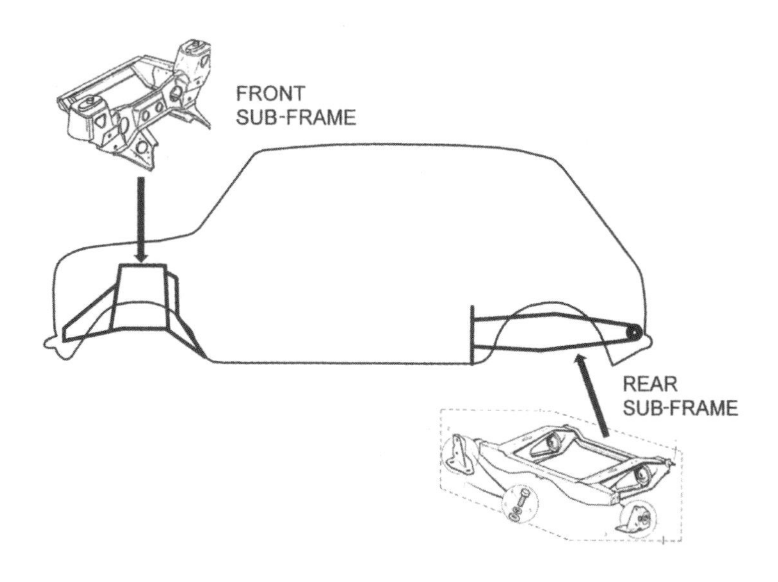

FIGURE 7.10

Diagram of the Mini and Approximate Positions of the Front and Rear Subframes [5]

FIGURE 7.11

Chassis Modules for the Chevy Volt [6]

There are other modular devices that have been incorporated, such as wheels that are driven by their own internal motors and batteries that can be charged from a variety of sources. Chassis modules are shown in Figure 7.11.

Modularization is used extensively in computers, buildings, railroad signaling, and many other products. The bed-lifting device shown in Figure 7.4 is also a modular construction. Standard parts can be bolted together using small modular components that are interchangeable. The drive unit, which fits against the wall, becomes one standard base module. The bed sizes, which may change, are manufactured as modules so that queen, king, double, and single beds can be accommodated for a particular order. This modularization greatly reduces the organization and energy required to manufacture while still preserving the flexibility to serve customers' requirements.

Under normal circumstances, high-volume manufacture reduces the individual cost per item. This is one of the great benefits of standardization. Modular-build systems effectively standardize larger systems rather than just smaller components, thus reaping cost advantages of large-volume production. In a similar fashion, modularization also reduces the embodied energy in each package. If modules are mass produced, they are more efficiently produced, therefore demanding less embodied energy input and consequently having a lower environmental impact.

One of the most heavily modularized products in the modern world is the computer, which is designed to accommodate modules. A standard base unit comprises power supply units, processes, motherboards, graphics cards, hard drives, optical drives, and so on. All these parts are interchangeable and can easily be replaced if, for instance, there is a damaged graphics card or an upgrade is needed to allow higher functions to be gained from a more basic computer.

A similar approach has been taken to software provision. A computer with an appropriate level of memory and speed can be "tuned" using appropriate software. For instance, the computer hardware may be exactly the same for professional use as for home use, but a computer intended for computer games will have a different set of software than the computer intended for use by a design engineer. Similarly, an engineer using his computer for internal combustion engine data acquisition will have a completely different set of software than that for home use or for engineering design.

Modularization has been used for many years and to great effect. The use of modules reduces cost and allows easy maintenance due to quick replacement of modules. Refurbishment of modules is therefore possible, prolonging the life of the main product. Thus we can see that the use of modular construction fits well within the model of sustainability and improves the SMV.

7.5.4 Manufacturing

Manufacturing is the most expensive part of the product development process. It involves the procurement of materials, their manipulation, and their finish, to various degrees. This process usually takes place in a factory of some kind using machine tools, labor, and energy. The cost of providing this facility can be enormous.

It is the designer's role to design components and products that the factory can manufacture, but too often the design and manufacturing model has been separated into pure design and pure manufacturing. This can lead to great inefficiencies in terms of cost and wasted energy.

Even if design is separate from manufacturing, the designer must consider manufacturing during the design process. For instance, the designer must decide whether to fabricate or cast or whether to mill or turn. Corbett and Dooner [4] suggested that 70% to 80% of the production costs are defined at the design stage. With this in mind, it is useful to define the elements that contribute to efficient manufacturing and which are often built into designs automatically. These elements are worth itemizing here:

- Minimize total number of parts
- Develop modular design

- Use standard components
- Design parts to be multifunctional
- Design parts for multiuse
- Design parts for ease of fabrication
- Avoid different fasteners
- Minimize assembly directions
- Maximize compliance
- Minimize handling

7.5.4.1 Minimizing the number of parts

A reduction in the number of parts normally leads to a reduction in handling, a reduction in inventory, reduced background paperwork, reduced number of drawings, and so on. Generally a reduction in parts creates a simpler product with much less energy expended and, normally, reduced costs. Even though numbers of parts may be reduced, the functions are still required within the product. This is normally achieved by creating more efficient single-piece parts. Care must be taken, however, since there is a breakeven point at which a reduction in parts leads to complex single components that may cost more than several smaller parts.

In general, a reduction in the number of parts in a product can lead to a reduction in cost, a reduction of embodied energy, and an increase in SMV for the product.

7.5.4.2 Developing modular designs

Developing modular designs, discussed in depth in Section 7.5.3, has several advantages:

- Design flexibility
- Efficiencies of quantity production
- Easier maintenance
- Refurbishment possibilities
- Reduction in embedded energy
- Offers easy customization using combinations of standard components
- Resists obsolescence
- Shortens the redesign cycle
- Offers new generation products often using old modules
- Changes provided with a minimum of design input
- Simplifies final assembly
- Reduces the number of parts to assemble

A modular design actually reduces parts, leading to reduced costs. If applied correctly, modularization can lead to a reduction of applied energy (embodied energy) and an increase in the SMV for the product.

7.5.4.3 Multifunctional parts

The cooperation of parts that can perform several functions can often reduce the overall number of components. For instance, a structural member can also be designed as a spring, or a structural member may also act as a conductor or perhaps a heat sink. A reduction in the number of parts is always a valuable contribution, since it reduces embodied energy and improves the SMV of the product.

7.5.4.4 Design parts for multiuse

A multiuse part is a component that can perform several functions, depending on the end use of the product. For example, a mounting plate on a machine may have several location holes and captive nuts that will accommodate several sizes of electric motor flange. This mounting plate can therefore be made in quantity, thus reducing costs and improving manufacturing efficiency. An excellent example of this idea is the worm-wheel drive casing shown in Plate 7.4. This casing is manufactured from die-cast aluminum alloy and houses a particular size of worm and wheel reduction components. Though these may be standard internal components, the casing is a multiuse device that can fit a range of sizes of electric motors and internal combustion engines. Furthermore, the casing can accommodate motors from various manufacturers.

Multiuse components offer the opportunity of reducing infrastructure costs and manufacturing components in quantity. The advantage of using such a device is that the casings can be manufactured in quantity, thus taking advantage of low cost and low embodied energy quantity production. This process greatly improves the sustainable source value (SSV) and the SMV.

7.5.4.5 Design parts for ease of fabrication and assembly:

The aim here is to reduce the time spent fabricating and assembling a product. Less time spent here will mean greater efficiencies and reduced energy input. The designer has ultimate control over the process, since he chooses the components that contribute to the overall design. When assembling a multipart product, the designer should be aware that the two major elements that need to be present for any component assembly are:

PLATE 7.4

Worm-Wheel Housing Multiuse Component [10]

1. Parts location in three dimensions
2. Fastening method in three dimensions

Each component should be examined to ensure that location and fastening method are present. If a component is located in two dimensions, it is often possible to secure the third dimension with an appropriate fastener. Location and fastening devices can therefore work together to fully secure and locate a component. Furthermore, the designer needs to consider the following points during the design process:

- Reduce the number of parts
- Keep parts simple
- Apply modular components where possible
- Consider options of simpler assembly procedures
- Improve parts access (one-direction assembly, space for access by hands, etc.)
- Consider ergonomics for assembly personnel (fit components at reasonable heights, reasonable arms reach, etc., to reduce tiredness)
- Reduce lengthy fastening processes (e.g., screws, welding, etc.)

The assembly of a gearbox, for example, can be achieved in several ways. Large gearboxes such as those used in ships' transmissions are often split along the line of the main shaft so that when the lid is lifted, all the gears and bearings are exposed. This means that the main shaft can be assembled with bearings, spacers, and gears prior to being hoisted into the main gearbox body. Though seemingly easy, this method has its difficulties. There is fine adjustment when tightening the lid so that bearings are not crushed, merely held in place. Precise machining and skilled fitting are required to achieve the workable result, which takes time and energy.

Another approach to assembling gearbox components is to load the shaft from one end. This method requires that one of the bearings needs to be large enough for the shaft and components to pass through its outer housing diameter. Other elements such as large gears can be loaded on the inside of the gearbox. This system requires a similar precision in machining but is much easier and quicker to assemble, thus requiring fewer parts such as shims and specialist seals. This approach not only reduces the parts inventory but also improves ease of maintenance.

Assembly is often a manual process. Any reduction in time spent on this process will reduce costs, but in a more global view it will also reduce the energy spent on infrastructure to keep workers comfortable, provide light with which to work, and all the other aspects required of a manufacturing plant. A reduction in energy here will improve the SMV.

Some assembly procedures are automated, which means that the designer cannot rely on human hand-eye coordination. When the human element is removed from the process, assembly becomes much more difficult to achieve. Consideration of assembly without human input is an excellent exercise since it focuses the mind toward the best method of assembly.

7.5.4.6 Fastening systems

Any product that comprises more than one part will require some form of fastening device. There are many fastening systems available to designers. Some of these are:

- Welding
- Brazing

- Screws
- Pins
- Rivets
- Adhesives
- Snap fasteners
- Spring fasteners

This list offers only a glimpse of available fastener types; it does serve to highlight the full range of possibilities.

The most used fastening device is probably the screw, available in many shapes and sizes. Screws require locating and rotating to be fitted properly. This is a relatively lengthy process and requires dexterity. Welding falls into the same category in that it is a dirty and lengthy process requiring much energy and time. Often these processes can be replaced by cleaner and easier-to-apply systems.

7.5.4.7 Case study: Car stacker

A three-dimensional car-parking device was to be designed according to the schematic shown in Figure 7.12. To facilitate ease of assembly on-site, the whole unit was to be supplied as a kit of parts and assembled using bolts, screws, and nuts. The deck was to be assembled from galvanized steel "C-section" units of approximately 250 mm width. The units stretched across the deck, fastening to side stringers via high-tensile screws.

When the stacker was assembled, the deck was found to flex, and although considered safe, this tended to unnerve users. The proposed solution was to stiffen the deck by fitting 0.5 mm galvanized sheet steel to the underside. Several options for fastening the galvanized sheet were proposed:

- *Self-tapping screws.* This would have required holes to be drilled and screws to be inserted. The time allocated for the assembly procedure was estimated at three hours.
- *Blind rivets.* Sometimes called *pop rivets,* this method still necessitated drilling holes and applying the rivets. The time taken to accomplish this task was estimated at two hours.

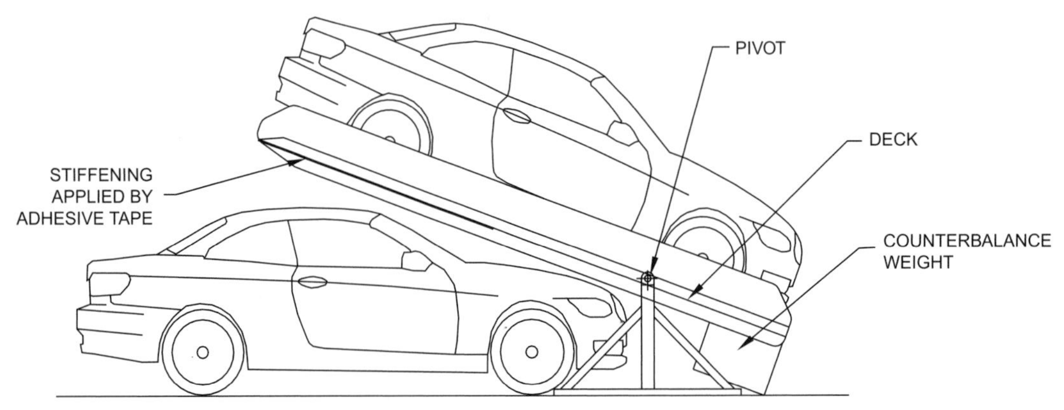

FIGURE 7.12

Schematic of the Car Stacker [5]

- *Double-sided adhesive tape.* This was found to be the most convenient option, with several advantages over the other methods that were considered. The time saving was enormous; it took only 20 minutes to apply the sheet steel. This was done by applying the adhesive tape from a roll and placing the sheet steel onto it, firmly pressing it in place. Not only was this method quicker, but it was much cheaper in terms of components and time and required much less energy, since no drilling was involved. Another added benefit was that the skill level required to accomplish the task was very low.

In practice, the designer had considered the flexure of the deck during the design process and had a solution ready when the flexure problem arose. He had considered screws and rivets and discarded those in favor of the double-sided adhesive tape.

When a designer considers the type of fastener to use, she may be led by current practices within her current company; however, the designer must be open to alternative thinking and reducing lengthy and expensive practices, especially if new practices can reduce the embedded energy. This approach improves the SMV.

7.5.4.8 Minimize assembly directions

Normally, smaller components are attached to a base component. Access for fitting staff and the use of automatic machines is essential for easy assembly. This can be done by minimizing the assembly directions. If personnel need to move around and assemble or have to keep turning the assembly to access assembly directions, time and energy are not being used efficiently. It is the designer's role to ensure that assembly can be completed in the most efficient way. Usually this means minimizing the number of directions from which parts can be fitted.

This approach minimizes assembly time and reduces embedded energy, thus improving the SMV.

7.5.4.9 Minimize handling

Handling and moving components and assemblies to different positions or from one station to another are effectively dead time. No work can be carried out on the product during these periods of manipulation. This movement must be reduced as much as possible. It is in the designer's purview to select procedures and methods to achieve this goal. Minimizing transport and handling has great benefits:

- Reducing time moving components from one operation to another
- Reducing energy
- Reducing staff time on the job
- Improving the speed of assembly or manufacture

In all of the benefits mentioned, speed of assembly is improved, cost is reduced, and, from a sustainability point of view, embedded energy is reduced. This can only improve the SMV.

7.5.5 Maintenance

A sustainable product is one whose life can be extended through maintenance and repair. During the normal process of design, maintenance should have been considered as an automatic adjunct to the creation of the design. This means that the product should be designed to be regularly maintained by replacing or refurbishing worn parts. To be fully sustainable, a product should be able to be maintained indefinitely. This latter statement is more of a wish than a reality, but there are excellent examples of products that have been maintained beyond what would be considered a normal lifespan.

PLATE 7.5

70cc Motorcycle [5]

The motorcycle shown in Plate 7.5 is typical of the personal transport found in countries such as India and Pakistan. It is unclear whether these motorcycles were designed and built to sustainability values; nevertheless, they are an excellent example of a product that can be maintained almost indefinitely and therefore has excellent sustainable credibility. These motorcycles are perfect for the environment in which they are used for a single rider, though it should be said that the author has personally witnessed up to five people riding on one of these machines.

The essence of the design for maintainability of the motorcycle is that parts can be accessed, removed, and replaced with a fairly simple toolkit. Another excellent example is the truck shown in Plate 7.6, which was originally built circa 1980.

It is typical for these trucks to be regularly maintained so that their life expectancy may be more than 50 years.

It is refreshing to note that the mindset of the mechanics who maintain such motorcycles and trucks is that of repairing or recycling components. For instance, if the aluminum crankcase of the motorcycle becomes cracked, the mechanic will recycle the crankcase by melting the material and casting another.

PLATE 7.6

Truck [5]

True sustainability has to start with the designer, who can design the appropriate elements into the product. In the case of the motorcycle and truck in Plates 7.5 and 7.6, a culture of sustainability has grown around the repair and maintenance of the machines.

As society moves away from the mindset of "throw away and replace," a new approach will gradually take hold, which will be to make products last as long as possible. This can only happen through regular maintenance and a clear mindset of the user.

7.5.6 Usage

Whenever materials are manipulated and formed to produce products, energy is expended. For the process to be sustainable, materials need to be regenerated, applying zero impact on the environment. Furthermore, the processes involved in forming the new product need to use naturally generated energy.

Energy is required to extract materials and to build products, but in some cases the energy required to bring a product to market is insignificant compared to the environmental impact of the device in use. This is true of most vehicles and items of plant. During its working life, a digger similar to that in Plate 7.3 (this example could be any make of digger worldwide) will use much more energy than that required to create it. These devices are usually powered by diesel engines, which burn fossil fuel that can be termed artificial energy. This artificial energy cannot be regenerated by the planet and therefore is not sustainable. If the digger could be powered by electric motors whose energy is captured from natural sources, the unsustainable energy use would be changed into a sustainable energy use, thus improving the SUV.

Wherever artificial energy (diesel, petrol, kerosene) is used, there will be a large unsustainable element to the whole life view of the product.

Not all devices are unsustainable in use. A large flywheel system that stores energy will probably run for 10 years without stopping. The only impact on energy resources is in its inefficiencies relating to friction in the bearings. In comparing the flywheel to the digger, the digger possesses a very low SUV, whereas the flywheel possesses a very high SUV.

Designers must select the lowest-impact devices wherever possible. Using current technology to provide power for the digger, the best drives available are likely to be lean-burn diesel engines, perhaps using biofuels, which can be grown and are therefore sustainable. Biofuels still impact the environment, however, since they are responsible for emitting carbon dioxide when burned in an engine. Carbon dioxide is not desirable and could be much reduced as natural energy becomes more widespread as the energy of choice for vehicle power.

7.5.7 Disposal

As with all other areas of product creation, designers can control the disposal of products. Disposal can be achieved by using the 4R rule, which this section discusses in detail. The main principles are as follows:

- Reduce
- Reuse
- Refurbish
- Recycle

7.5.7.1 Reduce

Reduce probably sits more comfortably in the early design stages when a reduction in mass, size, or the like could lead to reduction in product size and a further reduction in energy consumption in manufacture as well as in product use. At the end of a product's life, however, reduce refers to reduction in throwaway waste and therefore a reduction in waste disposal costs. It is therefore preferable to make use of materials derived from end-of-life products. The remainder of the 4Rs should therefore occur wherever possible.

7.5.7.2 Reuse

Components removed from an end-of-life product could be reused in other products. For instance, steam pipe, once used for supplying pressurized air throughout a factory, could be reused as handrail. Rolling element bearings that are removed from a product often still possess a substantial amount of life. These could be reused in less critical applications. These two examples are fairly superficial, but with a little inventiveness a designer can think beyond disposal to industries that could benefit from use components from the product.

7.5.7.3 Refurbish

Refurbishment is really a more intense form of maintenance. A well-used product is stripped down to repair major components, replace worn components, and make as new. The refurbished product exits the factory as though it were a brand-new item. Refurbishment takes a fraction of the cost and energy input of a new device and prolongs the life of the current product. This, then, has an impact on the sustainable life value (SLV) and the sustainable disposal value (SDV).

7.5.7.4 Recycle

When components are worn past any further use and it is not possible to refurbish or reuse them, recycling is the only sustainable option. The process takes end-of-life products, separates the materials, and recreates the materials as a raw material product. Steel is remelted and recast so that the steel can be used once again. Steel is actually the most recycled commodity on the planet and has a substantial recycling infrastructure.

Other commodities such as vehicle tires have to be dealt with in a different way. The remaining rubber is ground so that it can be supplied as a raw material as ground rubber pellets. Much of this material is recast as speed humps, children's playground floors, or firing range walls.

If the end-of-life disposal is conducted efficiently, very little of any product should be taken to landfill. The reuse of materials is not only cheaper than obtaining newly hewn raw materials, but it also requires much less energy and is therefore a very sustainable practice. The designer must plan for end-of-life disposal, since doing so improves the SLV of the product.

7.5.8 Conclusions to the sustainable design function

The designer is in a unique position. It is the designer alone, or at least the design function, which can influence the whole life of a product simply by designing elements into the product to enhance certain outcomes. It is true that a major goal of design is that of maintaining low cost while the design fulfils all the functions required of it.

It is proposed that the designer needs to approach his design work with a new eye so that he can apply sustainable engineering design techniques. Many of these techniques are already well known and often used by designers to reduce costs, but they are not always associated with sustainability engineering. The modern designer should therefore design to two major goals:

- Design to a low cost
- Design to achieve a high SLV

Referring again to the sustainable whole-life model of Figure 7.1, we can see that the application of the design function can improve the sustainable value of all the elements of this model. Sustainability can only be achieved for new products via the design function. This is the only element of the whole sourcing, design, manufacturing, and marketing system that can define the sustainability value of a product.

7.6 MANUFACTURING

The most costly and most energy-draining part of the product creation process is that of manufacturing. The classic design-and-manufacturing model is that designers design and manufacturers make the product. Communication between the two functions has often occurred purely through engineering drawings, but in any efficient production process the two factions work together to provide an efficient product development process. Design and manufacturing should be a simultaneous approach; however, it should be realized that this is not always possible. A lone consultant designer may be very experienced but cannot be expected to know and understand all the nuances of manufacturing.

It has been proposed that the design function should be in total control of the whole product life process in creating and disposing of a product. It is such a vast task that large products should be designed by teams of specialists who can decide the best course of action for a particular aspect of the product development cycle. The specialists should therefore be expert in cross-communication with other members in the team. In particular, manufacturing specialists should liaise closely with design specialists.

A lone design consultant might want to offer his services to a particular company, which then intends to build his design. In such a case, the design consultant must liaise with key people in the various departments within the company, from procurement through manufacturing to marketing. This is really the only way efficient product creation can evolve.

Designers must consider manufacturing during the design process; they are the only people who can create the best outcome of the design. For instance, a designer must decide whether to fabricate or cast or whether to mill or turn, though this decision could be made in consultation. Corbett and Dooner [4] suggested that 70% to 80% of the production costs are defined at the design stage. With this in mind, it is useful to define the elements that contribute to efficient manufacture and that are often built into designs automatically. These elements are worth itemizing here:

- Minimize total number of parts
- Develop modular design
- Use standard components
- Design parts to be multifunctional

- Design parts for multiuse
- Design parts for ease of fabrication
- Avoid separate fasteners
- Minimize assembly directions
- Maximize compliance
- Minimize handling

7.6.1 Minimizing the number of parts

Minimizing the number of parts means that less of everything is required to manufacture a product. This includes engineering time, drawings, part numbers, material, backup paperwork, accounting details, service parts and catalogues, number of items to inspect, complexity of assemblies, facilities, and training. These are merely a few of the items that cost the company. Fewer parts will reduce costs massively. The number of parts can be reduced and reduced, but care must be taken, since the process may actually lead to creating a large single component that incorporates all the small parts but is very expensive.

Integrated design, or the combining of two or more parts into one, often decreases weight and complexity and eliminates fasteners or joints and perhaps fewer points of stress concentration. The very fact that the number of parts has been reduced has the effect of decreasing the infrastructure and support cost and effort. If these effects can be gained in reducing cost, then it follows that a reduction in applied energy (*embodied energy*) can also be gained.

A reduction in the number of parts in a design leads to much lower energy costs, material costs, and time to complete the product. Reducing parts therefore greatly improves the SMV.

7.6.2 Developing modular designs

Developing modular designs was discussed in depth in Section 7.5.3 and has several advantages:

- Design flexibility
- Efficiencies of quantity production
- Easier maintenance
- Refurbishment possibilities
- Reduction in embedded energy
- Offers easy customization using combinations of standard components
- Resists obsolescence
- Shortens the redesign cycle
- Offers new-generation products, often using old modules
- Changes to the product accomplished with a minimum of design input
- Simplifies final assembly
- Reduces the number of parts to assemble

Using standard components is always less expensive than a custom made item. Standard components such as screws, nuts, bearings, and seals are usually made in quantity and are often available off the shelf. This means there is little or no lead time and they are very low in cost. Standard components are easy to replace and can often be carried in a maintenance toolkit or as a maintenance package.

Since these components are made in large batches, the individual embodied energy is also very low; hence their sustainability value is very high.

7.6.3 Multifunctional parts

Multifunctional parts can often reduce the overall number of components and improve a product's functional efficiency. Several authorities have suggested that the driver of a vehicle should not remove his hands from the steering wheel while driving, yet there are many functions that require drivers' hands to be removed from the wheel. Most vehicles have lighting-control stalks, which are multifunction components; the stalks control headlights, main beam, and indicators and often have incorporated windscreen wiper controls, windscreen wiper speed, windscreen wash, and sometimes cruise control. This is indeed a multifunctional part built into one single stick, placed in such a position that the driver can reach the stick without taking her hands off the steering wheel. The design of this multifunction part was probably very complex but possessed obvious safety elements. This single device replaces what was formerly a bank of switches and as such is probably cheaper to produce and embodies less energy than the original bank of switches.

Components such as the lighting-control stalk can be incorporated into most designs. This is often the case in design development, but a wily designer may consider multifunction components from the beginning of the design process.

7.6.4 Multiuse parts

Multiuse parts may be used on a variety of components. A mounting plate may be designed to mount a variety of components, or a spacer can serve as an axle, lever, or the like. The key is to identify multiuse parts by sorting all parts in the design into two main groups:

- Parts that are unique to a particular design, e.g., crankshafts, housings
- Parts that are generally needed in all products, e.g., flanges, webs, bushes, spacers, gears, handles

A company manufacturing a range of similar products would normally hold their own library of standard parts that can then be used within the product range. For instance, a fabrication company manufacturing chassis for trucks might keep an inventory of parts they use on most of their chassis. These parts could include cross-beam spacers, flanges, and webs. Another company might use steel piping for compressed air lines around a factory and could use the same steel piping as walkway hand rails.

Several advantages are gained by employing multiuse components:

- Components can be manufactured in quantity, thus reducing cost.
- Designers can draw on basic components without having to redesign.
- Raw material stocks can be reduced.
- Raw material stocks can be purchased in larger quantities, forcing the supplier to give cost advantages.

In a similar way to multifunction parts, multiuse parts offer all the benefits of a reduction in components, giving ongoing benefits of reduced costs, reduced embodied energy, and reduced time to market.

7.6.5 Design parts for ease of fabrication

This principal suggests that parts should be designed using the least costly material that "just satisfies" functional requirements. If the material more than satisfies the requirements of the duty, material is wasted. For example, Formula One racing car engines are designed for a single race. If the engines could be used for, say, two races, then overdesign is present and material and processing have been wasted. A Formula One engine is an extreme example of designing so that the function "just satisfies" requirements. Often designers need to build in some form of safety margin, or factor of safety, to compensate for various factors, from operating environment to misuse by operators. (Factors of safety are discussed in Chapter 6.)

Major principles of design for ease of fabrication also include ease of access for welders or to merely fit screws. There is no point in designing a fabrication where welds are inside an enclosed box, making it impossible for the welder to access the joint. The same is true of access for screws and bolts. These are usually applied with tools such as wrenches, Allen keys, and hexagon sockets. All require access and an area around the fastener to swing the tool.

Many modern products are welded. Some are welded using spot-welds, but large fabrications are often continuously welded by some form of electric arc process. This process is expensive in terms of both welder's time and electrical power. The designer should always review the welding regime and perhaps apply intermittent welds rather than continuous welds, thus reducing welding time and power input to the weld. Excessive heat input to a weld is always dangerous in that heat from welding may warp a structure due to the release of internal stresses within the material. Minimal welds, and therefore minimal heat input, will reduce this problem. Figure 7.13 shows a typical intermittent fillet weld.

In many applications, a continuous weld is required in such places as where the joint needs to be water or gas-tight. It is the designer's prerogative to select appropriate welds for appropriate conditions, but designers should be mindful of the cost and environmental implications of applying welds.

To simplify fabrication time, effort, and complication, it is useful to break down the costs and energy input associated with a normal welding process. Barckhoff, Kerluke, and Lynn [8] suggested that labor was the major cost of welding at 85% of the total cost. The pie chart in Figure 7.14 shows the percentage costs of the major elements of a welding application.

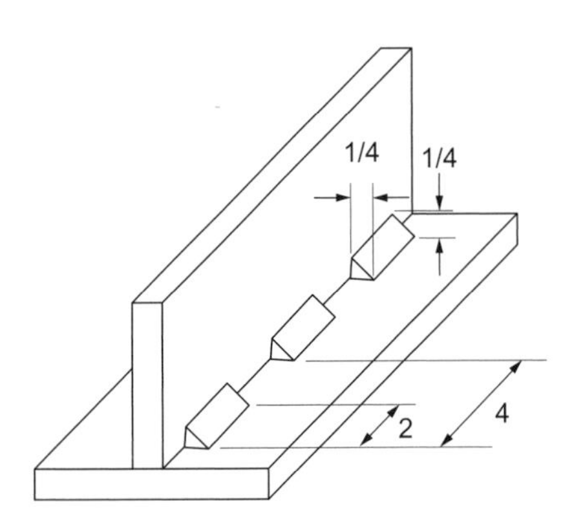

FIGURE 7.13

Typical Intermittent Fillet Weld [5]

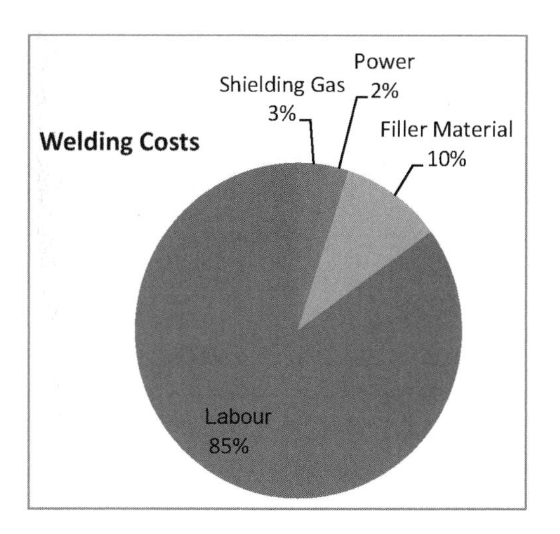

FIGURE 7.14

Percentage Costs of the Major Elements of a Welding Application [5]

We can see that power consumption is only 2% of the cost, but it should be noted that power equates to embedded energy, and any reduction in power will lead to a reduction in the embedded energy within a fabrication. It is also true that labor contributes 85% toward the cost of the welding process. The application of labor equates to time spent, and a reduction here would give a large cost saving and, by implication, a saving in power applied.

It is often the case that fabrications are over-welded. In practice this relates to the human condition mentioned previously where a specified 5 mm throat weld in practice will receive a 7 mm throat weld, "just to be safe." The application of a weld that is too large for its duty means that unnecessary time, unnecessary energy, and unnecessary materials have been expended. From a sustainability point of view, this is wasted energy and effort and increases the embodied energy within the fabrication.

7.6.6 Reevaluation of welds

In an exercise in reducing costs on a mobile plant chassis with a mass of 2 tonnes, a designer reevaluated the strengths of the chassis but also of the welds.

Welding costs can vary enormously with the thickness of material, the size of welding rod, and the power required. For this example, the chassis was fabricated from a 100×80 mm rectangular hollow section. Calculations are based on the mass of the deposition rate and the total mass of the welding rod deposited. The example uses a medium-duty weld type with medium power.

The objective of the exercise is to reduce the amount of welding within the chassis. The chassis comprises a rectangular hollow section with corners strengthened by webs. The average throat thickness (weld size) has been set at 6 mm. The values are based on a medium electrode size and the mass of electrode deposited during the welding process.

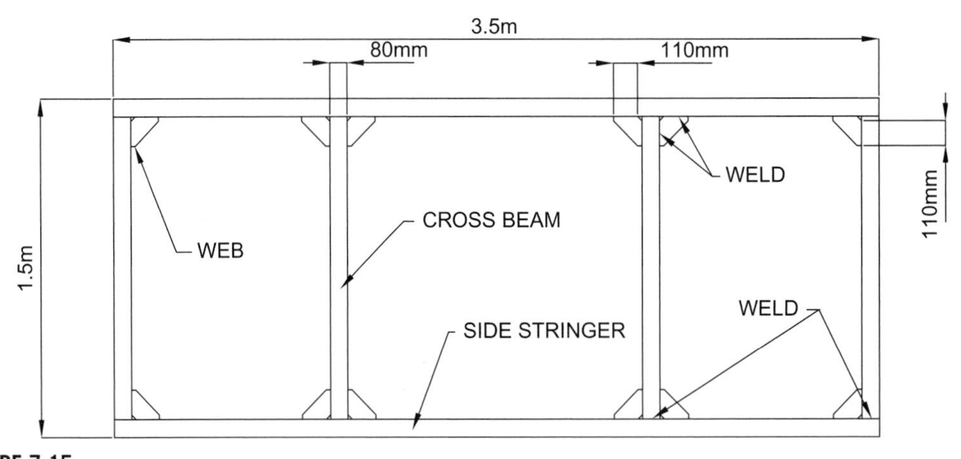

FIGURE 7.15

Item of Plant Chassis Showing Welded Joints [5]

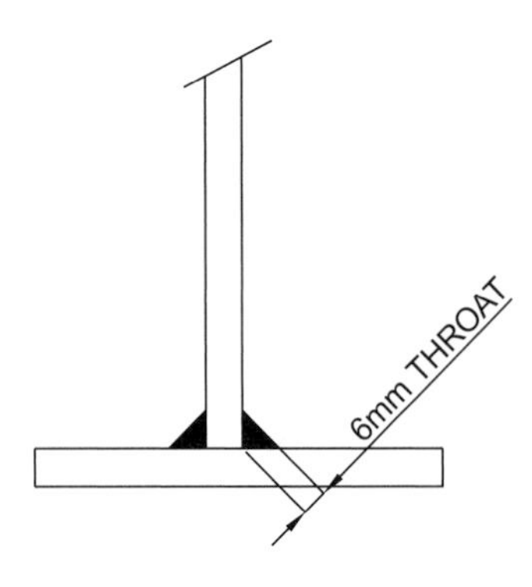

FIGURE 7.16

Throat Dimensions of a Typical Fillet Weld [5]

The chassis set out in Figure 7.15 indicates the welded joints, which were dimensioned at 6 mm throat in the original chassis. Figure 7.16 shows the throat dimensions of a typical fillet weld.

Determine the mass of the deposition in the welded joints of the original chassis. These are as follows:

Main chassis welded joints:

Mass = **0.814 kg**

Fillet weld joints for the web:

Mass = 1.49 kg

Total mass of weld in the original fabrication: 2.03 kg = 5.82 lbs

New chassis weld details:

New weld regime:

- Reduce the general weld throat to 3 mm
- Use continuous welds in fabricating the rectangular hollow section
- Use intermittent welds: 50/50; say, 25 mm weld/25 mm gap

Main chassis new welded joints:
Find mass of welds:

Mass = **0.2 kg**

New fillet weld joints for the web:

Mass = 0.187 kg
Total mass of weld in the new fabrication: 0.387 kg = 0.85 lbs

The charts in Figures 7.17 and 7.18 give estimates of the time taken to lay down a mass of weld and an indication of cost implications.

Time Needed (mins) to Deposit 1 lb of Weld Filler					
Rod Size	Operating Factor				
	60%	50%	40%	30%	20%
1/8"	30.4	36.5	45.6	60.8	91.2
5.32"	23.4	28.1	35.1	46.8	70.2
3/16"	16.8	20.2	25.2	33.6	50.4
7/32"	14.9	17.9	22.4	29.8	44.7
1/4"	12.2	14.6	18.2	24.3	36.5

FIGURE 7.17

Time Taken to Deposit One Pound of Weld Metal [9]

Total Cost ($/lb) of Deposition with $50 Labour & O/Head rate					
Rod Size	Operating Factor				
	60%	50%	40%	30%	20%
1/8"	$29.91	$34.97	$42.97	$55.24	$80.57
5.32"	$24.03	$27.02	$33.77	$43.52	$62.73
3/16"	$18.63	$21.43	$25.63	$32.63	$46.35
7/32"	$17.05	$19.54	$23.26	$29.47	$41.61
1/4"	£14.80	$16.83	$19.87	$24.94	$34.80

FIGURE 7.18

Approximate Cost Estimation Pound of Weld Metal Deposited [9]

Original fabrication times and costs:

Using 1/8" welding rod diameter
Using 40% operating factor

Time taken:

Time = 265.4 mins = 4.42 hrs

Total cost:

Cost = $247.76

New Fabrication Times and Costs
Time taken:

Time = 38.76 mins = 0.65 hrs

Total cost:

Cost = $36.18

Percentage reductions:

Time reduction = 85.4% reduction in time
Cost reduction = 85.4% reduction in cost

7.6.7 Assembly methods: conclusion

Fabrication using various methods of welding is an extremely important and well-used assembly process. This exercise shows how a little thought by the designer in specifying welds could lead to an enormous saving, both in terms of cost and, more important from a sustainability point of view, power consumed. It highlights the fact that cost savings and sustainability improvement are symbiotic. An improvement in the area of cost often leads to a reduction in energy input.

The efficiencies have been based on costs, but it is reasonable to conclude that a reduction in time spent welding is, of course, a reduction in time spent applying power. The new fabrication with the improved welding regime will benefit from a much lower embodied energy value.

7.6.8 Fastening systems

In any assembly process, components need to be fastened together. The natural fastener for any designer to use is the humble screw fastening. These components, however, can prove difficult to feed, may jam, and require monitoring for presence and tightening torque. Fasteners often need expensive feeders, extra labor, and extra assembly stations. Corbett and Dooner [4] estimate that the driving of screws can be 6 to 10 times more costly than the fasteners themselves. A little thought by the designer as to how to fasten the components may eliminate screwed fasteners, perhaps instead using tabs or snap-fits.

It is inevitable that fasteners must be used in some circumstances. The following checklist highlights some elements of best-practice principles for selecting fasteners.

Fasteners in general:

- Reduce the quantity
- Use standard fasteners
- Use the same strength of fastener throughout the assembly
- Reduce the size of fastener
- Use one size of fastener

Receiving components:

- Avoid tapped holes.
- Employ captive washers and nuts, thus reducing the number of parts to assemble.
- Consider automated assembly requirements, e.g., the shape of screws required for vacuum pickup or using chamfered points for improved placement success.

7.6.9 Minimize assembly directions

All parts should be assembled from a single direction. If the main assembly has different access points, the whole assembly may have to be turned to allow access. This involves transfer stations and inspection stations and eventually leads to increased cost and energy input.

7.6.10 Maximize compliance

Components are not always identical. Misalignment, out-of-shape components, and component flexure are all problems that may reduce assembly efficiency. This was a major issue during the authors early days as an apprentice machie-tool fitter: covers rarely fitted the mating location on a casing. Major factors that can affect the mating parts include:

- Accuracy of component geometry
- Consistency of component geometry
- Stiffness of the component
- Stiffness of the assembly tool
- Rigidity of jigs and fixtures
- Friction between parts

To avoid assembly problems, compliance must be built into the assembly process. It is the designer's responsibility to ensure that the preceding list of possible problems is avoided. Most can be circumvented through forethought; however, consistency of component geometry (accuracy of parts) may often be difficult to achieve, especially if parts are manufactured by an external source.

If automated assembly is being employed, compliance may be improved by the use of robot assembly systems employing elements such as tactile sensing, vision systems, quick change end effectors, and tools.

The designer can apply compliance features to the design, such as generous tapers or chamfers for easy insertion and other guidance features. An effective method is to use one of the major parts as a base unit. In the case of assembly of a passenger vehicle, it is conceivable that the car body would be the base unit to which all other parts can be fitted. This part base should be designed to be as stable and as rigid as possible, which will inevitably improve accurate insertion and simplify handling. This base unit may also be provided with accurate location points, generous tapers, and other features that can be used to guide insertion of additive components.

7.6.11 Minimize handling

Whenever components are moved to effect access or are moved from one location to another, costs and energy are involved. The greater number of times an assembly is moved, the greater the risk of:

- Damage to the assembly
- Slower cycle times
- Risk of diminished quality
- Increased complexity of equipment
- Increased cost of equipment

To minimize the handling and movement of assemblies, the designer should not only be aware of improved access but also should build in symmetry and orientation features that will guide and locate parts. Most products require some form of packaging, and it is with the assembly in mind that the designer could consider how the packaging should be applied. The packaging should be designed for easy handling and easy application and, when applied to the product, should also allow for easy handling and delivery. This can be achieved by understanding how the packaging could be fitted, including:

- Product flows within the facility
- Workspace facilities
- Packaging supply flow

For each element in this list, the designer should consider how the product, subassembly, or component can be designed to simplify flow to the packaging facility and away from it.

7.6.12 Design constraints and total design team formation

The designer must always perform tasks under several constraints. Costs may be considered the severest of the constraints and may consist of:

- Manufacturing costs
- Development costs
- Product support costs
- Warranty costs

Other constraints may be listed as follows:

- Timing and entry into the market
- Product technology
- Materials availability
- Manufacturing technology
- Volume, which dictates the method and rate of production; small batches would be treated very differently from mass production

- Ability to develop the product
- Projected life cycle
- Sustainability and environmental impact

This list is driven entirely on the type of product and has to be treated as a priority list from which elements may be added or subtracted to suit the design.

Traditional design methods complete the design process before manufacture has even been considered, but as the preceding list of considerations shows, an essential part of the design process is that of manufacturing method (among many other elements), which has to be built into the product at the design stage.

The design function can be polarized with a single designer working alone, perhaps as a consultant. His skills and knowledge would need to span design techniques, the product, the industry, and the manufacturing technique. As a consultant he would probably be commissioned by a company, in which case he would naturally look at the machine tools and processes available within the company so that he could familiarize himself with skills and techniques that are already in use. The aim here would be to reduce the costs by performing essential manufacturing in-house. To understand the intricate workings and systems of a particular company, the designer would need to liaise with key personnel and base designs on their advice and input.

At the other end of the polarized scale is a multidisciplinary design team that could proffer many specialist skills for the procurement, design, and manufacturing processes. Multidisciplinary teams offer an integrated and balanced approach and are an important factor in improving communications and the ability to consider many concepts simultaneously.

Corbett and Dooner [4] suggest that 70–80% of the production costs are defined at the design stage. The great advantage of multidisciplinary teams is that multiple aspects of the product can be considered simultaneously. Principal among these is the correct method of manufacturing. Specialist manufacturing staff, along with marketing staff, purchasing staff, and so on, work alongside the designers to assure product success by ensuring that the following elements apply:

- The product must work efficiently.
- The product can be manufactured efficiently.
- It is saleable.
- It is profitable.
- It has a low environmental impact.

It is the designer's responsibility to ensure the total success of the product. Since 70–80% of the cost is also prescribed at the design stage, it can be seen that the designer has a huge responsibility, and it is suggested that the design function within a company requires *total control* in order to fully respond to this responsibility.

Some of the skills required within this multidisciplinary design team can be described by the particular tasks that are required, as described in the following discussion.

7.6.12.1 The value engineer

The value engineer considers the manufacturing process, associated costs, and material costs and traditionally sees the new design after it has left the design office. He may suggest modifications and improvements, but often these suggestions would prove too costly when manufacturing has already been implemented. It seems reasonable that the value engineer should be incorporated into the total design team during the design process.

7.6.12.2 Manufacturing engineer

The skills of a manufacturing engineer are very wide and varied. His role is to develop and create physical artifacts, but his knowledge must span production processes and the technology required to manipulate materials into a product. The manufacturing engineer will possess design, organizational, and communication skills, along with statistical and other mathematical skills. It seems appropriate that the manufacturing engineer should be one of the major contributors to the total design team.

7.6.12.3 Engineering purchaser

The engineering purchaser will need a fundamental knowledge of engineering, but her skills will lie in negotiating prices for commodities as well as understanding the fundamentals of the components required in the manufacture of a product. The role of the engineering purchaser within the total design team would be to advise on component costs, availability of products, the timing for procurement of components, and finally, the sustainability value of proposed components.

7.6.12.4 Product designer

The term product designer may have several meanings. The product designer's role is to combine art, science, and technology to create new products. Within the total design team, it is the function of the product designer to create a product that is appealing and that will assist the marketing and sales function. Engineering designers and manufacturers may be able to create a product, but their skills are not normally in creating aesthetically pleasing articles. The product designer requires technical engineering skills, but her role is in applying aesthetic principles, along with her technical skills, in providing a product that is acceptable to the consumer. A food mixer, for example, may have the best technical specification on the market, but it would not be purchased and placed in people's homes if it did not look good.

7.6.12.5 Marketing and sales

Through market surveys, the marketing function will provide the need and the brief for the design of a product. It is not normally commercially viable to design and develop a product when there is no market and no perceived need for it. The marketing department provides the market analysis that informs the development company that there is a need to be fulfilled. For example, the Apple iPad was a new introduction to the market. The Apple marketing team identified a market need. The foresight of Apple as a company developed the iPad into a very marketable product.

The sales function is often allied to the marketing function and is sometimes combined. The role of the salesperson used to integrate the product into the market during the development process, largely after the product had been produced.

Both marketing and sales personnel require technical insight, but this is not their primary function. However, they do need a place on the total design team to advise on several elements, including the accuracy of functions compared to market requirements, packaging, manufacturing quantities, and timing into the market.

7.7 LIFETIME USAGE

Lifetime usage relates to the environmental impact and hence the energy used and pollution created during the life of the product. For some products this element of the life cycle is the dominant feature of the whole embodied energy of the product.

7.7.1 Case study: item of plant: water-well rock drill

The water-well rock drill shown in Plate 7.7 comprises many different and diverse components, from steel fabrications such as chassis, fuel tanks, mast, and gearboxes through bought-in components such as diesel engines, compressor, wheels, and bearings. Clearly, the calculation of embodied energy in this product is complex, but using CES EduPack [3] software, we can analyze the major components and formulate an overview of the embodied energy applied at each stage of the whole life of the product.

For this particular exercise, it has been assumed that the whole rock drill has been manufactured through a fabrication process from low-carbon steel. This was the major manufacturing technique used in producing the rig. The rock drill was destined to be used in remote villages in the Himalayas and was transported via rail over 5,000 miles. The engine powering the rig was 800 hp and the rig life was assumed to be five years. It has been assumed that most of the components can be reused, refurbished, or recycled. The full breakdown of details is shown in the "Eco-Audit Report" in Section A7.3 of the Appendix.

The eco-summary for embodied energy is shown in Figure 7.19; the eco-summary for the carbon footprint appears in Figure 7.20.

We can see that the carbon footprint and the embodied energy are very closely related. This is because it is assumed that all the energy applied to the rig in extracting materials, through manufacturing, use, and disposal, is also applied by fossil fuels, which can be termed artificial energy. The use of natural energy such as solar or wind power will inevitably reduce the carbon footprint, but it is unlikely to reduce the embodied energy value. If the drill rig is to function as normal, no matter what the energy

PLATE 7.7

Water-Well Rock Drill [5]

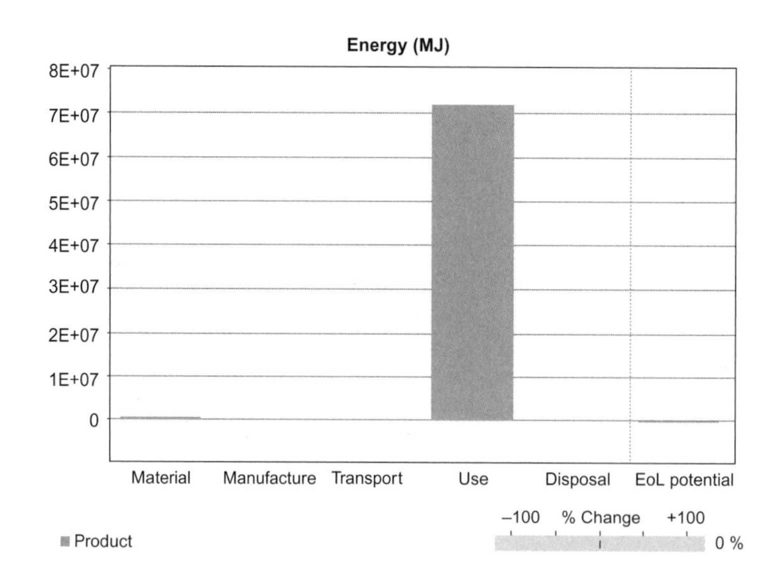

FIGURE 7.19

Eco-Summary for Embodied Energy: Rock Drill Life Cycle [3, 5]

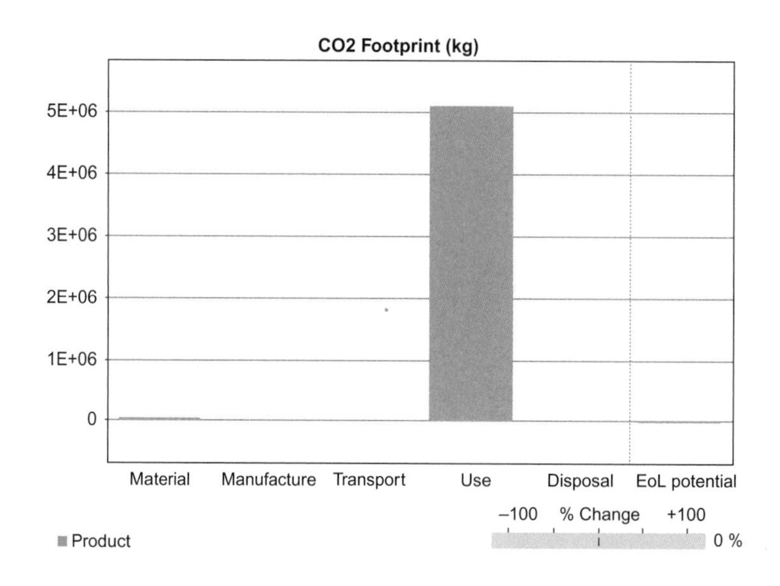

FIGURE 7.20

Eco-Summary for Carbon Footprint: Rock Drill Life Cycle [3, 5]

source it will still need the same value of embodied energy. If natural energy is used, the rig function will become more sustainable and the carbon footprint will be reduced.

Both Figures 7.19 and 7.20 indicate that sourcing the materials, manufacturing, and transportation requires minimal energy, but this energy requirement is totally overshadowed by the energy used during the normal operation of the rig. If the designer is to reduce embodied energy in the life cycle of the drill, he has to carefully consider the energy needed to power it and how that energy is derived.

7.7.2 Case study: steel fabricated footbridge

The fabricated footbridge shown in Plate 7.8 requires a large element of energy during material sourcing and manufacturing. After installation, however, the footbridge is passive and requires no energy for it to function, apart from occasional maintenance. The life span was expected to be approximately 50 years. An eco-audit using CES EduPack [3] software determined the embodied energy value of the bridge. The results are shown in Figures 7.21 and 7.22, with a full detailed analysis in Section A7.4 of the Appendix. For the purpose of the exercise, the material was assumed to be virgin material that could only be reused or recycled at the end of its life. Furthermore, the deck was assumed to be timber from a certified source. This material was also to be reused at the end of the bridge's life.

The footbridge requires no energy for it to function properly; therefore the embedded energy in its life of use is zero, which is quite a contrast to the rock drill in Plate 7.7. It is clear that the bigger part of the embodied energy, over 1 MJ, is used in obtaining the raw material, with approximately 100 kJ used in manufacturing. At the end of its life it was assumed that the steel and wooden components of the bridge would be reused or recycled, thus giving an enormous end-of-life potential of 1 MJ. This almost completely negates the energy used in sourcing material.

We previously stated that true sustainability is almost impossible to achieve. The data displayed in Section A8.4 of the Appendix suggests that there will be a 94% end-of-life potential for the bridge's steel components due to the reuse of the material. The end-of-life component of the timber is only 6.5%, but this seemingly low figure is misleading, since the timber originated from a sustainable source, so the end-of-life potential will be much less. However, this is a figure that should be a major goal. There will always be losses in the system, but this bridge offers a glimpse of the sustainable efficiency that could be achieved.

The CES EduPack [3] software gives designers the opportunity to quantify products' environmental impacts in terms of energy expended and carbon footprint. This software is an essential tool for designers and design teams who are attempting to reduce products' embodied energy.

PLATE 7.8

Steel Fabricated Footbridge

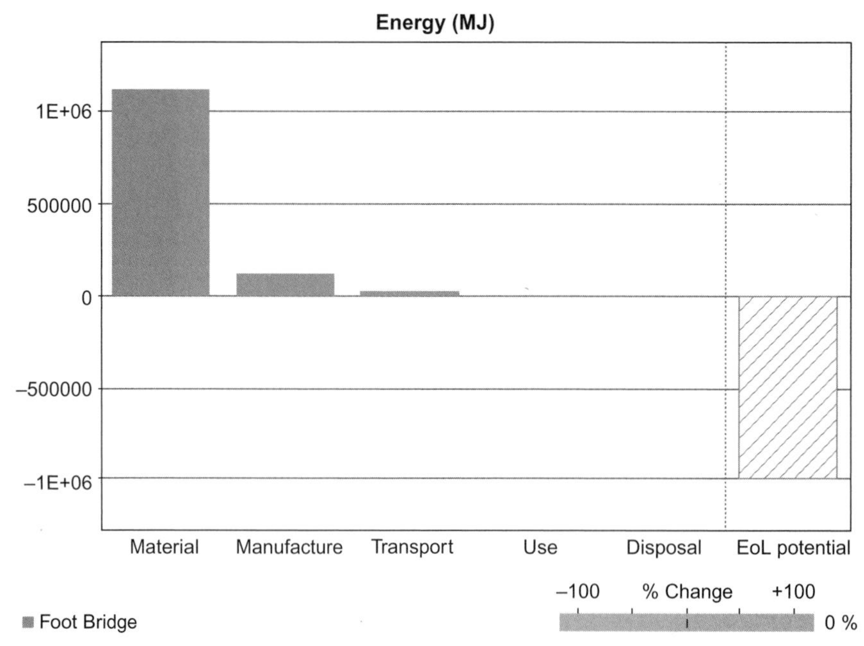

FIGURE 7.21

Eco-Summary for Embodied Energy: Footbridge Life Cycle [3, 5]

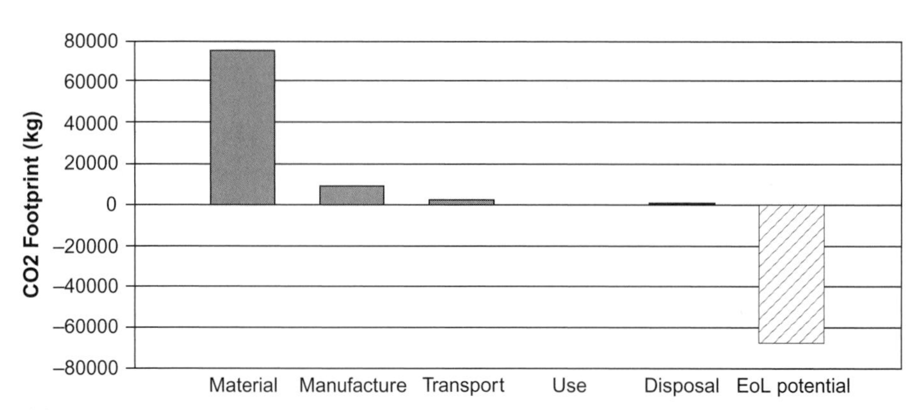

FIGURE 7.22

Eco-Summary for Carbon Footprint: Footbridge Life Cycle [3, 5]

7.7.3 Designer's duty

The duties of the designer or the design team are many and varied, but most consider that the primary objectives of any design are to:

- Reduce the cost
- Reduce the embedded energy

Cost reduction has been a primary exercise for designers since the design function came into being. The duty of evaluating embedded energy reduction is a more recent feature of the role and should be explored a little further.

The embedded energy within a product can be manipulated and reduced by concentrating on those areas that are shown to be the biggest contributors. In the case of the rock drill, the biggest contributor to the embodied energy is during its useful life, where the greatest environmental impact is derived directly from the use of the diesel engine from which it gleans its power. The designer must naturally concentrate on this element of the product's life in order to reduce the power use. The reduction in value may be a tall order, since the drill still has to function, but the critical point here is to reduce artificial energy (energy derived from diesel engine) to natural energy or energy derived from natural sources such as wind or solar power, thus employing a sustainable power source.

The bridge's embodied energy comes from the sourcing of the virgin material. The designer's attention should therefore be drawn to sourcing the material, since this will repay the design in reduced embodied energy and can be achieved by using more recycled or reused materials.

As designers embrace sustainable engineering concepts, they not only assist many environmental conservancy issues, but they also reduce the costs of the product to the manufacturer.

7.8 MAINTENANCE

In designing a product such as a motorcycle, the designer needs to consider the following points:

- Life of product
- Life prediction of components and design for scheduled component replacement
- Simplicity of components and standardization
- Accessibility for ease of removal of components
- Downtime: Dual components (redundancy) for continuous working while maintenance is being carried out
- Detail design for easy removal and component replacement, easy access, easy removal of fasteners, e.g., cap head screw, M 12 hexagon heads external use: easy fitting, bearings, seals
- Maintenance location: In the field, at the factory
- Modular build
- Lubrication and lubricant delivery

The approach to maintenance presupposes that there will inevitably be components that wear and need to be replaced. Any engineer can appreciate that some parts will wear more quickly than others simply because there is a part of the system that is more highly used. Bearings and seals fall into this category and can be regularly replaced.

7.8.1 Life of the product

The design life of a product relates to the period of time expected by the designers to function normally. The life of the product depends entirely on the product and its intended use. A short design life could relate to a disposable camera, which fulfills its useful life once its roll of film has been exposed.

It should be noted that once the camera's film has been extracted and converted into photographs, the casing is reused, repackaged, and marketed once again.

A longer design life could be attributed to a digital camera, which is more expensive than the disposable camera and which would be expected have a design life measured in years or even decades. Design life is often related to obsolescence but is distinct from that particular concept. Many products are still very useful and work perfectly to their original design parameters after a number of years but may become obsolete in the eyes of the user simply because they do not possess the latest technological advances.

7.8.2 Component life prediction

Whenever a design is created, the designer must have in mind the lifespan of the product, or at least the life expectancy of some of its components. The designer must also appreciate that the failure of some components, when in use, may lead to fatal consequences. The bearings carrying the rotors for a jet engine, for example, need a high level of reliability, since their failure in flight could lead to an aircraft crash.

It is a useful concept to design components into a product so that their lifespans can be predicted. This is often achieved by calculating the number of cycles and replacing the component when its end of life is predictably close. One such component is the humble rolling element bearing, normally rated with an L10 life. The life of rolling element bearings is defined as the number of revolutions at a given constant speed and load that the bearing is capable of enduring before the first sign of fatigue occurs in one of its rolling elements. It should be noted that fatigue is not the only mode of failure of a bearing. The L10 life value relates to the expected survival of 90% (10% failure) of the bearings under prescribed loads and speeds cycling for 1 million revolutions. Statistically, bearing companies expect a median life of approximately five times the calculated basic L10 life.

To put this into perspective, the designer can relate the load and speed applied to a bearing, enabling the calculation of the expected life through a number of cycles. A rule of thumb here suggests that half the maximum load will lead to roughly twice the length of life of the bearing. Armed with this information, the designer can predict the life expectancy of a particular bearing and suggest a period of use prior to essential maintenance. This is normal practice with passenger vehicles, which have particular maintenance periods for certain components. Typically these maintenance checks could include the following:

- Weekly checks: Minor elements, e.g., coolant level, tire pressure, engine oil level
- Six months: Renew engine oil and filter
- Twelve months: Drive belts, road wheels, brake calipers and discs, front suspension components, rear suspension components, drive shaft joint gaiters
- Two years: Renew brake and clutch fluid, renew coolant
- Four years: Renew spark plugs, renew timing belt, renew fuel filter

This list is indicative but shows that vehicle designers have considered the wear rate of various components and have suggested a maintenance program to the owner. Such a maintenance program would be absolutely essential for aircraft, since failure in the air could lead to a substantial number of fatalities.

7.8.3 Simplicity of components and standardization

Simple components are normally low-cost and require low energy to manufacture. They should also be easy to fit, especially if the design considered the location and fixing technique. This could be as simple as a hole in the receiving component and a spigot on the replacement component with a simple snap fitting. A simple vehicle headlight bulb is an excellent example. The replacement simply plugs into the headlamp socket, which normally then fits into the headlamp receiving hole from behind, requiring only a quarter turn to fix it in place. The unit is located and fixed in one simple movement.

Many standard components can be purchased from most engineering stockists. These components can be incorporated into the design and include bearings, seals, screws, rivets, nuts, pins—the list is endless. These components are very low-cost and readily available. It is normally good practice to replace bearings as well as seals, screws, washers, and the like whenever maintenance is carried out. When maintenance is done in the field, the maintenance engineer would probably carry with him a standard set of parts intended for the maintenance of a particular device.

Standardization may also apply to manufactured components within the product. A range of gearboxes may require spur gears in different combinations. Gears could be standardized to benefit from high-volume production, but careful design would reduce the number of different gears required to be kept in stock for replacement in the range of gearboxes.

7.8.4 Accessibility for ease of removal of components

During the design process the designer must consider the assembly procedure as well as the disassembly procedure of a device. It is poor practice to weld a plate in place when the bearing behind it requires removal during the next maintenance operation.

It is important that the device's design gives appropriate physical access to accommodate tools, extraction devices, maintenance personnel, or at least the hands of maintenance personnel, with enough room to rotate wrenches, sockets, and screwdrivers. Components requiring regular replacement should be located to the outside of the product rather than buried internally and creating access difficulties, thus extending the time to maintain the device.

7.8.5 Downtime

The time it takes to maintain a component is critical, since often production will come to a halt. Companies plan scheduled shutdowns so that maintenance can take place on critical equipment. If equipment breaks down during a production cycle, millions could be forfeited in lost production when the fault could merely be a rolling element bearing that costs £20 (US$30).

Designers need to build in simplicity of maintenance so that speedy maintenance can be accommodated. Sometimes it is necessary for a critical unit to function while maintenance is being carried out. In such a case, redundancy is built in in the form of dual components. Here one set of components can perform the function adequately while the other is being replaced. A typical such application can be seen in the oil recirculation unit shown in Plate 7.9. Twin motor/pump units can be seen placed on top of the tank. This is typical of built-in redundancy where maintenance on one motor/pump unit can take place while the other motor/pump unit performs the normal function of the unit.

PLATE 7.9

Oil Recirculation Unit Showing Twin Motor Pump Units [13]

An unscheduled shutdown caused by a parts failure can be extremely expensive for a company. Scheduled maintenance is usually planned to avoid this issue, but more and more companies and factories are using modern technology to detect obscure vibrations and noise in things such as gearbox drive units so that they are forewarned of an impending failure and can plan for an ad hoc maintenance exercise on, say, a drive unit. This could be done between shift changeovers and may involve planning that a new gearbox be inserted while the failing gearbox is removed.

7.8.6 Detail design for quick-and-easy maintenance

The detail designer, when providing a product's manufacturing drawings, needs to consider easy access for tools and extractors. For instance, a shoulder may be provided for a bearing to be located against, but the detail designer will need to increase the diameter of the shoulder to give access for a drift to be inserted to knock out the bearing, as shown in Figure 7.23. Here there is ample access for the drift to purchase on the outer and inner race of the bearing so that it can be knocked out of the housing. The same method needs to be applied to other replaceable components such as seals.

Consideration also needs to be given to removal of screws. Designers will specify which screws and where they should be applied, but a golden rule is to apply screws so that they can be easily removed. This often depends on the end use of the product and the environment in which it will be used. If the device is an item of plant used outside in the open, it would seem ridiculous to insert small cross-head screws. These screws will easily rust and be impossible to remove. Similarly, cap-head screws sporting an Allen socket hole in the head will easily fill with debris if used in outside environment. This will make it difficult to insert the Allen key. These screws are also high tensile and have been heat-treated. They are usually supplied covered in oil film but will easily rust when exposed to weather. These screws are clearly intended for internal use on such devices as machine tools.

Good practice for items of plant used outside and subject to normal weather conditions is to apply a minimum of M12 (1/2" UNC) hexagon headed screws (bolts). The hexagon head can easily be cleaned of debris with a wire brush and will give enough purchase for a socket or a ring spanner, thus allowing the maintenance personnel to easily remove the screws.

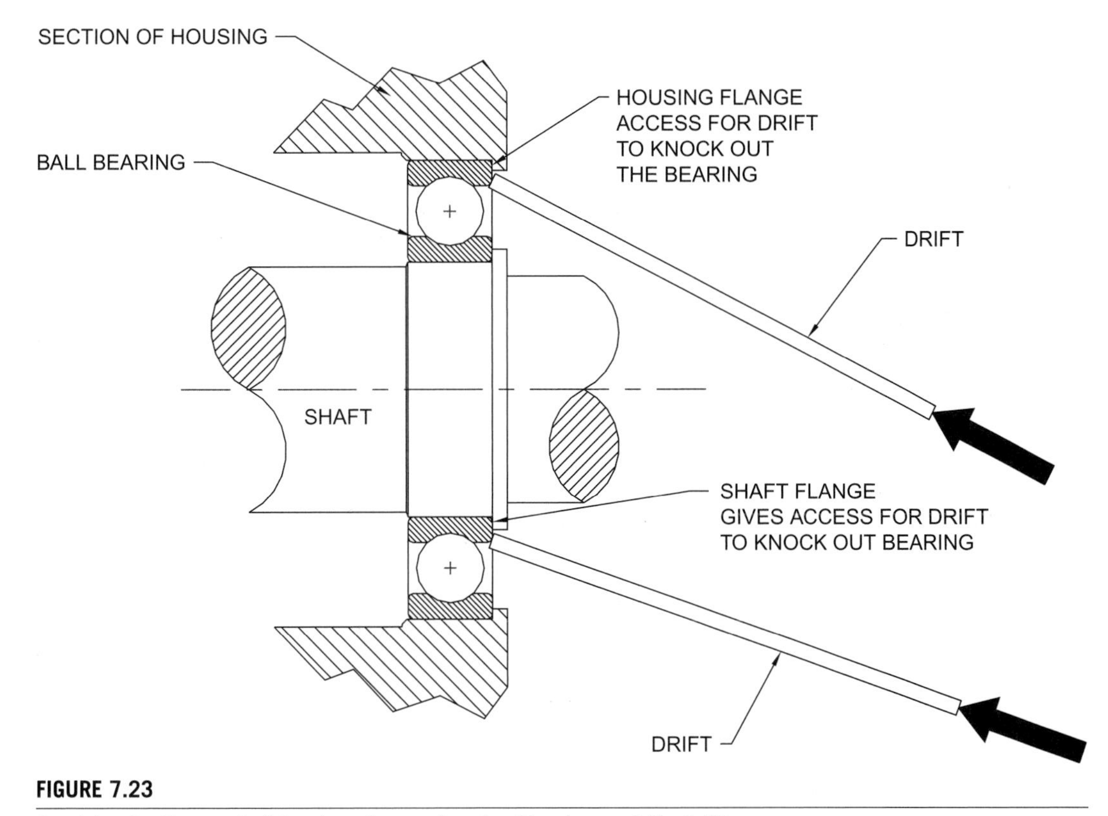

FIGURE 7.23

Provision for Removal of Bearings from a Bearing Housing and Shaft [5]

These are just a few of the elements an experienced designer must consider for the maintenance process when providing detailed drawings prior to manufacture of the components.

7.8.7 Maintenance location

The location of the normal maintenance practice has an enormous effect on the design and selection of components. If the maintenance can be done in comfortable surroundings of a workshop where there are available tools, cleanliness can be observed, and temperatures are reasonable for maintenance personnel, then the maintenance practice is fairly straightforward. Consideration can therefore be given to more standard components that can be removed, perhaps with a bench press or with available power tools. Maintenance in the field is quite a different matter.

Maintenance in the field could mean on the factory floor, where production is taking place, or it could mean that the maintenance personnel have to travel to the machine, which could be in a quarry or in a mountainous region hundreds of miles from the nearest workshop. If maintenance is to be done *in situ* on the factory floor, then access to power tools may be limited and, depending on the process, might prove dirty and uncomfortable for the maintenance personnel. A dirty environment is also the enemy of clean assembly. Should dirt ingress the assembly, premature wear could take place. It is particularly important for hydraulic assembly to take place in a clean environment, since small dirt particles (smaller than the naked eye can see) could infiltrate the very close-fitting components in, say, hydraulic pumps and cause premature wear and premature failure. In such cases, design elements and good practice, as discussed, should be built into the design to promote quick, easy, and clean maintenance, thus reducing downtime and easing the load of the maintenance fitter.

In extreme cases, items of plant may be used in a very hostile environment located many, many miles from the nearest service depot. In such cases, service engineers need to travel to the device that requires maintenance. An excellent example of such a case is explained in the following case study.

7.8.8 Case study: water-well rock drill

A water-well rock drill was designed to operate independently in the bush in Africa. The specification for the rock drill was to drill 100 mm diameter (4") holes, approximately 400 m (1,200 feet) deep, to access the water table. When the hole had been drilled, a down-the-whole water pump could be inserted to bring water to the surface. This would then be the water supply for local villagers.

The drill was based on a trailer and towed by a long wheelbase Land Rover. The whole rig was designed for maintenance in the field and independent use. This included the application of maintenance procedures in the field. One particular component on the rig that had the most arduous use was the rotation gearbox, which is highlighted in the schematic of the drill rig in Figure 7.24.

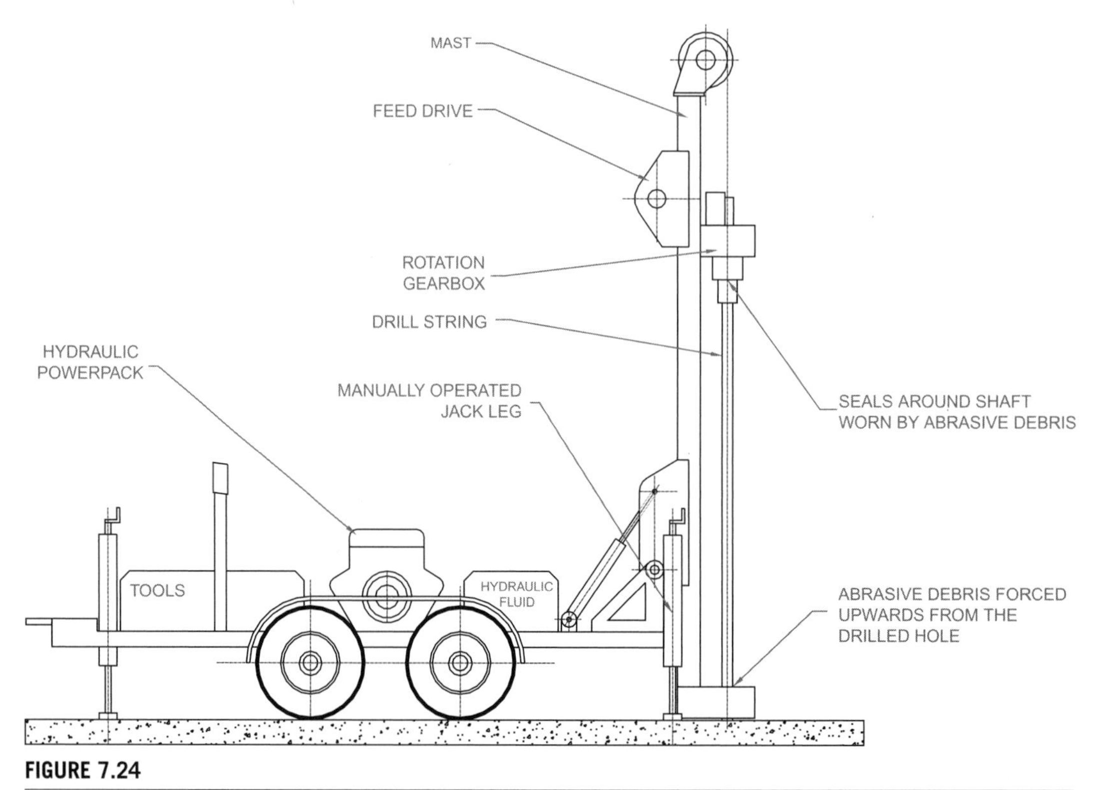

FIGURE 7.24

Trailer-Mounted Water-Well Rock Drill Used in Remote Areas [5]

The rotation gearbox rotates the drill string and supports several tonnes of drill pipe. Compressed air is passed down the drill pipe to operate a hammer bit at the bottom of the hole. The remaining energy within the compressed air brings highly abrasive dust to the surface, hitting the underside of the gearbox. Seals in this location operate under extreme wear conditions. The seal retains the oil in the gearbox and prevents ingress of the abrasive debris. Inevitably oil escapes through the seal and forms a paste with the abrasive debris, which tends to wear a substantial groove in the shaft under the seal. Under normal circumstances the shaft would be removed from the gearbox and sent to a workshop, where the seal/shaft contact surface would be renewed by a deposition process and machined to accept the seal once again. The shaft could then be replaced in the gearbox.

This process is impossible when the rig is operating over 600 miles from the nearest factory. During the design process, a solution had to be found for the wear problem or risk weeks of downtime and lost production in the field. The solution was to use a "cassette seal," a cross-section of which is shown in Figure 7.25.

The seal provides a sacrificial sleeve on which the lip seal rides. The sacrificial sleeve is a tight fit on the shaft forming a static seal with the shaft. This means that the sacrificial sleeve wears rather than the shaft. Maintenance is now possible in the field, since an extractor can remove the sacrificial sleeve and a new cassette seal can be applied to the shaft. The shaft does not suffer any wear at all and does not have to be returned to the factory. Downtime in the field is now measured in hours rather than weeks.

The service engineers suggested that gearbox lid gaskets often get damaged in transit. A solution was found in PTFE cord. This is pure PTFE and can be purchased from seal providers in diameters of 3 mm, 6 mm, and up to 15 mm. Rather than a gasket, the designer specified 3 mm diameter PTFE cord, which could then be applied to the interface between the gearbox lid and the gearbox case, providing a perfect seal when the gearbox lid was fitted. The problems of protecting gaskets in transit did not exist with a length of PTFE cord, since it could be transported in a roll. This feature also promoted maintenance in the field and was applied at the design stage.

7.8.9 Modular build

The advantages of modular build were dealt with earlier in this chapter. From a maintenance point of view, the replacement of a module is quick and easy. If the module has stopped functioning, it could be replaced by a working module and removed back to a workshop where appropriate facilities and maintenance time can be allocated. Perhaps the biggest advantage of using modules is that downtime is minimized on the main product.

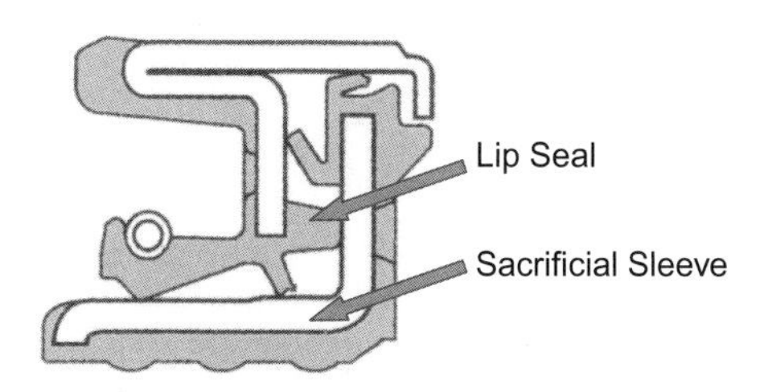

FIGURE 7.25

Example of Cassette Seal Showing Sacrificial Sleeve [11]

7.8.10 Lubrication and lubricant delivery

Lubrication is often the last thing considered in any design, but it should be treated like any other major component. Most lubricants after all can be considered liquid engineering.

A *lubricant* is a device introduced between two components that slide relative to each other. The task of the lubricant is to prevent the two components from coming into contact and therefore abrading each other. Dry steel rubbing on dry steel will tend to "pick up." This means that material from one component will be deposited on the other component, and vice versa. It does not take long before both surfaces become rough and unusable. This can be demonstrated by rubbing together two glass bottles. Both will appear burned and both will "pick up." When lubricant is introduced between the two adjacent surfaces, the lubricant film will shear, effectively allowing the surfaces to ride on the sacrificial film. The end result is that they will slide relative to each other without touching and without wear.

Lubricants are available in many forms, from hard, highly viscous lubricants that are the consistency of frozen butter to lubricants with a low viscosity, an example of which is paraffin. Lubricants are normally considered oils, but these, along with greases, are the more common types. More specialist lubricants could be any material that can be inserted between the two sliding components. Graphite is among these, as are copper and other specialist materials, including engineering plastics, though the use of these tends toward "self-lubricating bearings" rather than pure lubrication.

Perhaps the most obscure "lubricant" is that of *lignum vitea*. This slow-growing timber of the genus *Guaiacum* has been used for centuries as a self-lubricating bearing, especially in underwater applications. The timber is the only wood that will sink in water due to its extreme density, caused by the saturation of lignum (tree sap). The classic use of the timber is between a ship's hull and the slipway prior to launch. This use is in the true sense of a lubricant in that it is a sacrificial element between two sliding surfaces.

Oil can be used in various applications and can vary in viscosity. Oils can be "designed" for a particular application and may have additives such as antifoaming agents for high-shear applications such as in gearboxes. Oils can also be designed for high-pressure applications for such components as vehicle differential gearboxes.

Greases are effectively oils that are mixed with soap. The soap, perhaps silicon soap or lithium soap, is a carrying agent that acts like a sponge, absorbing the lubricating oil. Greases can be regarded as plastic solids and differ from normal oils in that they will not flow unless subjected to shear stress. Also when subjected to shear stresses, greases become increasingly mobile, that is, they become more liquid. A very useful property of a grease is that it will stick to the component, whereas oil needs to be contained and physically kept in contact with the component.

In designing a component with moving parts, the designer needs to consider not only the type of lubricant but also the delivery system. Moving components will require a lubrication system and a lubricant designed for that system only. An internal combustion engine requires an engine oil of a particular viscosity, where the oil is contained in the sump and pumped to vital components in the engine and gearbox. The gearbox bearings and gears in the rotation gearbox in Figure 7.24 are lubricated by being partially submerged in an oil bath. Grease lubrication is useful for standalone bearings such as those used on the rolls of a paper mill. A plumber block (sometimes called a *pillow block*) bearing, shown in Plate 7.10, is usually packed with grease simply because grease is sticky enough to stay inside the bearing. Note the grease nipple on the top of the housing, through which grease can be injected via a grease gun or other grease pressurization device.

PLATE 7.10

Typical Plumber (Pillow) Block Bearing [12]

Fast-moving components tend to discard the lubricant by flinging it away. Such is the case with motorcycle chains. Specialist lubricants have therefore been developed for these kinds of situations. The lubricant needs to be sticky enough to coat the moving parts without being flung away, but in the case of a motorcycle chain, it should also avoid picking up dirt and debris. These conflicting requirements can be met by combining the oil with various additives.

7.8.11 Lubricant delivery systems

Delivery of lubricants depends on the lubricant and the object being lubricated. Perhaps the simplest method of injecting (usually) grease into a bearing is via a grease gun. A selection of grease guns is shown in Figure 7.26.

Grease guns deliver grease through a grease nipple at high pressure up to 50 bar, forcing grease into a bearing. Bearings can take several forms and can be used in many varied situations. Plumber blocks

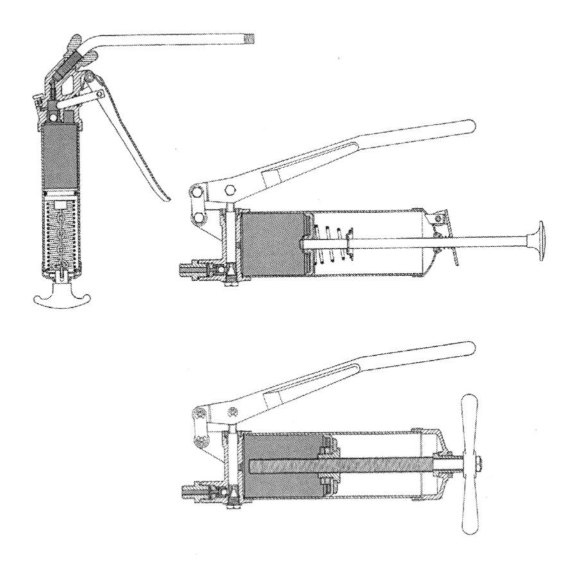

FIGURE 7.26

Selection of Typical Grease Guns [13]

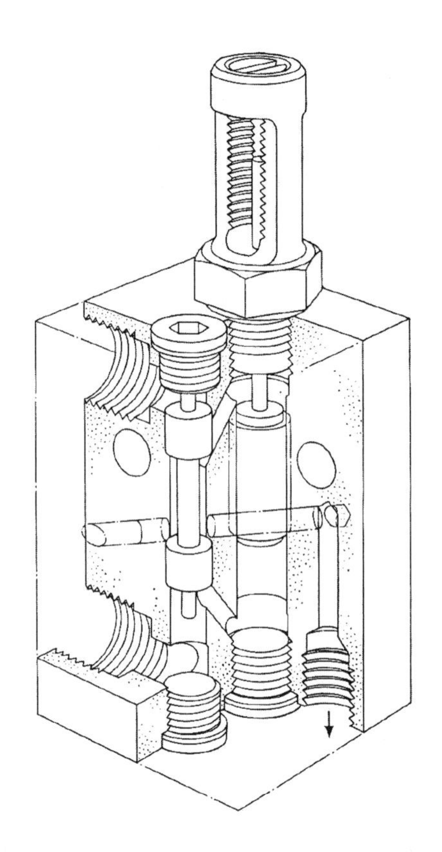

FIGURE 7.27

A Dual-Line-Injection Measurement Valve [13]

are shown in Plate 7.10, but there are many other styles and types, such as rod ends on hydraulic rams or, indeed, in any situation where there is individual metal-to-metal contact.

There is an element of guesswork in dumping grease into bearings, since quantity cannot be measured and there is uncertainty in some cases with the grease reaching the bearing. When more certainty is required, we can use automatic lubrication systems, which feed measured quantities of grease or oil through connecting pipes directly into bearings. A typical dual-line-injection measurement valve system is shown in Figure 7.27.

This type of valve is equipped with an indicator that can sound alarms if there is a blockage in the line. These valves are placed in blocks of up to 12 valves. The outlets from each valve are then piped directly to the bearings through smaller-bore pipes. Pipe fittings carrying the small-bore pipe then replace grease nipples. A typical dual-line lubrication system application is shown in Plate 7.11, where there are several dual-line-injection valves piped directly to bearings.

This typical dual-line lubrication system is fitted to a mobile machine and is therefore fed by a hand pump. Pressurized grease is forced through the injection valves, which in turn feed the pressurized grease through smaller-bore pipes directly to the bearings through pipe fittings.

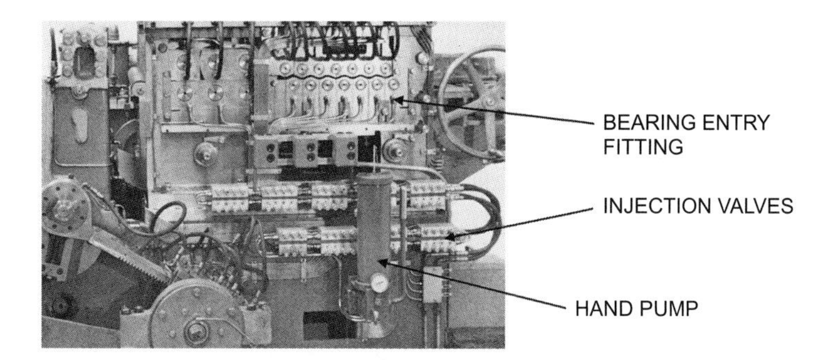

PLATE 7.11

A Typical Dual-Line Lubrication System Application [13]

PLATE 7.12

A Typical Lubrication Pump Room [13]

In stationary applications—for instance, in rolling mills—there may be a specific room set aside for the pumping systems. A typical pump room is shown in Plate 7.12. Several grease reservoirs can be seen. Beneath the reservoir is the pumping system, driven by electric motors. Typically these pump units will feed high-pressure grease or oil to a dual-line injection system such as that shown in Plate 7.11.

7.8.12 Engineering plastics

Engineering plastics have become revolutionary new bearing materials that, under some conditions, have many more benefits than standard bearing systems.

Engineering plastics are group of plastic materials that exhibit superior mechanical and thermal properties in a wide range of environmental conditions. These materials can be manufactured to suit a particular application and are excellent when encountering conditions in which rolling element bearings or other stationery bearings are unsuitable. Such conditions are listed here:

- Low speed
- High load
- Wet or submerged conditions
- Dry conditions
- Impossible-to-lubricate conditions
- Oscillating shafts
- Linear bearings
- High-wear situations
- Self-lubricating situations

Engineering plastics can be manufactured to be porous and thus be impregnated with extra lubricant. The basic materials listed here can also be mixed with strengtheners such as fiberglass, Kevlar, carbon fiber, or the like. During the manufacturing process, engineering plastics may also have included chemicals and other additives that improve properties such as brass particles, which enhance the high-temperature performance.

Engineering plastics are based on several popular plastics such as polyethylene, nylon, PTFE, ABS, polyamides, and many more. Specialist companies are very willing to give advice on particular applications.

Engineering plastics can be found almost everywhere, but for the layman the most obvious place to find these plastics is in the passenger vehicle. Engineering plastics are used extensively in fenders, facia, steering wheels, engine manifolds, and rocker box covers. There are many other applications, such as food pumps, skateboards, motorcycle crash helmets, coffee pots, kitchen knife handles, and toolboxes.

Plastic bearings offer the opportunity of almost zero maintenance. In many instances they can be considered a "fit and forget" bearing. The drive shaft for the bed-lifting device in Figure 7.4 rotates in plastic bearings. These units are placed in domestic bedrooms and therefore cannot be regularly maintained. Plastic bearings are an ideal choice and are expected to last longer than the life of the drive unit. When drive units are thus replaced, bearings will also be replaced.

Let's look again at the water-well rock drill in Figure 7.24, which uses linear (flat) engineering plastic bearings on the slides, thus enabling the rotation gearbox to slide up and down the mast. Furthermore, the drill string requires guidance before entering the ground. The engineering plastic guide bush at the base of the mast guides the drill string. Both the linear bearings and the guide bush operate in an extremely harsh and abrasive environment. Practice shows that debris tends to embed itself in the plastic and polishes the metal on which it slides. It would be difficult to find other bearing materials that could fulfill the same function so efficiently.

Plastic bearings are also used on the pintle bearings of ships' rudders. When ships are loaded with cargo, the pintle bearings are submerged. When the cargo ships are empty and riding high in the water, the upper bearing is often dry. Engineering plastic bearings cope with this situation extremely well.

7.9 END-OF-LIFE DISPOSAL

End-of-life disposal was covered at length from a practical point of view in Chapter 3. It is now necessary to revisit the methods by which a product can be sustainably disposed of from the designer's point of view.

Every product eventually comes to the end of its life. This may be due to a number of factors:

- Overtaken by technology
- Reduced popularity in the market
- Worn but still functions
- Worn beyond usefulness
- Functioning at less than full capacity
- Broken, beyond repair

When a product suffers any of these fates, it may be deemed that it is at the end of its life. It is the role of the designer to recognize these modes of failure and design into the product easy, low-cost, and low-energy methods of disposal. Sustainable thinking at the design stage will greatly reduce the embedded energy and cost in the disposal process.

The general approach to end-of-life disposal is the familiar 4R approach:

- Reduce
- Reuse
- Refurbish
- Recycle

These are almost universally accepted methods of disposal at the end of life of a product, but when they are cross-referenced to the end-of-life failure modes, the designer may see some clarity in design approach methods. A table cross-referencing failure modes to end-of-life disposal methods is shown in Figure 7.28.

7.9.1 Reduce

It is immediately obvious that the term reduce does not seem to have any significance. In the context of end-of-life solutions, reduce merely means to consign as little as possible to landfill. It also means that the remaining three options of reuse, refurbish, and recycle should be applied as much as possible so that in theory, 100% of a product is reapplied in some way, with nothing being just "thrown away." This may be an idealistic number, but through proper design the end of life of a product may come close. Vehicle designers and manufacturers, for instance, aim for 95% of their product being reapplied in some way.

	Reduce	Re-use	Refurbish	Recycle
Overtaken by technology		#	#	#
Reduced popularity		#	#	#
Worn but still functions		#	#	
Worn beyond usefulness		#		#
Low function capacity			#	#
Broken beyond repair				#

FIGURE 7.28

Failure Modes and Disposal Methods [5]

The application of the idea of reducing can be more usefully employed during the design stage by reducing mass. If the product requires energy for it to be moved, a reduction in mass would require less energy, which would have a great impact on the sustainable use value (SUV). For example, in the United Kingdom, fuel stations are a maximum of 25 miles (42 km) apart, yet most vehicles have a range of 300 to 400 miles (500 to 666 km) on a tank of fuel. It seems reasonable to reduce the size of fuel tanks so that fuel will not be carried unnecessarily.

Reduction at the design stage also means “do not overdesign.” Overdesign generally means more weight, more parts, and more energy required to process the material into a component. Careful analytical work can be employed here so that the only material applied to a product is that which is needed. The designer may also be tempted to reduce the design to such an extent that it loses its market personality. A computer would still function with all the components arrayed on an appropriate base, but it would not sell in the marketplace because the market would expect it to have an aesthetic casing.

The golden rule, therefore, is to reduce the design to what it needs to function as a product, but it may need much more than just mere function, in which case all the needs of the product must be considered, including those required to entice the market.

7.9.2 Reuse/Refurbish

The reuse of articles most appropriately applies to products that are still functioning but for many reasons might not be useful. Some products may be overtaken by technology, where advances have made the product obsolete. There are several excellent examples, described here.

In the early 20th century, sheet music was replaced by vinyl singles and long-playing (LP) records. These were subsequently replaced by eight-track and then cassette tapes, and as technology advanced, these were overtaken by compact discs (CDs). The CD format was in turn replaced by Internet downloads and digital music. Significantly, the pace of technological change has increased almost exponentially.

Photographic technology developed in the latter portion of the 19th century using glass plates and silver halide as a light-sensitive material. In the early part of the 20th century, the glass was replaced by celluloid film. In the second half of the 20th century, photographic slides were used and viewed with a slide projector similar to the Aldis projector shown in Plate 7.13.

In a short space of time since then, digital technology has made obsolete celluloid film cameras and their viewing equipment. Digital technology is also changing the way photographs are taken. Today almost every mobile phone is equipped with an excellent camera. Photographs can be converted into files and shared via the Internet or merely through the mobile phone service.

The dilemma for today’s designers is that of understanding where technology will advance so that they can build into their designs an element of “future proofing.” To some extent, this has been done with computing equipment and software, for which the technological advances have improved at an exceedingly fast pace. Some computer manufacturers have kept pace by offering modular construction, in particular toB2B clients that have skilled IT professionals to specify and upgrade units in-house using new modules. The challenge is now to change the habits of B2C customers, who tend to be more fashion driven and will often discard units well before the end of their useful working life.

During the design of new products, the designer should be aware that this product may come to the end of its life before its useful life has ended. Design for dismantling should therefore be a feature of the design process, enabling not only maintenance but also end-of-life disposal. Dismantled parts may be reused, refurbished, or, as a last resort, eventually recycled.

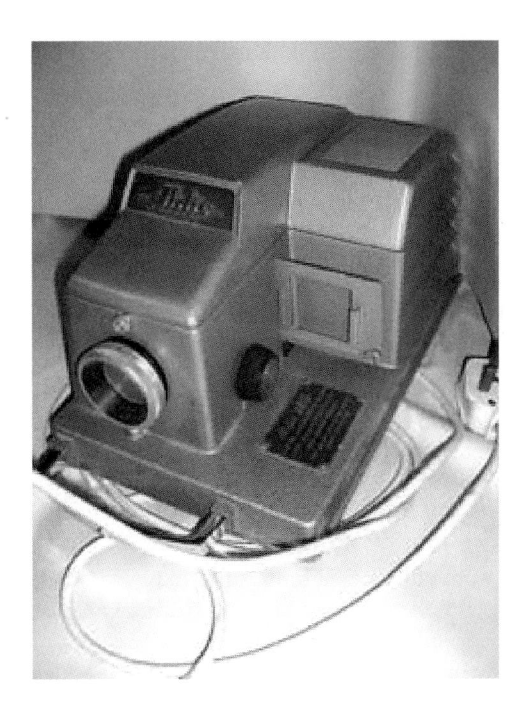

PLATE 7.13

Slide Projector, Circa 1955 [5]

Industrial products do not suffer so much from market fickleness. Industry relies on equipment functioning efficiently, but as the equipment becomes worn and less efficient, companies normally prefer to refurbish rather than replace, since replacement is more costly than refurbishment. A piece of refurbished industrial equipment can be returned to full use and 100% efficiency for a fraction of the cost of a new device.

When a piece of equipment is worn beyond usefulness or repair, the only option is to either reuse useful components or recycle the individual components. Within this scenario could be included vehicles that have crashed and are beyond repair or industrial components that have been left in the environment to rust and therefore cannot be usefully repaired.

7.9.3 Recycle

Eventually all components and products will end their useful lives and suffer dismantling and disposal. Designers' aim is to provide the wherewithal to separate and recycle all the different components and materials that originally made up the product.

Inevitably, components that are to be reused or recycled need to be separated into their unique materials. As materials are sorted for recycling purposes, the most difficult process is that of defining individual materials.

For example, glass cullet ready for remelting is often made up of multiple glass colors such as green, brown, blue, and clear. There are uses for colored glass, but often reused glass needs to be pure and clear. Glass recycling agencies run into difficulty in separating colors.

Furthermore, there are multiple types of plastic, such as polypropylene, polyethylene acrylic, polycarbonate, and polyurethane. Recyclers encounter great difficulty in separating the different plastics when they all look very similar and there is no usable detection method.

7.9.4 The designer's role

During the process of design for manufacturing, including ease of assembly, it is often the case that the same techniques can be used to disassemble a product that were used to assemble it. However, elements such as snap fasteners and adhesives can slow disassembly. As the design progresses through the concept into the detail design stage, the designer may consider dismantling techniques alongside the more standard assembly methods. Since the separation of materials is one of the most difficult tasks, distinct joints should be made in the product between different materials so that it becomes easy to separate the various components at those joints.

The process of separating individual materials is not likely to be careful dismemberment of the product, as one might expect with an assembly process. It is most likely that powered machines will roughly separate components. Joints should therefore be designed into the product that will function properly when the product is in use but that will easily separate under the right conditions at disassembly.

7.9.5 Case study: reuse of floor panels

Floor panels of 600 × 600 × 40 mm thick are manufactured from fiberboard encased in a thin galvanized steel skin. The floor panels are used for suspended flooring and are often faced in carpet or plywood, such as maple, then varnished. The refurbishment process involves separating the 10-mm-thick plywood from the floor panel, but the adhesive used between the floor panel and the plywood was a thermo-set and could not be melted or reduced by a solvent. The separation process involved designing and building a hydraulically driven knife blade that separated the plywood from the floor panel. A great deal of cost and energy were applied in developing and using the machine tool that separated the two materials. If at the floor-panel design stage thought had been applied as to the end of life dismantling method, many different adhesives could have been specified that would have been easier to separate.

It can be increasingly seen that the designer or design team has an enormous task in order to design a product that meets market requirements and considers the requirements of sustainability. Though many engineers consider sustainability an additional burden, it is clear that to reduce costs, many sustainability principles are already being applied. Some elements of the sustainable design process are perhaps not as obvious as others. One of these elements is designing for disposal. An easy disposal process designed into the component or product will return a low embedded energy, especially when components are recycled into raw, usable materials. It is therefore the designers' remit to reduce the embedded energy and return an improved sustainable disposal value (SDV).

References

[1] E. McLamb, Ecology Global Network, 2011. www.ecology.com.
[2] Alternative Energy Group, 2012. www.altenergy.com.
[3] Granta Design CES EduPack, Granta Design Ltd.: Rustat House, 62 Clifton Road, Cambridge, CB1 7EG, United Kingdom.
[4] Corbett, Dooner, Meleka, Pym, Design for Manufacture, Addison-Wesley, Pearson Education, Harlow UK, 1991.
[5] Author supplied.
[6] S. Lampkin, U.S. Air Force, 2012. www.defenseimagery.mil. (This image or file is a work of a U.S. Air Force Airman or employee, taken or made during the course of the person's official duties. As a work of the U.S. federal government, the image or file is in the public domain.).
[7] T.B. Christiensen, Modularized Eco-Innovation in the Auto Industry, J. Clean. Prod. 19 212–220.
[8] J.R. Barckhoff, K.M. Kerluke, D.L. Lynn, Certified Welding Supervisor Manual for Quality and Productivity Improvement, American Welding Society, Miami, 2005.
[9] R. Geisler III., Specials Certification Engineer and AWS Certified Welding Inspector, regis_geisler@lincolnelectric.com, The Lincoln Electric Company: www.lincolnelectric.com.
[10] Brown Europe Ltd: Gleaming Wood Drive, Unit 33-34, Lordswood, Chatham, Kent, ME5 8RZ, United Kingdom; mail@browneuropeltd.com.
[11] Freudenberg Simrit GmbH & Co. KG, D-69465 Weinheim, Germany.
[12] SZR Bearing Co.: szr-bearing.com.
[13] The Application of Lubricants, Shell International Petroleum Co., London, UK, 1965.
[14] M.F. Ashby, Materials Selection in Mechanical Design, Elsevier, Oxford, 1999.

Bibliography

[1] C.L. Dym, P. Little, Engineering Design, John Wiley & Sons, Hoboken, NJ, 2009.
[2] R. Sheard, The Stadium: Architecture for the New Global Culture, Periplus Editions, Hong Kong, 2005.
[3] M.F. Ashby, Materials Selection in Mechanical Design, Elsevier, Oxford, 1999.

GLOSSARY

Embodied energy. Additive energy required to create a product. Depending on the processes used in its design and manufacture, this embodied energy will consist of elements of artificial energy and natural energy. *Artificial energy* embodies energy derived from such sources as diesel, petrol, and gas. *Natural energy* is derived from sources such as wind power, solar power, and wave power.

Artificial energy. Notionally, the energy derived from fossil fuels used in the creation of a product.

Natural energy. Notionally, the energy derived from natural means such as hydroelectric, wind, or solar used in the creation of a product.

Carbon footprint. The value of carbon emissions released during the conversion of energy from fossil fuels.

Depth of section. In beam theory, the *depth of section* is the distance between the upper surface of the beam and the lower surface of the beam. An increase in depth of section generally improves the load-carrying capacity of the beam.

USEFUL ADDRESSES

FSC
Rainforest Alliance
259-269 Old Marylebone Road
London NW1 5RA, U.K.
Tel: +44 (0)207 170 4130

U.K. Media Inquiries
Stuart Singleton-White
+44 (0)771 040 3092
Washington, D.C. Office
Rainforest Alliance
2101 L Street, NW
Suite 800
Washington, D.C. 20037 U.S.A.
Tel: (202) 903-0720

CHAPTER

Drivers of Sustainability in Design: Legislation and Perceptions of Consumers and Buyers

8

A number of studies conducted in the first decade of the twentieth century concluded that current growth rates in both human population and resource consumption were unsustainable and that "unfettered" or natural capitalism would not curb that growth. It followed that legislation should be introduced nationally and internationally to address the issues identified. Indeed, the Stockholm Conference of 1972 proclaimed that "The protection and improvement of human environment is a major issue which affects the well-being of people and economic development throughout the world and it is the duty of all government and people to exert common effort for the preservation and improvement of human environment, for the benefit of all people and their posterity" [1].

8.1 LEGISLATION

Legislation to date has followed the "trends" of political pressure and popular science and has hitherto largely reflected the clear and present impacts on society of the extraction and use of natural, indeed, largely mineral resources. In the latter part of the twentieth century, legislation was focused first on pollution, then on fuel consumption and use, and finally on carbon emissions.

As the global economy began to grow, with a shift of the production of white goods and other domestic consumable goods together with some industrial machinery and equipment to cheaper Far Eastern producers, we also saw an increase in the use of nonrecyclable packaging materials and systems, designed to protect and preserve the product *en route* to market. At the same time, marketers were developing more sophisticated packaging to make their products more convenient, appealing, and attractive to the consumer. Under pressure from the environmental lobby, and led in the EU by Germany, some countries started to put pressure on the suppliers to take responsibility for the disposal of this packaging, in an effort to reduce the use of nonessential packaging and increase recycling of such packaging as was used.

Similarly, the developed world, driven by labor cost increases, has moved to a situation where

(a) domestic appliances have become uneconomic to repair and are therefore discarded after a relatively short working life;
(b) mobile technologies such as mobile phones and I-players are increasingly seen as a fashion item and discarded well before the end of their useful life;
(c) governments in the developed world put pressure on the automotive and other mobile equipment industries to discard old, less efficient machinery rather than repair it, as the older machinery is seen to be more "polluting."

Sustainability in Engineering Design. http://dx.doi.org/10.1016/B978-0-08-099369-0.00008-X

Thus, the reaction to the issue of carbon and other emissions and to the shift from disposal to recycling has been led by legislation on the one hand and by the adoption of positive reinforcement programs on the other: leading brands are demanding a "greener" appearance, and the richer market segments in both developed and developing world are often prepared to pay a premium for what are perceived to be greener products.

Although the broader arguments can be said to have been won, legislation has not been consistent across the globe. The subsections later in this chapter are intended to give a brief examination of the overall thrust of environmental and sustainability legislation in some of the world's key economies.

8.1.1 The United States

In the United States, the role of environmental protection falls to the EPA, which produces regulations based on the laws written by congress and presidential orders (EOs), issued by the president's office.

Thus, the basic laws include sweeping legislation such as the clean air and clean water, and environmental policy acts and the EOs include EO13211 Actions Concerning Regulations That Significantly Affect Energy Supply Distribution or Use.

In addition to the specific regulations on calculating and reporting fuel economy and exhaust gas emissions including the "gas guzzler tax" for vehicles, there have been a number of regulations passed on stationary power sources, including requirements for existing sources for pollutants to use best practice and prevent significant deterioration in their operational effects. More recent legislation has focused on water pollution and water resources development, including Underground Injection Control for Geologic Sequestration (or CCS)—the issue of using underground facilities (e.g., oil and gas wells) to store captured carbon as a carbon offsetting measure [2].

In the built environment, the emphasis is on energy use reduction, and the US department of energy has issued guidelines on such issues as insulation, and local or state building codes often exceed those basic requirements in terms of *R* values or insulation thickness [3]. Following a number of studies and surveys, including US Homebuilder survey in 2006 [4], observers have concluded that the demand for "greener" homes has increased, not the least because of potential cost savings. There has been some adoption of the UK's BREEAM and LEED certification systems via US Green Build Council, and corporate buyers are apparently prepared to pay up to 4% more for a LEED-certified building [5].

We will examine the BREEAM and LEED certification system and principles later in this section to draw any lessons that can be learned in applying these principles from the built environment into the mechanical engineering design environment.

Academics and engineering bodies such as ASME are lobbying hard to persuade industry in general of the economics of sustainable design [6] and to indicate that legislation in the EU and elsewhere is ahead of the United States in this field. Some work has been published in the United States on the use of whole life cycle analysis in engineering design and the benefit to the triple bottom line [7], and the concepts have been adopted by forward-thinking manufacturers.

There is, however, relatively little legislation affecting design and material choices, as focus has been on the effects of the production and use of products rather than the embedded issues engendered by design practice.

8.1.2 Canada

Environment Canada uses a number of acts, regulations, and agreements to fulfill its mandate to preserve and enhance the quality of Canada's natural environment. The key instruments are the regulations covering issues such as emissions and waste disposal on land and at sea and the use of toxic substances in manufacture and processing.

In addition, guidelines and codes are issued for environmentally sound practices and objectives for desirable levels of environmental quality. These provide a scientific basis for the development of environmental quality objectives and for performance measures for strategic options and risk management initiatives [8].

These guidelines cover emissions, waste, environmental quality, etc., but, as with the United States, do not yet cover the concept of sustainability by design.

8.1.3 Europe

As with other states and groups of states, the EU led first on pollution issues including SO_2, NO_2, and similar pollutants (1999/30/EC) and then considered waste issues (including WEE directives 2002/95/EC and 2002/96/EC, now revised as 2012/19/EU). This directive was specifically aimed at restricting the waste in electronic equipment, often exacerbated by the short life cycle and rapid obsolescence of this equipment. It also had the effect of driving a reduction in the use of hazardous substances in electrical and electronic equipment, promoting recycling and reuse of equipment and reducing the use of heavy-metal elements. The broader policy was aimed at reducing waste going to landfill and shifted the emphasis from waste as a burden to waste as a resource. Targets across domestic and industrial waste are to recycle 50% of municipal waste and 70% of building waste by 2020 [9].

More recently, the *EU* has recognized that "Over 80% of all product-related environmental impacts are determined during the design phase of a product" and in response created a directive (2009/125/EC), which is to form the framework for the setting of ecodesign requirements for energy-related products. Under this directive, it is argued that the production, distribution, use, and end-of-life management of energy-related products (ErPs) are associated with a considerable number of important impacts on the environment, namely, the consequences of energy consumption, consumption of other materials/resources, waste generation, and release of hazardous substances to the environment. It uses the directive to define conditions for setting requirements for key properties and characteristics of products and will be followed by "implementing measures," which will act *in lieu* of firm legislation [10]. Compliance with this directive is nonetheless required for CE marking.

The *Ecodesign Directive* foresees two types of mandatory product requirements:

A Specific requirements, which
- set limit values, such as maximum energy consumption and minimum quantities of recycled material.

B Generic requirements, which
- do not set limit values;
- may require, for example, that a product is "energy-efficient" or "recyclable" (compliance with the relevant harmonized European standard gives presumption of conformity with the requirement);

- may entail information requirements, such as material provided by the manufacturer about best practices to use and maintain the product in order to minimize its environmental impact;
- may require that the manufacturer perform a life cycle analysis of the product in order to identify alternative design options and solutions for improvement.

The regulations are to be phased in based on specific groups of products: the ecodesign committee has identified a list of "high"-energy-consuming machinery and equipment that will form the first focus of design improvement regulations.

The following is an indicative list of product groups covered by this working plan:

- Air-conditioning and ventilation systems;
- Electric and fossil-fueled heating equipment;
- Food-preparing equipment;
- Industrial and laboratory furnaces and ovens;
- Machine tools;
- Network, data processing, and data storing equipment;
- Refrigerating and freezing equipment;
- Sound and imaging equipment;
- Transformers;
- Water-using equipment [11].

The EU has also been addressing the issue of waste reduction by considering the waste hierarchy. Directive 20008/98/EC sets out five steps for dealing with waste, ranked according to environmental impact—the "waste hierarchy" [12].

Prevention, which offers the best outcomes for the environment, is at the top of the priority order, followed by preparing for reuse, recycling, other recovery, and disposal, in descending order of environmental preference (Table 8.1).

In addition to considering emissions and waste, EU directives have also reflected the EU's increasing attention on energy use and efficiency, covering efficiency of boilers, fluorescent lighting, household appliances, etc., and of course efficiency and emissions of passenger vehicles and construction equipment. In addition, there have been directives aimed at promoting sustainable and renewable energy generation.

Table 8.1 Waste Hierarchy [12]

Stages	Include
Prevention	Using less material in design and manufacture. Keeping products for longer reuse. Using less hazardous materials
Preparing for reuse	Checking, cleaning, repairing, refurbishing whole items or spare parts
Recycling	Turning waste into a new substance or product. Includes composting if it meets quality protocols
Other recovery	Includes anaerobic digestion, incineration with energy recovery, gasification, and pyrolysis that produce energy (fuels, heat, and power) and materials from waste; some backfilling
Disposal	Landfill and incineration without energy recovery

The EU directives are adopted by the EU member states and enshrined in their local laws. The room for maneuver to transpose directives, especially the so-called framework directives into national law, is often substantial and can lead to large interpretation differences and subsequently in different environmental ambition levels.

It pays, therefore, to look at two key examples of legislation within EU countries.

8.1.4 The United Kingdom

The UK legislation is based on EU legislation and directives, and the Environment Act 1995 covers air, land, and water pollution and empowers the Environment Agency to intervene to ensure quality of air, water, or land. It also outlines a national waste strategy and defines responsibility for waste disposal between producer and consumer. The Waste and Emissions Trading Act 2003 was modified by the Waste (England and Wales) (Amendment) Regulations 2012, which came into force on 1 October 2012. The amended regulations relate to the separate collection of waste. From that date, it also imposes a duty on waste collection authorities, when making arrangements for the collection of such waste, to ensure that those arrangements are by way of separate collection. These duties apply where separate collection is "necessary" to ensure that waste undergoes recovery operations in accordance with the directive and to facilitate or improve recovery and where it is "technically, environmentally and economically practicable" [13].

DEFRA has also recently consulted on the waste hierarchy. The resulting report will amend the current guidance, which is based on life cycle assessment, by taking into account ecological footprinting and other evidence-based views, including

- high- and low-efficiency energy recovery;
- open-loop recycling, for example, of glass and plastics;
- plastics energy recovery versus landfill; and
- paper energy recovery versus composting.

The waste hierarchy has been transposed into UK law through the Waste (England and Wales) Regulations 208. The regulations came into force on 29 March 2008. The provisions relating to the hierarchy (set out in Regulations 12, 15, and 35) came into force on 28 September 2008. The regulations place a responsibility on businesses that produce or handle waste (including local authorities on behalf of the public consumer) to take all reasonable measures initially to prevent waste and then to reuse, prepare for reuse, recycle, recover, or dispose waste. The definitions of these categories are based on EU directive 2008/98/EC. Schedule I of the regulations sets out a nonexhaustive list of disposal operations [14] and includes the policy interpretation of the waste hierarchy.

8.1.5 Germany

According to the EU's Europe Web site, Germany has no specific legislative act covering sustainability, "but rather a sustainability strategy."

This strategy is the responsibility of the federal government and contains 21 topics mapped to 35 goals [15].

The strategy is supported on four key pillars:

- Responsibility to future generations
- Quality of life
- Social cohesion
- International responsibility

The key priorities are as outlined in the succeeding text:

1. Sustainable business and commerce—today's business taking responsibility for tomorrow at an individual company level. Correctly formulated, such policies should lead to a competitive advantage.
2. Climate and energy—the key driver is the 50% reduction in 1990 level greenhouse emissions by 2050, to mitigate the predicted global temperature rise of 2C.
3. Sustainable water—access to clean water is a key basis of life on earth. The standard of German groundwater has improved markedly over the past years due to strict regulation of and high investment in wastewater treatment.

One of the key implementation rules is that the producer is responsible for environmental impacts of its processes and products. This includes issues relating to material selection and use. The remaining rules cover issues such as the (restricted) use of nonrenewable materials and energy sources, the need to build an inclusive society, and the need for public organizations to embrace sustainability in all its aspects [16].

The resultant legislation has included energy legislation, such as the Combined Heat and Power Act (KWKG) and Energy Conservation Act (EnEg) in 2008. Much of the legislative effort has been in the form of support and empowerment, rather than punitive. Thus, schemes such as *Bürger initiieren Nachhaltigkeit* ("Citizens Initiate Sustainability") were launched to ensure the involvement of the population as a whole.

The Germans were early leader in the use and recycling of packaging and the introduction of the green dot (*grüner Punkt*) label. The legislation (*Verpackungsordnung*) is aimed squarely at the producer. In a B2C (business to consumer) transaction, the business (producer) has an obligation to participate in one of the nine recognized recycling schemes. The packaging in this case includes anything that is "unpacked by the consumer," including transport packaging. The legislation applies to all B2C suppliers into the German market, regardless of country of origin.

In B2B sales, there is no such obligation, although companies are often obliged to account for their total packaging material usage and shipping pallet accounts are generally held between logistics companies and their clients, where the pallet has a fairly high monetary value, to encourage both the adoption of the standard "Euro"-pallet (1200 × 1000 mm 4-way entry) and their repair and reuse rather than wastage.

The result of this general approach to packaging has been a perception by the public and the producer that unnecessary packaging is to be avoided and the use of nonrecyclable packaging materials has declined sharply [17].

The German government also reports regularly on the effectiveness of their legislative approach. For instance, from the 2012 report by the Bundesamt für Statistik,

- indications are that although overall usage is down due to the world downturn, energy efficiency savings at just under 30% are not quite on target;
- in resource usage efficiency, however, they are much closer to target at 83% of 1990 rates of use;
- similarly, greenhouse gas emissions are reducing in line with target at just under 75% of 1990 levels.

It is clear from the internal reports that financial issues have been dominating the legislative mind over the last few years: the rate of introduction of specifically "green" measures has declined during that time [16] (Figure 8.1).

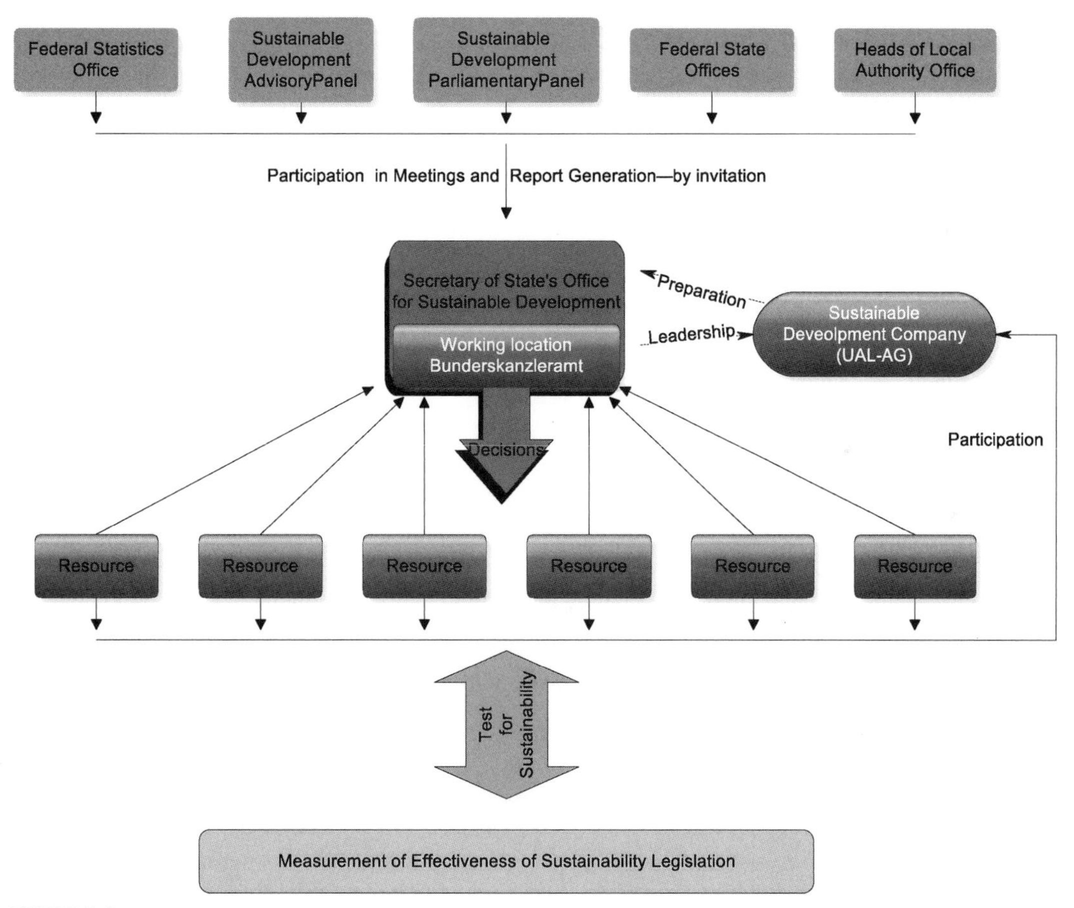

FIGURE 8.1

Implementation of German Environmental Legislation [18]. (For color version of this figure, the reader is referred to the online version of this chapter.)

8.1.6 BRICS Nations

8.1.6.1 Brazil

Despite Brazil's emergence as a global economic power, its massive environmental resources, and its environmentally conscious people, many Brazilian companies and politicians have still not fully taken onboard sustainable development as a strategic priority. The ongoing debate about a forestry law that could encourage deforestation has exposed deep divisions in Brazilian society. To date, there have been less emphasis on issues of energy efficiency and recycling and more on the protection of the natural environment. Recycling rates are only at around 26%, but about 50% of Brazil's energy supplies are from renewable resources (hydro and biomass).

Since 1981, Brazilian legislation has focused on environmental impact of industrial activity. There is no longer any environmental damage that is released from proper remediation: strictly speaking, pollutant emissions are no longer tolerated. The bottom line of this new legislation is that even a polluting waste tolerated under the established standards may cause environmental damage, thus making the polluter liable for redress. This entails the concept of strict liability, by which an industry undertakes all risks inherent to its activity: damages are no longer to be shouldered jointly with the community [19].

8.1.6.2 Russia

Russia does not currently have a well-developed system of environmental regulations and standards, and most notably, it lacks a procedural framework. Therefore, Russian environmental specialists, despite their deep professional knowledge, are not fully experienced with implementing international regulations for environmental assessments and environmental management. Without developing new methods adapted to national and regional characteristics, specifically addressing regional environmental issues on the one hand and a detailed approach incorporating international environmental procedures on the other, the sustainable development of the Russian economy appears difficult, if not impossible [20].

According to Russian official sources, the environmental situation typical of densely populated urban and industrial areas (constituting 10-15% of the country's territory) is "alarming." Examples are as follows:

- Systems of industrial and domestic waste handling are inadequate in Moscow and other cities. 1.8 billion tons of toxic waste has accumulated and this increases annually by 108 million tons.
- The potential for energy savings in Russia is vast. For example, the potential for energy efficiency gains is estimated at 400 mtoe (compared with annual natural gas production of 490 mtoe).
- Industrial and vehicle emissions cause severe air pollution in many cities.

The current position is one of the unclear legislations applied inconsistently across the Russian federation. As the Russian economy grows, we expect new and tighter legislation on emissions and waste disposal [21].

Pressures on the state budget have, however, led to constraints on funding for environmental policy, institutions, and enforcement. Sovereign guarantees for environmental loans have been unavailable for several years. The rate of public and private investment in environmental projects is critically low. Massive technological and management modernization is needed—including in environmental technology, much of which will need to come from abroad [22].

8.1.6.3 India

Article 48-A of the constitution lays down the fundamental approach to environmental protection, namely, "The State shall endeavour to protect and improve the environment and safeguard the forests and wildlife of the country." The constitution then goes on under article 51-G to indicate that all citizens are responsible for the protection of the environment. The 1974 42nd amendment of the constitution makes it the state's responsibility to actively protect and improve the environment [23].

The Union and State *lists*, under which Union and States, are empowered to make laws and both encompass the same subjects (industry, mines and minerals, and rivers and fisheries), with land and colonization an additional subject under the State list. The *common list* under which both may legislate includes wildlife and forest protection and population and social planning [24].

Consequently, laws have been enacted on some of the subjects such as the following:

- The Factories Act 1948
- The Insecticides Act 1968
- The Water (Prevention and Control of Pollution) Act 1974
- The Air (Prevention and Control of Pollution) Act 1981
- The Forest (Conservation) Act 1980
- The Wildlife (Protection) Act 1986
- The Environment (Protection) Act 1986

As with many other developed and developing economies, to date, there has been no specific legislation enacted dealing with sustainability in product design and manufacture. The emphasis is on the protection of the environment in terms of pollution and emission control and public safety.

Legislation alone does not guarantee sustainable development. The situation is made more difficult in India as enforcement and implementation are not straightforward. Rules can work as a driver toward conservation: for instance, those that are inbuilt in the permission system force industries to take environmental issues into consideration [25].

Also, India is a signatory to many of the international treaties bearing directly on environmental protection and sustainable development and takes an active role in prosecuting polluters.

8.1.6.4 China

The Chinese government recognizes that the current patterns of economic growth are not environmentally sustainable. As a result, they built a comprehensive regulatory and institutional framework for environmental management in the late 1990s. The set of five-year plans (FYPs) contains a number of quantitative, time-bound targets and is linked to project programming designed to meet these targets. The environmental regulatory framework expanded particularly rapidly between 2000 and 2004 when a number of new environment laws were enacted, such as the law on environmental impact assessment (EIA) and the law on cleaner production. In this period, a number of other legal acts were also amended, including the Environmental Protection Law (EPL) as well as the Air, Water, and Waste Prevention and Control Laws. The legal system is supported by a number of specific regulatory instruments for industrial pollution control, including environmental quality and emission/discharge standards, the discharge permit system (DPS), "three synchronizations" ("3S"), and EIA.

The national environmental agency has been strengthened since the turn of the century, leading to the creation of a State Environmental Protection Administration (SEPA) placed directly under the State Council.

As with India, sustainability in the built environment is beginning to gain ground as an issue, but there is still no impulse to drive sustainability in engineered products, except conformity with the legislation of the target markets for any exported product [26].

However, since 2007, there has been a joint initiative with the Environmental Protection Agency of the United States to strengthen the application of the law in China. The resultant pressures have improved compliance measures and allowed an information and best practice exchange and led to a joint US-Chinese statement on climate change in April 2013 [27].

It has also helped stir the debate in 2010 into the "unacceptable" pollution levels from Chinese industry, leading to operational activity such as China's *Green Fence* initiative, designed to halt dangerous goods scrap dumping in China by applying the existing rules under Article 12 more rigorously. Article 12 states "In the process of importing solid waste, measures shall be taken to prevent it from spread[ing] seepage and leakage or other measures to prevent pollution of [the] environment" [28].

8.1.7 Japan

Japan's approach to environmental legislation is to control pollution levels in

- air;
- water;
- soil and land, including ground subsidence;
- noise and vibration.

The Basic Environment Plan is based on the basic plan for environmental conservation Article 15, the Basic Environment Law, and was decided by the cabinet while consulting the Central Environment Council. It was approved on December 1994. It declared the basic idea for environment policy based on the principles of the Basic Environment Law and four long-term objectives of cycle, harmonious coexistence, participation, and international activities and looking to the mid-twenty-first century. It also indicated the direction for measures to be taken for the early twenty-first century, to be developed comprehensively and systematically [29].

The introduction of the basic laws in 1993 was backed by sizeable public support, and implementation of the basic plan has helped Japan tread a careful balance between economic growth and the maintenance of the environment. Unlike in Europe and the United States, the Japanese government has traditionally taken on the role of guardian of the environment, and lobby groups and NGO have enjoyed scant support. The law requires EIA for any major project, and local ordinances can often require EIA even for smaller projects. Waste management is also a key issue, and regulations are strict.

Following the early days of cleaning up ground and water pollution, there was a focus on regulating vehicle emissions and noise, and this has helped drive the growth in research into hybrid and electric vehicles.

Since the disaster at Fukushima, the government has been wrestling with balancing the issue of increased safety standards and increased use of fossil fuels.

There is, as yet, no legislation or guidance on design impacts for sustainability.

Nonetheless, the 2010 OECD review of Japanese policy remained broadly positive in terms of the effect of the legislation, but recommended that such legislation be updated, and also recommended the increased use of financial and commercial instruments to reduce emissions [29].

8.2 EFFECTIVENESS OF INTERNATIONAL ENVIRONMENTAL REGIMES AND LEGISLATION

Measurement of efficiency or effectiveness of environmental legislation requires either empirical observation of alternative regimes for comparison or calculations based on theoretical models that do not abstract away determinants of efficiency that are critical in practice. According to OR Young [30], the former are generally absent, while the latter too restrictive to generate reliable guides. Indeed, there are still issues regarding the effectiveness of environmental impact assessment work in terms of

- the quality and veracity of data,
- the inability to compare with "control" or nonintervention situations,
- the lack of follow-up on EIAs.

All of which tends to lead to a "fit-and-forget" approach to expert interventions for specific projects [31].

Even the European Environment Agency (EEA) declared in 2001 that although there was no shortage of reporting obligations imposed on member state governments, there are often complaints about "reporting fatigue" from the more than 100 pieces of environmental legislation in force, and the majority of European Union measures require greater effort to demonstrate their effects on the environment [32]. "Evaluation should not be an afterthought," said the report, outlining a number of measures, including explicit quantified objectives combined with timetables for their achievement:

- A preference for "evaluation-friendly" instruments over less quantifiable policies, for example, economic instruments rather than information and awareness campaigns
- A requirement for baseline monitoring prior to the implementation of measures, facilitating "before and after" comparisons
- Greater provision for pilot projects in order to facilitate fine-tuning

The UK DEFRA commissioned a report on enforcement of environmental legislation in the United Kingdom in 2006 and concluded that 70% of regulatory breaches were dealt with by the issuance of cautions and warning letters as a means of obtaining compliance, followed where required by enforcement notices. It is difficult to assess whether fines imposed are proportionate to the seriousness of the breach or effective as a deterrent. Similar views were expressed by Emma Bethell in the Plymouth law review of 2009, who also indicated that prosecutions had been made under the "polluter pays" principle, and reflected that the rules of strict liability allow for a relatively straightforward approach to prosecution. The downside, however, was that there was little differentiation under law between negligent and unintentional pollution [33,34].

According to the DEFRA 208 report, the push for higher recycling of domestic waste has resulted in over 40% recycling in 2010/208 compared to 8% a decade earlier. The improvement in commercial waste recycling has been less dramatic: from 42% to 52% over the same period. Of the 8.3 million tons of waste produced in the United Kingdom annually, currently, a vanishingly small proportion is used to generate power, whereas the potential is of the order of 3-5 TW h. Around 1 TW h is generated from anaerobic digestion of sewage sludge; however, more could be done by diverting waste from landfill to AD. Landfill rates at 50% are still higher than the EU average of 42%. Direct emissions from the waste management greenhouse gas inventory sector in the United Kingdom accounted for

3.2% of the UK's total estimated emissions of greenhouse gases in 2009 or 17.9 million tons CO_2 compared to 59 million tons CO_2 in 1990. Of the 2008 total, 89% arises from landfill, 10% from wastewater handling, and 2% from waste incineration [35].

The Chinese government has identified inadequate enforcement as one of the key factors in China's deteriorating environmental situation. The 9th, 10th, and 8th FYPs emphasized the need to strengthen environmental enforcement and compliance assurance. In this context, a number of enforcement activities have been carried out, closing down or penalizing most polluting industries. The legal system has been built up to fight against noncompliant firms, and the courts have been brought in to support the prosecution of environmental offenders. The public has been gradually engaged in noncompliance detection and compliance promotion [26].

In general, then, the effectiveness of legislation, local or international, is a subject much studied by social scientists and lobby groups but little understood by the lawyers and lawmakers [27].

Such studies, as are conducted, are often

- limited in scope to one effect of one piece of legislation (e.g., CO_2 emissions as a result of legislation on vehicle exhaust emissions);
- conducted by public lobby groups to demonstrate a lack of progress in their specific area or to gain a specific and limited end;
- undertaken and reported by social scientists, whose papers and symposia are not necessarily followed by the legal and legislative professionals;
- undertaken by governmental bodies to demonstrate the success of their apparent or real adherence to international standards and laws;
- undertaken by industry-based specialists and lobby groups to militate against perceived adverse economic effects of following specific regulations and legislation;
- interpreted by specialists in legal or social fields rather than by those schooled in the science under consideration.

And cherry-picking results from specific surveys and studies are a potential hazard resulting from the plethora of such publications.

It is clear that the effectiveness of international standards and laws on environmental protection in any given country is also determined to some extent by the stage of development that country has reached. Empirical evidence is clear that there is a certain turning point in economic development, where increased economic welfare goes hand in hand with environmental improvements [36]. This raises important questions in relation to the level of development of the BRICs countries, whose potential to pollute far outweighs that of the currently developed world, due to demand-side pressures coupled with the fact that they may not have reached the critical point mentioned earlier.

In parallel with this argument runs the "race-to-the-bottom" argument, where heavy polluters choose deliberately to relocate their "worst" activities in those countries and states where economic development and technical development are not so advanced, and there is therefore less attention paid to the results of heavily polluting production and extraction processes. For example, Kolstad and Xing argued that the laxity of environmental regulations in a host country can represent a significant element in foreign direct investment decisions from the US chemical industry. The more lax a country's regulations are, the more likely the country is to attract foreign investment [37].

The conclusion we may draw from the wealth of literature available on the subject is that the trend toward the adoption of a positive and preventative approach to sustainable production and use is driven by a combination of the following:

- Reactions to existing legislation both at home and in other "target" markets
- Anticipation of further impending legislation or regulation
- Reaction to public perception and media pressure
- The desire to be seen as a leading exponent of environmentally sustainable production
- The extent to which sustainability can be built into a "total" product or service without making a significant impact on profitability

Thus, the next logical step is to consider these nonlegislative pressures on producers of goods and services and examine at a high level some of the measurement and guidance tools.

8.3 NONLEGISLATIVE MEASUREMENT AND GUIDANCE TOOLS

Much of the regulation and legislation since the 1960s have involved determining compliance levels for pollution emissions. However, other environmental management tools have also been developed. These include environmental auditing, environmental accounting, environmental reporting, life cycle assessment, environmental management systems, risk assessment, and environmental impact assessment.

Most of these tools have been developed by research and educational institutions and are now used commercially to assist in the measurement by commercial enterprises of their environmental impact or that of their products. Some are focused on carbon fuel consumption or carbon emissions' equivalence, and others take a broader view of other resource depletions or a qualitative view of environmental impact assessment.

Current focus is largely on resource consumption or energy-based and carbon-based models, including EIA, Eco-Rucksack, and carbon footprint.

Some other systems, such as EAI and IDEA abacus, are more qualitative in nature and are intended to guide the design process and thinking, rather than to produce measurable outcomes on which to base material selection manufacturing process or other key design decisions.

The growth industry in the development of "eco" tools in the engineering sector has been the development of software applications that can measure or indicate the relative merits of materials and manufacturing methods. Allied to a database of such materials, to which carbon, toxicity, resource depletion, or other values have been assigned, and to a second database of production methods, to which typical carbon emission data, etc., have also been assigned, the resulting algorithm can give an indication of the relative merits of alternative components and design approaches when designing quite complex systems. One such algorithm has been developed by Granta at Cambridge University in the UK, and is described in 8.3.5.

There are a number of detailed reviews and critiques of the currently available models. A recurrent criticism is that the model needs to be simple in application in order to allow ease of use. At the same time, it must be robust enough to encompass data and information from many sources and produce a "typical" value for the component or item under consideration while being flexible enough to differentiate between embedded energy obtained in a sustainable way and that of nonsustainable energy. We also find that some of the models lean toward a simplistic calculation that does not take account of all stages of the total life of the product, from sourcing to reuse and disposal.

A further detailed review and comment on efficacy of the leading models, databases, and algorithms will form the basis of ongoing research and a future publication from the authors of this book and will lead to further research and the development of our own sustainability model, the SLV (sustainable life value) that is outlined in concept elsewhere in this book.

For the present, we will focus on the early models, such as EIA (environmental impact assessment), Eco-Rucksack, and carbon footprinting. We will then give a brief outline of outline some of the current LCA programs.

8.3.1 EIA: Environmental Impact Assessment

Environmental impact assessment (EIA) was introduced under the US Environmental Policy Act in 1969 to identify the possible impacts of a proposed construction project on the environment. A similar approach has been adopted by most developed and many developing countries, often embedded into their legislation program.

It is intended to provide decision makers with an understanding of the probable environmental consequences of a proposed project and thereby facilitate the making of more environmentally sound decisions [38].

The standard steps in a typical EIA process are shown in Figure 8.2.

Since EIA is now widely used, there has been some debate as to its effectiveness as a mitigation tool. It is certainly seen as cost-effective on large projects, typically at less than 1% of total costs [38]. Initial concerns are centered on the fact that the institutional framework preceded any scientific basis for establishing EIA, leading to potential inaccuracies and poor decision. Overall, while EIA was seen to be an effective consultation tool, there was some concern about a "fit-and-forget" approach, focusing on the decision-making process and largely ignoring follow-up and the measurement of actual impacts; indeed, research conducted by C. Harmer concluded that greater and better use of follow-up is required to improve effectiveness [39]. The other main drawback to EIA is that it is generally impossible to make true comparison to either the status quo or an alternative strategy, although this may improve as predictive models are tested more robustly and improved in accuracy.

The built environment has been a key focus of carbon emission reductions along with (clearly) power generation and transport systems for goods, products and people.

What we have not seen is a serious review of the impact of broader engineering and product design on sustainability.

8.3.2 Eco-Rucksack

Eco-Rucksack is a tool based on the concept of NRE (natural resource efficiency), which seeks to decouple energy and material consumption in making a product from the economic performance of the product. The effect of NRE-based decisions has been to prolong in some cases the resource bank of finite resources and to slow consumption rates. Newer technologies, for example, tend to use resources more efficiently. However, the increasing demand created by and for these new technologies may result in the sheer volume of consumption outstripping any savings made in energy or resource terms, and this has informed much of the policymaking at national and international level. The finite

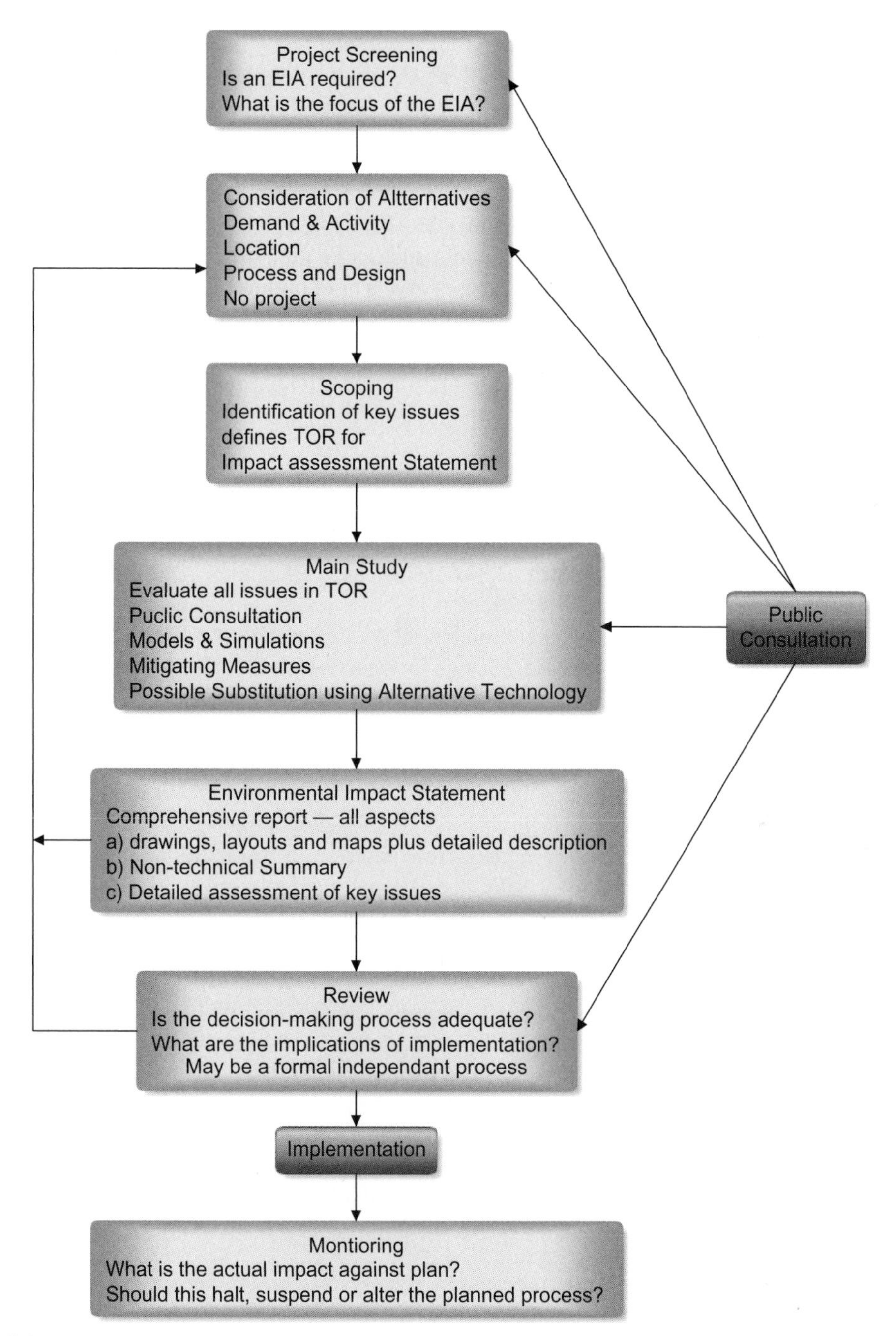

FIGURE 8.2

Flow Diagram for an Environmental Impact Assessment. (For color version of this figure, the reader is referred to the online version of this chapter.)

levels of natural resources were addressed as long ago as already in 1972 at the UNCHE, albeit in broad terms. It was not until 2002 that the World Summit on Sustainable development (WSSD) sought to obtain general acceptance that current consumption patterns were unsustainable and that we should formally delink growth patterns from environmental degradation. However, it was also adopted without concrete targets, both from the lack of international will to agree to such targets and from a lack of usable scale of measure for any targets.

The Eco-Rucksack was then created as a measurement concept to address this lack. It seeks to measure the amount of materials used in relation to a products' entire life cycle, making some provision for issues of recycling and for "hidden flow," that is, the indirect consumption associated with manufacture or production. A more accurate, less catchy term for this measurement is material input per unit of service. This is computed using a material flow analysis to give the total mass of material directly and indirectly used in making and using the product or item divided by its own mass. The inverse of this figure is the Eco-Rucksack ratio.

This can be a useful comparator when selecting, say, raw materials but must be used in conjunction with the other properties of those materials such as density, tensile strength, and fatigue resistance. It may well be that a material with a lower Eco-Rucksack ratio (RS factor) may be inferior in other ways. Nor can the RS factor take into account the importance of a product or service to human sustenance (using, for example, Maslow's hierarchy of human needs), which again may disallow judgments and decisions made simply on the basis of RS factor.

The RS factor will increase when one component of the materials used in its production process becomes less easily extractable or accessible, and this factor will decrease with enhanced material recycling. Sometimes, a simplified methodology is used without full life cycle approach, that is, for the assessment of the hidden material flows only for the production of a commodity without taking into account the extra materials (and energies) used for the extraction or disposal.

The term *virtual material* (VM) is also used in specific context, and it means the total amount of a particular material or substance used directly or indirectly for producing a commodity. The most frequent application of the latter concept can be met in relation to water resources, especially for the demonstration of virtual water transfer coupled with international trade of food, etc. It also applies to other "hidden" resource use such as overburden removed to allow access to mineral deposits or residuals from harvest in agriculture.

There are international aspects of this problem as well. When new technologies are installed and the older, less efficient (but still usable) technologies are transferred to less affluent partners, it may lead to an international rebound effect, where larger volumes of cheaper product are produced with greater ecological inefficiency and consequent adverse environmental impacts of the "country of origin." Thus, the make-or-buy decisions to be sustainable must be decoupled from simple economics. This may also apply to old vehicles and equipment exported from the developed world due to fiscal and other pressures and legislation. Such technologies or products can contribute to well-being, relatively higher standards of living, or even poverty reduction of the recipient community (e.g., in terms of energy poverty), but in a global sense, such a transfer will increase the overall ecopressure. At least, this secondary RB effect should also be taken into account together with relative RS-factor gain achieved in the country of origin [40].

Many examples could be mentioned for the Eco-Rucksack, including any form of mining (e.g., digging large rock tunnels in order to mine coal, where the rest of rocks do not directly enter the production process and the product itself), bottled or canned mineral water, timber or other furniture, rare earth, and modern infocommunication technologies, to name a few obvious examples. One of the more frequently quoted cases is that of platinum and the catalytic converter: "In order to extract one gram of platinum from a platinum mine, for example, we must displace and modify 300,000 grams of rock. Without platinum we would not have the catalytic converter in our automobiles. Two to three grams of platinum are found in one such catalytic converter, in addition to high-quality steels, ceramics and other materials. Thus, the ecological rucksack of the catalytic converter, i.e. the total amount of material translocated for the purpose of constructing it, amounts to about one metric ton of environment. This means in effect that the catalytic converter burdens the automobile with as much matter as the car itself weighs" [41].

8.3.3 Carbon footprint

Carbon footprint is a measure of the impact that our activities (by a person, organization, event, or product) have on the environment and in particular climate change. By creating a relative index, consumers can make an informed choice as to what products and services they may buy and consume, mindful of the environmental impact of the choice. A carbon footprint considers all six of the Kyoto Protocol greenhouse gases: carbon dioxide (CO_2), methane (CH_4), nitrous oxide (N_2O), hydrofluorocarbons (HFCs), perfluorocarbons (PFCs), and sulfur hexafluoride (SF_6).

The main types of carbon footprint for organizations are the following:

- Organizational carbon footprint—emissions from all the activities across an organization, including buildings' energy use, industrial processes, and company vehicles.
- Value chain carbon footprint—includes emissions that are outside an organization's own operations (also known as Scope 3 emissions). This represents emissions from both suppliers and consumers, including all use and end-of-life emissions.
- Product carbon footprint—emissions over the whole life of a product or service, from the extraction of raw materials and manufacturing through its use and final reuse, recycling, or disposal.

Life cycle assessment provides the existing and internationally agreed basis for calculating carbon footprints of goods and services in a robust way.

Data are generally collected throughout the entire life cycle at a consistent level of detail as in an ISO LCA, although not all emissions, resources consumed, and impact categories are evaluated. This limitation in scope raises the prospect of burden shifting—solving one problem while creating another. This can unfairly promote products that do not necessarily have a better overall environmental performance or environmental footprint.

In recent years, the carbon footprint has gained recognition as an indicator of the contribution of goods and services to climate change. It is often based on a life cycle approach but focuses only on the emissions linked to a product that contribute to climate change [42].

8.3.3.1 Carbon footprint initiatives

Retailers are becoming increasingly aware of the opportunities to improve the environmental performance of products and influence purchasing decisions. There are currently various activities to capture and record life cycle data on a range of household products and pass this information on to customers. This offers consumers a better understanding of the environmental impacts of their purchasing choices.

Some initiatives display carbon footprinting information on a wide range of products from potatoes to light bulbs. Retailers are beginning to present this kind of information on product labels. Some carbon labels are directly linked to the commitment of actively working to reduce greenhouse gas emissions. In addition to reducing their own carbon footprint, the benefit to retailers is that they are being seen as national and international leaders in engaging consumers on climate issues and helping them reduce their carbon footprint. In the case of one multinational retailer, more than 500 product lines have been or are in the process of being footprinted [42].

A number of private consultancy companies have been set up to offer a carbon footprinting calculation services or software to industry, indicating that the concept has a strong position as a unit of measure in the B2B and B2C market place. Some of these, including www.carbonfootprint.com, also offer offsetting programs, largely in the form of tree planting [43].

8.3.3.2 Pros and Cons of carbon footprint

- It allows measurement for services as well as goods and product.
- If computed accurately, it can allow make-or-buy and grow-or-import decisions to be decoupled from purely economic decisions.
- It tends to be the preserve of the well-meaning consumer in developed economies and does not inform choices of the developing world.
- It can lead to half-understanding the problem and incomplete systems measurement.

For example, the happy concert goers, who travel miles to a venue in their cars to attend an ecoevent where the sound and lights are solar or even pedal-powered, without realizing that the on-site energy savings are dwarfed by the energy consumed in travel, preparations, cleaning, etc.

- Carbon footprint calculation is based on an arbitrary or at least nominal sequestration rate and may lead to miscalculation of the effectiveness of offsetting measures.
- Different calculation tools often produce wide variations in the results [44].
- Carbon footprinting is only one measure of impact.

For example, some environmental impacts, notably those related to emissions of toxic substances, often do not covary with climate change impacts. In such situations, carbon footprint is a poor representative of the environmental burden of products, and environmental management focused exclusively on CFP runs the risk of inadvertently shifting the problem to other environmental impacts when products are optimized to become more "green" [44].

8.3.4 ECO-IT

ECO-IT as a simple tool concept was developed by Mark Goedkoop an industrial designer specializing in "ecodesign." It came out of work undertaken on developing SimaPro life cycle analysis software in the 1990s. It is now managed by Pré Consulting. According to their Web site, "ECO-it allows you to model a complex product and most of its life cycle in a few minutes. ECO-it calculates the environmental load and shows which parts of the product's life cycle contribute most. With this information you can target your creativity to improve the environmental performance of your product."

The program includes over 500 environmental impact and carbon footprint (CO_2) scores for commonly used materials such as metals, plastics, paper, board, and glass as well as production, transport, energy, and waste treatment processes. These scores are like predefined building blocks to model product life cycle. The database can be edited and expanded to include client-specific materials [45].

This is one of several similar LCA programs. The keys to their use and efficacy appear to be the following:

(a) Ease of use and simplicity—if the application is overcomplex, designers will be less likely to use it.
(b) Veracity of database—the software needs to access well-established and creditable data on "environmental load."
(c) The environmental load must be broader than carbon rating and should include issues such as resource scarcity and efficiency of primary resource production, including overburden and ore percentage issues in mining.
(d) The data for each item may not or indeed cannot easily differentiate between individual sources—for example, the difference between sourcing a drop forged component in Germany, India, or China would require and intimate knowledge of the specific equipment employed by the specific potential suppliers in each case.
(e) Their use should be seen as a guide or indicator rather than a definitive impact statement. The objective is then to compare alternative materials and methods of production to guide the designer in their choices at an early stage of the design process.

8.3.5 Granta

From detailed and extensive research applied to a good understanding of the issue facing production engineers, Granta has developed a comprehensive and interactive "material intelligence" design aid tool at Cambridge University. Granta was formed by Professors Mike Ashby and David Cebon at the Cambridge University Engineering Department.

The Granta research and development team became aware of the limitations of many LCA systems and specifically five key issues:

a. Sustainability improvement in design requires the designer to understand when and where the environmental impacts occur across the product life cycle.
b. Many of the commercially available LCA software packages have been developed with LCA practitioners in mind, requiring an expert knowledge of the methodology for both analysis and interpretation. Most designers and engineers do not possess this knowledge.
c. LCA software is unsuitable for use during the early stages of the design process as the requisite detailed knowledge of manufacture and use is not yet available to designers at that stage.
d. Designers' time is limited, and LCA can become another extraneous demand on that time.
e. The use of in-company or external consultants and specialists divorces the LCA analysis from reality in the designers' minds, preventing it from being integral to the design process.

In consequence, they have developed a series of simple tools that can be used early on in the design process, on the basis that "imprecise data can guide good decisions" during the design process. By allowing early evaluation of the alternatives, the systems enable changes to be made at a point in the design process where the positive impact of those changes on product cost and performance outweighs the negative impact of additional redesign and process engineering costs (see Fig. 8.3).

Their tools are available to both academic users and private companies and corporations.

Their Eco Audit Tool produces a graphical representation of the key issues (such as CO_2 footprint or embedded energy) across the life cycle stages of the item under consideration, with a focus on speed

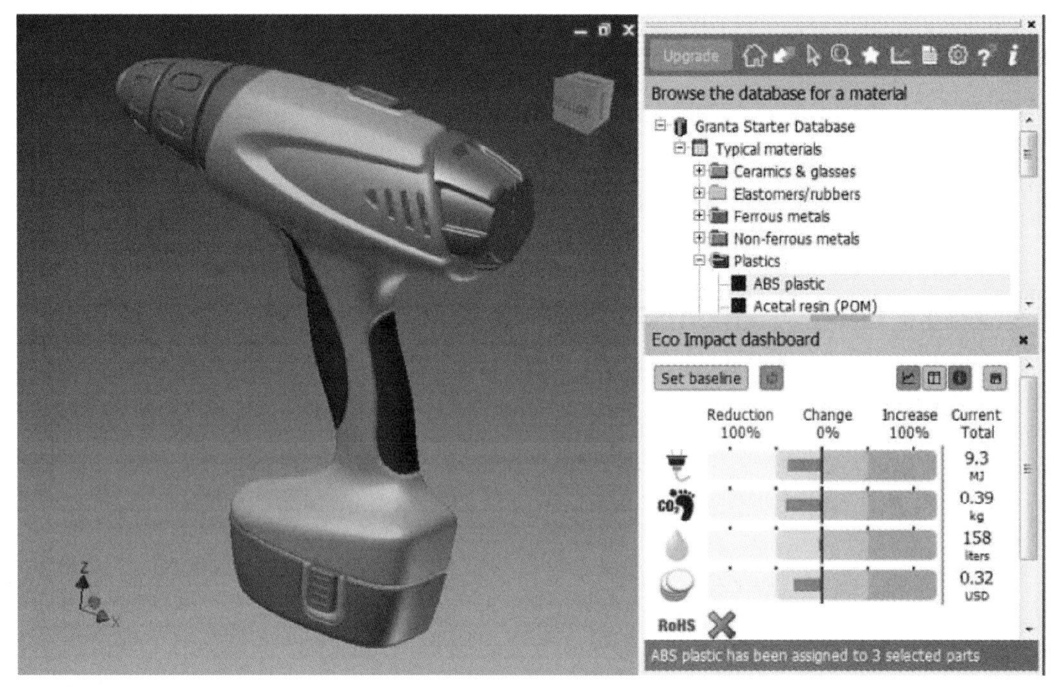

FIGURE 8.3

Ecoimpact Dashboard [46]. (For color version of this figure, the reader is referred to the online version of this chapter.)

and ease of use, setting the confidence interval for the data accuracy at "sufficient" to guide decision making rather than pinpoint.

In addition, they have a CES Selector that allows the designer to compare, for example, Young's modulus against embodied energy or material mass for a range of materials, which, together with the restricted materials database, aids initial material selection for critical components.

The software has been designed to integrate with the more common CAD software programs, where the user can search and browse a database of materials and processes, apply these two parts within their model, and instantly generate an ecoimpact "dashboard"—see Figure 8.3. This allows interactive exploration of the effect of material selection changes [46].

8.3.5.1 EMIT consortium

The University of Cambridge has launched a collaborative project building on Granta's ecosoftware and data, with a goal of embedding ecodesign into routine engineering workflows.

The consortium includes many internationally renowned engineering companies and organizations, including Rolls-Royce, Boeing, NASA, Emerson, Honeywell, and Eurocopter. Two key focus areas are as follows:

- Restricted substance regulations (such as REACH)
- Ecodesign to meet objectives such as reduced energy or carbon footprint

The members of the consortium ensure feedback to Granta to improve and prioritize development of the design audit tools and to promote the use of these and similar tools throughout industry.

As these tools develop in both ease of use and integration with commonly used design software packages, their impact on design decisions will become more pronounced, and most forward-thinking companies will use them as standard practice.

8.4 OTHER DRIVERS OF SUSTAINABLE DESIGN

Apart from legislative drivers and the increasingly ubiquitous measurement tools, the chief drivers to sustainable design are financial and social.

8.4.1 Social drivers

In social terms, there is a segment of the market in the more affluent, developed world that has identified global temperature rise, growing demand on limited resource from a growing global population, and the need reduce waste to landfill as key problems facing their and their children's future. Since the majority recognize the benefits offered by the technologies of the developed world, including increased life expectancy, increased leisure and wealth, and improved communication, they are unwilling to reject technology altogether as the route to a reduced footprint and seek out instead those producers and manufacturers who have demonstrably engaged with the process of improving sustainability.

These aware consumers are led by the young, the "millennials" (18-32 year olds), and this group is as strong in Asia as in Europe or in the United States. According to Keeble, from Accenture's research, they are less interested in the view that sustainability in a product is gained by sacrificing performance or value for money and are more interested in positive impacts claimed by manufacturers and marketers. There is some indication from sources, such as TerraChoice, that these consumers are not given reliable information on which to base those decisions, and later on in the book, we examine some of the pro and con arguments raised by the debate on these issues [47].

Indications are that there is a good public awareness of energy savings, also spurred on by high energy costs. This has led, for instance, to the widespread adoption of low-energy lighting products and the phasing out of filament bulbs or to changed habits of switching equipment off rather than leaving it on standby and conservation of water and other resources by fitting lower-use equipment.

In Europe, at least, and driven by German activism and the legislation that followed, there has been a reasonable degree of success in tackling some of the problems and issues associated with packaging, especially second- and third-tier packaging.

Some consumer product suppliers have taken the opportunity to supply refill packs for product ranging from coffee to printer inks, reducing the cost of production and/or transport and increasing margins into the bargain.

Concern for carbon emission reduction has fueled a drive toward more efficient energy production. This is evident not just in large-scale power generation such as the move from coal to gas and renewables, but in the consumer goods arena from domestic boilers and space heating systems to mobile equipment and vehicles.

In the same vein, there is concern for energy efficiency, as this benefits the consumer not only in the longer-term more abstract manner by reducing carbon emissions and thereby impacting on climate change but also in the shorter term by reducing the cost of usage. As an example, we have selected the Dyson Airblade™ hand dryer. The hand dryer uses a highly efficient motor and high-velocity air "blades" to shear water from the hands, eliminating the need for air heating. The short cycle time also reduces energy use [48]. The unit has been designed to be robust in use, giving it a long working life, and is easy to dismantle to allow repair and recycling at the end of its normal working life, thereby allowing it to rank well at all points along the life cycle (Figure 8.4).

FIGURE 8.4

Dyson Airblade Hand Dryer [48]. (For color version of this figure, the reader is referred to the online version of this chapter.)

The brand of JCB, established in 1945, stood originally for robustness and durability. Like many responsible and growing global businesses, JCB has embraced the need for it to "realign its responsibilities in line with the changing world" [49]. These responsibilities cover the three elements of the triple bottom line:

- To respond to the threat of climate change
- To improve workplace health and safety to an even higher level
- To rise to the challenge of rebuilding global prosperity in the wake of the financial crisis

As an organization, JCB decided to make energy efficiency part of its strategic planning process. Working with the Carbon Trust, they have reduced their direct emissions by 27% since 2007, reduced their accident potential rate by 50%, and improved fuel consumption on their backhoe loaders by 16% [49].

The direct emission reductions have been obtained through a mixture of measures including the following:

- Increasing the recycled content of their water usage in their Indian facility to 40%
- Tracking energy consumption in real time and identifying waste in lighting and heating
- Monitoring air leaks and thus reducing compressor demand and output
- Raising staff awareness and integrating energy-saving measures into shutdown and PMS maintenance procedures and processes

The following are savings at a glance:

- Total project cost: £300,000
- Projected annual cost savings: £1,500,000
- Projected annual CO_2 savings: 7800 tons [50]

Apart from looking at their own emission levels as an organization, JCB has focused on meeting and exceeding the increasingly stringent Stage IIIB/Tier 4i emission regulations for their own off-road vehicle engines. These emission regulations limit emissions from off-road vehicle engines and cover the EU and Japan and North America, respectively.

Using sophisticated design, including electronic fuel injection, variable geometry turbocharger, diesel particulate filtration (DPF), and selective catalytic reduction (SCR) in their engines, makes them fully compliant with Stage IIIB/Tier 4i and fully in tune with a cleaner world. The exact technology used on each product range is matched to the machine and application using sophisticated modeling techniques [51] (Figure 8.5).

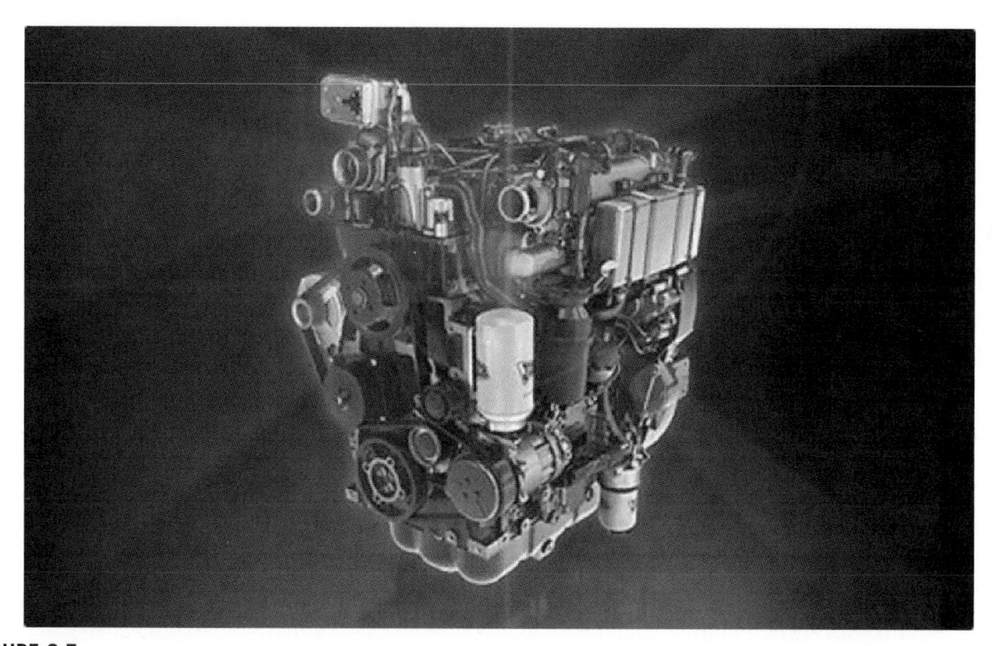

FIGURE 8.5

JCB Ecomax Engine [50]. (For color version of this figure, the reader is referred to the online version of this chapter.)

8.4.2 Financial drivers

Basically, energy cost is a driver, so any low-energy product or "free" energy source looks attractive. Nonetheless, the high investment cost has driven government subsidies on renewable energy sources at an industrial and domestic level.

A reduction in material consumed (optimized component weights, bottle weights, etc.) should also lead to a cost reduction, as does a simplification of the design. Furthermore, the careful selection of material may also allow a more efficient manufacturing process, moving increasingly away from traditional machine removal-based manufacturing of milling and turning to the increased use of precision casting, stamping, cold forming, and most recently 3D printing, coupled with microremoval technology such as spark erosion and polishing. All of these developments have reduced embedded energy in a number of components at the same time as reducing production cost.

As we have mentioned earlier, the move from skilled bespoke fitting process to less skilled assembly has also saved cost and energy. Few machine tools these days require hours of scraping-in on bedways, and fewer engine and gearbox systems require shimming and the extensive use of marking blue much beloved of the author's time in the fitting shop as an apprentice. These changes have reduced assembly times and hence production cost. Furthermore, as part of a package including advanced seal technology and the use of sacrificial sleeves, they improve the sustainable life value (SLV) of a product by increasing operating life and improving ease of maintenance and repair. This approach was taken in the design of the rock drill gearbox as discussed in Chapter 3.11 earlier in the book.

Packaging improvements have been shown to benefit costs. These improvements are often about better design of packaging to allow reuse, reduction in weight thickness, etc., particularly of secondary and tertiary packaging, and the intelligent integration of the supply chain system to standardize the use of multiway secondary and tertiary packaging.

Examples of how this can be achieved include the following:

(a) Reduce packaging via elimination of a layer—see example in the succeeding text.
(b) Redesign to reduce volume, thickness, and hence weight of material, or substitution of one outer packaging type for another.
(c) Reuse.

Examples

(a) Elimination

A small spinning/weaving company in Lancashire used to pack a number of individually wrapped cloth "pieces" in a single polypropylene bale. Pieces are now only wrapped individually in polyethylene. The benefits of this change include the following:

- Polypropylene is no longer purchased and has been eliminated as a waste stream.
- Wrapped pieces can now be carried by hand (the bales had to be moved using a crane).
- Bales do not have to be sewn, thus reducing both labor costs and expenditure on bale hooks.
- More efficient use is made of container and aircraft hold space.
- Reducing transport costs and permitting a lower export price.

This process shows a cost and ecological benefit.

(b) Redesign/substitution:

banding for shrink-wrap. This not only reduces the amount of material used but also replaces a nonrecyclable material with a recyclable one.

(c) Using standardized reusable transport crates, sized to interlock on standard Euro-pallets, has allowed a major supermarket chain to eliminate one-way packaging altogether from its fresh produce supply chain [52].

In the author's direct experience, a pallet account was held by a company he managed in Germany with their main supplier of logistics services. By putting a monetary value on the standardized reusable Euro-pallets, their reuse and refurbishment were assured.

8.5 CONCLUSION

There are both positive and negative drivers pushing designers into embracing the concept of sustainability in design.

The negative drivers or in the classic "stick-and-carrot" approach are those coming from current and impending legislation, which drives manufacturers and producers to seek to produce ever more efficient, quieter, and sustainably produced equipment, machinery, and consumer goods. In the past, this has been seen as a burden on industry, where meeting increasingly stringent regulation would add cost, reduce efficiency, and ultimately reduce consumption of energy and resources by pricing them out of the market [53].

However, it appears that more recent trends in marketing and consumer demand are leading to positive drivers, the "carrot" where environmentally aware customers are increasingly demanding a more sustainable approach, not only to the design and production but also to the business model and the whole pattern of human consumption.

The challenge for the designer is to embrace the concept of truly sustainable design while ensuring that profitability is improved. Remember, we are seeking to improve the triple bottom line.

The second challenge is to avoid damage shift, where improvements in the developed world are offset by unfettered unsustainable growth in the developing world.

In our view, these challenges are to be met using one of the increasingly well-thought-out models measuring sustainability impact at an early stage in design to help guide decisions on make or buy, optimize material use, and focus on the greatest impact saving across the full life cycle.

Sources and references

[1] Declaration of the United Nations Conference on the Human Environment (UNEP), 1972.
[2] EPA United States 2012, www2.epa.gov/laws and regulations, accessed July 2013.
[3] Green building advisor. www.greenbuildingadvior.com. Taunton Press Inc.
[4] Staying Competitive in Today's Homebuilding Industry, McGraw Hill & Deloitte & Touche, US Homebuilder Survey, 2006.
[5] Green Building Drivers, Green Building Insulation a 2009 white paper by Greenguard.org.
[6] Sustainable Manufacturing the Road Forward, ASME article by Marion Hart March 2012, www.asme.org.
[7] Symposium on Sustainability and Product Development IIT, Chicago, August 7 and 8, 2008, Beyond the 3R's: 6R Concepts for Next Generation Manufacturing, Dr. I.S. Jawahir, University of Kentucky, 2008.
[8] Environment Canada, 18/07/2012. www.ec.gc.ca.
[9] Being Wise with Waste, Europe's approach to waste management ISBN 978-92-79-14297-0.

[10] EU Commission Energy Web Site 2012.
[11] Source EU Commission Ecodesign, your future brochure, 2012.
[12] Directive on Waste 2008/98/EC summaries of European legislation. www.europa.eu/legilsation.
[13] DEFRA 2012 "Legislation and Regulation Frameworks."
[14] The Waste (England and Wales) Regulations 208, UK Government legislation. www.legislation.gov.uk.
[15] Sustainability in Germany. www.europa.eu/yourbusiness/.
[16] Nachhaltigkeitsstrategie. www.bundesregierung.de.
[17] German Packaging Regulation (Verpackungsordnung) 2009 IHK Siegen.
[18] Nationale Nachhaltigkeitsstrategie Fortschrittberich 2012, Presse -und Informationsamt der Bundesregierung, Berlin 31 Oct 2012, translated by A.G. Gibson.
[19] Pinheironeto Advogados, Legislação Ambiental\2007\leg ambiental-ing2007.doc São Paulo, January 2007.
[20] Review of Russian Environmental Legislation ICF International 2012. http://www.icfinternational.ru/english/environmental-regulations.asp.
[21] State Report "On Condition and Protection of Environment of Russian Federation in 2003."
[22] Communication from the Commission on EU-Russia environmental cooperation of December 17, 2001 [COM (2001) 772 final—not published in the Official Journal].
[23] Environmental Regulations And Legal Framework In India SJVN web publication 2007. http://sjvn.nic.in/projects/environmental-regulations.pdf.
[24] Comparative Study of the Environmental Laws of India and the UK with Special Reference to Their Enforcement, Govind Narayan Sinha, 2003.
[25] Environmental legislation in India, FinPro Public Document, January 2008.
[26] Krzyztov Mychalak, OECD report, 2006.
[27] EPA-China Environmental Law Initiative, United States Environmental Protection Agency. http://www.epa.gov/ogc/china/initiative_home.htm.
[28] ISRI Convention, China's Green Fence Recycling Today, April 23, 2013.
[29] Information site from the Ministry of the Environment, Japan. http://www.env.go.jp/en/coop/pollution.html (accessed July 2013).
[30] Effectiveness of International Environmental Regimes, OR Young, MIT, 1999.
[31] Is Improving the Effectiveness of Environmental Impact Assessment in the UK Dependent on the Use of Follow-up? Views of Environmental Consultants Clare Harmer University of East Anglia, August 2005.
[32] Reporting on environmental measures—towards more 'sound and effective' EU environmental policies European Environment Agency report, 2001.
[33] Departmental Report, Department for Environment Food and Rural Affairs 2006. http://archive.defra.gov.uk/corporate/about/reports/documents/2006deptreport.pdf.
[34] Environmental Regulation: Effective or Defective? Assessing whether criminal sanctions provide adequate protection of the Environment, Emma Bethel, Plymouth Law Review, vol. 2, Autumn, 2009.
[35] Departmental Report, Department for Environment Food and Rural Affairs 2008. http://webarchive.nationalarchives.gov.uk/20130123162956, http://archive.defra.gov.uk/corporate/about/reports/annual.htm.
[36] Edward B. Barbier, Introduction to the Environmental Kuznets Curve Special Issue, Environ. Dev. Econ. 369 (1997) 369–370.
[37] Yuqing Xing, Charles D. Kolstad, Do lax environmental regulations attract foreign investment? Environ. Resour. Econ. 21 (1) (2002) 1–22.
[38] Peter Nelson/International Institute for Environment and Development (IIED), London, Autumn, Profiles of Tools and Tactics for Environmental Mainstreaming No. 1 EIA, 2007. www.environmental-mainstreaming.org.
[39] Claire Harmer, Is improving the effectiveness of Environmental Impact Assessment in the UK Dependent on the use of follow-up? Views of Environmental Consultants, Thesis paper, School of Environmental Sciences, University of East Anglia, 2007.

[40] Dr. Tibor Faragó, hon. prof. of St. István University (Hungary), former state secretary for environmental and climate policies, Critical factors and processes counteracting to resource efficiency enhancement efforts, ESDN Conference, 28 June 2011.
[41] Prof. Dr. Schmidt-Bleek, Wuppertal Institute, MIPS, The Fossil Makers, 1993.
[42] What is carbon-footprinting, Carbon-Trust Web site: www.carbontrust.com/resources (accessed June 2013).
[43] Laurent, Olsen, and Hauschild, Technical University of Denmark, Paper 2012 Environmental Science and technology, Limitations of Carbon Footprint as Indicator of Environmental Sustainability.
[44] Soil Association Producer Support Sheet, Soil Association, South Marlborough St., Bristol.
[45] Mark Goedkoop, SimaPro web-site, operated by Pré Consultants bv Amersfort, NL, June 2013. http://www.pre-sustainability.com/ (accessed June 2013).
[46] J. O'Hare, E. Cope, S. Warde, 208, FIVE STEPS TO ECO DESIGN. Improving the Environmental Performance of Products through Design. Granta Design, Cambridge, UK.
[47] Justin Keeble, Young Consumers Hold the Key to Sustainable Brands, MD of Accenture writing in the Guardian, April 18, 2013.
[48] Dyson Airblade "why it's better", Dyson Web site. http://www.dysonairblade.co.uk/handdryers (accessed June 21, 2013).
[49] JCB main Web site: http://www.jcb.co.uk/About.aspx (accessed June and July 2013).
[50] Carbon Trust Website: Our clients JCB Case Study downloaded pdf file Simon Appleby JCB Environmental Advisor & carbon Trust 2009 published http://www.carbontrust.com/our-clients/j/jcb.
[51] JCB Product Information site about tier4 emissions control. http://www.jcbtier4.com/en-gb/about-tier-4 (accessed July 2013).
[52] Examples from Packaging reduction saves money: industry examples Envirowise brochure April 2004 Harwell International Business Centre, Oxford, UK.
[53] Effectiveness of Environmental Law: What Does the Evidence Tell Us? Michael G. Faure, University of Maastricht—Faculty of Law, Erasmus University Rotterdam (EUR), October 23, 2012.

CHAPTER 9

Strategic Sustainable Design

9.1 TRIPLE BOTTOM LINE—THE 3P APPROACH

Early industrialists, the architects of the industrial revolution, were driven by the need for increased output to satisfy the demands of the growing markets within the Empire. Their focus was thus on improving the methods of production, designing machinery that could harness the emerging power sources to automate and speed up the manufacture of goods, and designing products that could more easily be made on this increasingly automated machinery. As a result, they were able to derive a greater return on investment in fixed and current assets—and became rich.

This was the first bottom line—*profit.*

Some of these industrialists such as Titus Salt and George Cadbury had or developed a social conscience, seeking to improve the "lot" of their workers, housing them in model villages, educating their children, and even investing money in improving safety in the manufacturing process.

More recently, leading companies have strived to distinguish themselves from their competitors as being more people friendly, either in terms of how they treat their employees, suppliers, and customers or in terms of how they interact with their external stakeholders.

This became the second bottom line—*people or society.*

As we have become aware of the depletion of the earth's resources and the challenges set by increased use of fossil fuels and consequent production of potentially harmful by-products of combustion, Western society has identified the need for a third element to be accounted within the industries of extraction, production, manufacturing and distribution, namely the environmental impact.

This is the third bottom line—*planet or environment.*

So where early capitalists focused on return on investment, which continues to be essential to the long-term prosperity of any nation and its citizens, their successors sought increasingly to benefit not just themselves but their wider "stakeholders" through a socially responsible approach to their company. Modern successful entrepreneurs and industrialist are increasingly mindful of the impact their activity will have on their environment.

9.1.1 What It Is and What It Does

This was called by Elkington and others the triple bottom line [1].

At first sight, it would appear counterintuitive for a company to focus on their environmental impact, particularly where this could have an apparently detrimental effect on their bottom line or profit, which was the first and still most important of the "3Ps."

However, changes in public attitudes certainly in the developed world have begun to manifest themselves in the form of legislation and market pressure. Thus governments, adopting a typical stick and carrot approach, have penalized high-energy users through high tax takes on fossil fuels, while

Sustainability in Engineering Design. http://dx.doi.org/10.1016/B978-0-08-099369-0.00009-1

allowing carbon credit trading to benefit the more efficient and effective production processes. Consumers increasingly review energy efficiency ratings for domestic appliances or fuel efficiency for vehicles and make purchasing decisions based on recycled material content or end-of-life destination for product packaging. It therefore behooves any good company to ensure that it is seen to follow or exceed these demands, not the least as a way of capturing a share of the most lucrative market segments.

The premise of triple bottom line approach to management of companies assumes that in order to manage an aspect of the business it must be measured. Thus, the financial health and profitability of an organization is measured using P&L accounts, balance sheet, and cash-flow forecasts. Similarly, an organization seeking to demonstrate its commitment to its key stakeholders, such as its employees, its community, customers, and suppliers will proudly display its attainment of ISO 9000 quality systems, Investors in People, Fair Trade, and other awards and demonstrations of good practice. It therefore seems logical to extend this idea to the third element of a successful company's output, namely the accountability of a company in its interaction with and effect on the immediate and broader environment.

In *Cannibals with Forks*, Elkington argues that the move from "pure" capitalism, in which companies instinctively seek to consume or devour their competitors, is giving rise to a more civilized form of business where enlightened self-interest demands a more sustainable, cooperative approach [1]. In *Enter the Triple Bottom Line* [2] published over 10 years later, Elkington argues that the process of "silo busting" to allow integration of 3BL into the corporate structure is already taking place, although not universally and not without some internal problems and issues.

Savitz and Weber [3] also look at how really successful companies are increasingly those embracing best practice in all three elements of the triple bottom line. They contend that those companies that embrace and adapt to society's emerging trends will tend to prosper: those that do not may have short-term success, but not survive in the long term. They also use a number of case studies to provide concrete examples of how corporations might turn the idea into profitable reality.

Not everyone agrees. Norman and MacDonald [4] challenge the notion, declaring that much of what is behind the concept is not novel, and that what is novel is unsound. Further, they argue that to attenuate the demands and claims of the concept in order to rescue the paradigm is pointless. They are supportive of the aspirations of 3BL, but unconvinced of its practical application and use.

Moses L. Pava in his article [5] challenges their argument, indicating that there is no simple single measure of a company's financial performance: P&L "bottom line" on its own is not an adequate measure, and needs to be viewed in context with balance sheet, cash-flow, and a risk analysis of assets and liabilities. In the same way, one cannot expect a single simple catch-all measurement of sustainability bottom line. He is, however, wary of the misuse of the term 3BL by many to promote their own company's interests.

Nonetheless, empirical evidence suggests that the nay-sayers are wrong. A sustainable business it seems is more likely to remain successful tomorrow and on into future generations. It is interesting, for instance, that companies applying for the prestigious Queens Awards for Innovation and Export in the United Kingdom are expected to show a sustainable responsible and profitable growth curve over a period of 3-6 years. Sales growth or bright innovation is not enough; we are demanding that companies act responsibly both socially and environmentally. Thus increasingly, companies are expected to become part of the solution rather than part of the problem, and where the interests of all three elements of society, environment, and shareholders can be made to overlap, we have a win-win-win situation as shown in Figure 9.1.

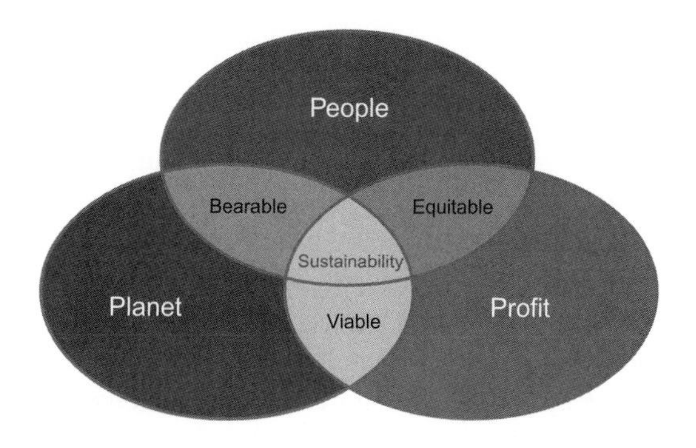

FIGURE 9.1

Venn-Diagram, Mapping 3BL [6]. (For color version of this figure, the reader is referred to the online version of this chapter.)

9.1.2 Measurement systems for 3BL

So the key issue is now that having defined the triple bottom line or 3BL, we need to measure it.

The three elements of the 3BL—people, planet, and profit—do not share either a common system or a unit of measure.

One approach would be to assign financial value to the environmental and social elements of the 3BL. This gives rise to issues such as how to assign agreed values to environmental or social impact. Embedded energy or carbon can be measured or estimated in manufactured objects or manufacturing processes, but how does one set a cash value to this? Equally, the cost of restorative work or contribution to social projects can also be measured, but is this a true measure of a company's social responsibility? Ethics in dealing with customers and suppliers must surely also pay a big role.

Another approach might be to set up an index system, which compares the performance of any given company with that of an idealized organization on a series of indices. This system, while being simpler to devise and operate, is potentially heavily subjective, and would require a good deal of thought into the weighting of the key elements, or indeed subcategories with the 3Ps.

Certainly, a 3BL calculation could include such elements as:

- The classic elements of profit calculation referred to above (P&L, B/S cash-flow)
- Environmental sustainability—percentage of end-of-life to recycle as opposed to landfill
- Durability and repair—life cycle prolongation by repair or replacement of wearing or failing elements
- Packaging efficiency and recyclability
- Relative consumption of energy over the working life cycle, using product or industry norms or standards
- Stakeholder satisfaction rates
- Transparency levels in dealings with stakeholders
- Impact on employment levels and rates
- Application of equality of opportunity to internal and external stakeholders [7]

This would also be in accordance with the EU Accounts Modernization Directive [6], which introduces requirements for (large) companies to include a balanced and comprehensive analysis of the performance of the business in their Directors' Report. This analysis must include "both financial and, where appropriate, non-financial key performance indicators relevant to the particular business, including information relating to environmental and employee matters."

There have been a number of recent studies outlining potential measurement concepts. These include the "3-E" concept of Isaksson and Garvare, representing a modified version of 3BL, where *ethics* replaces *social responsibility* as a better and broader criterion. The project is in early stages of development, and intends to use indicators of organizational performance split into four "drivers": input, enablers, output, and outcome. The concept is to develop similar approaches to those used already in quality audit systems such as ISO 9000 [8].

Foran *et al.* use output units of measure per dollar of GDP to review the performance of specific sectors of the Australian economy. Their argument is that this gives a measurement system that can be understood in terms of current accounting practices and can allow a comparison performance of any sector against the norm. By extension, a similar approach could be used to compare individual organization performance against (international) industry norm. Thus, financial aspects of performance can be expressed, for example, as dollars of export earnings per dollar of GDP. Social aspects such as employment can be portrayed as minutes of employment generated per dollar. Greenhouse issues can be portrayed as kilograms of carbon dioxide emitted per dollar.

The outcome can be displayed as a spider diagram, where the industry standard is represented as an average line. Thus, positions inside the line represent a better than average performance, and those outside the line represent worse than average (see Figure 9.2). This has been further developed as the basis of a software program by the ISA group at the University of Sydney, termed BottomLine3

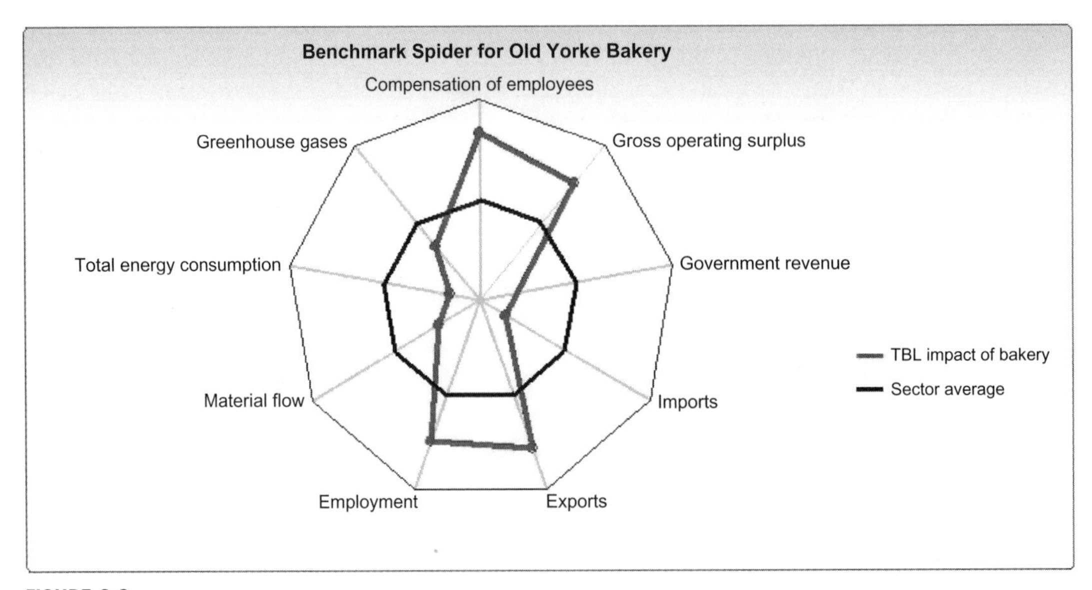

FIGURE 9.2

Spider Diagram Mapping 3BL Using BL3 Software [9]. (For color version of this figure, the reader is referred to the online version of this chapter.)

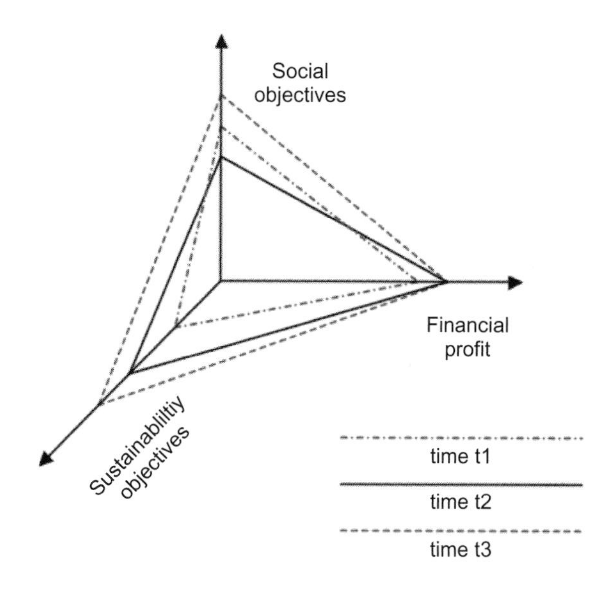

FIGURE 9.3

Three-Axis Representation of 3BL Influences. (For color version of this figure, the reader is referred to the online version of this chapter.)

Modified from Munasinghe [11].

("BL-cubed"). Outputs are rankings and breakdowns of economic social and environmental indicators, together with sector benchmarking and supply chain analysis [9,10].

Yet another way of thinking about the triple bottom line framework is as three-dimensional triangular planes where each axis reflects economic, social, and environmental objectives. This was first outlined by Munasinghe [11] and developed by Roberts and Cohen (2002).

Plotting the three key values of the triple bottom line on the three axes of Figure 9.3. At time t1 in this example, indicated by the chain-dotted green line, the organization appears to place about equal value on profit and social objectives and a lower value on the environment.

Sometime later, at *t*2, indicated by the solid black line, the profit and sustainability motives have increased in importance, while the social objectives lag behind. This could, for instance, be due to recessionary pressures, where restructuring and cost saving have had a negative impact on social objectives relating to the organizations stakeholders, but have helped maintain profitability and coincidentally a focus on waste and efficiency have improved the sustainability aspects.

At time *t*3, indicated by the dotted red line, it is clear that the financial performance has remained unchanged but, possibly now in a growth phase, the organization has focused more clearly on social and sustainability issues [11].

9.1.3 Are manufacturing companies embracing the triple bottom line?

Early exemplars of the 3BL approach include Unilever, General Electric, Proctor & Gamble, Pepsico, and Cascade Engineering. The last of these has issued 3BL (or TBL in their parlance) reports since 2003 (see below). In the United Kingdom, Southern Energy have embraced the concept, and according to Ric Parker, Director of Research and Technology at Rolls Royce, all products are assessed for environmental impact at the development stage a part of their commitment to 3BL [12]. EON also now produce a "Sustainability Report" rather than the previously titled CSR report. The new report covers all the three bottom line ingredients. The environmental aspects are measured under the GRI index, with detailed reporting on emissions, waste, and water management. The social issues cover gender, labor, and accident and incident statistics, and the financial reporting is covered only by a few KPI (key performance indicator) figures [13].

JCB do not refer to the triple bottom line as such, but recognize the potential impact they have on:

- The welfare of the communities in which we operate
- The health and safety of our employees
- The environmental footprint we generate

According to a survey in 2005 by KPMG, as reported in Professional Engineering, engineering companies were not the quickest to sign up to 3BL reporting. At the time, just 25% of companies were formally reporting performance against sustainability targets, ranking 9th out of 14 sectors, and well behind the utilities with 61%. James Stacey, KPMG Global Sustainability Advisor, did recognize the difficulties associated with 3BL reporting for complex organizations, but insisted that increasing professionalism in reporting standards was necessary to ensure that CSR was "fully integrated into the strategy and operations of complex organisations operating in a global economy" [14].

9.1.4 Case study example: Cascade Engineering Inc

Cascade Engineering manufactures a range of products for industrial use throughout the world. Their products include injections molded components and interior and exterior frame and body parts for the automotive industry, as well as materials handling, office furniture, and waste handling and recycling equipment. Since 2003, they have been reporting to their stakeholders and shareholders on their performance in all three elements of the triple bottom line. These reports cover key elements that make up the 3BL for cascade, including recycling and waste levels both in production and end-of-life or their products, improvements in production and product use efficiencies, corporate and local community investment projects, health and safety measures, and of course company profitability and shareholder returns [15].

In the foreword, CEO Fred Keller confirms that while some may think that a "focus on people and planet takes away from profits. . .if you do it right the opposite is true." As "B" is a registered company, by committing to and gaining commitment from internal and external stakeholders, there is a common goal and a proven motivation to work harder and with greater innovation [15].

The scorecard at the end of the report measures:

- Sales year on year (in $M)
- Energy consumption (in sales$/kWh consumed) year on year
- Greenhouse gas emissions (in tons of CO_2 equivalent)
- Use of recycled material (by weight)
- Landfill costs (as a percentage of sales)
- Employee training (average h/employee)
- Incident rate (per 200,000 h worked)
- Lost or restricted workday rate (per 200,000 h worked)
- Welfare to career retention rate

9.1.5 A critical review: is it an unacceptable, sensible, or imperative measure of a company's success?

There have been a number of papers examining the triple bottom line. The general consensus is that the concept is sound, but that since an acceptable standardized method of calculation has yet to be found, the use of TBL as a comparator or absolute indication of an organization's performance for public decision-making is not yet feasible.

Skouloudis studied the Global Initiative guidelines and concluded that they were more qualitative than quantitative, and that their study had revealed gaps in the reporting practices. The result was that an organization could improve its score by including issues for discussion in its reports rather than taking firm action [16].

A similar criticism is leveled in a report by Trucost for DEFRA into the FTSE 100 companies' reporting in the United Kingdom. They concluded that there was a lack of quantification in most reporting, certainly not any "single, quantifiable measure that companies can use as a KPI (key performance indicator) for the effect of their upstream supply chain on the environment" [17].

Tullberg concurs, stating that the triple bottom line is potentially a missed priority model, and suggesting another "sketchy" set of measurement criteria [18]. Similarly, Hubbard's paper of 2009 also proposes a scorecard system, noting that as sustainability concepts change this makes performance measurement even more difficult [19].

The key to triple bottom line is, it seems to us, that the lack of an agreed or standardized unit of measure should not detract from the validity of the overall concept. As with the development of other units of measure and systems of measurement, it will take time for one or two to be devised that gain universal currency.

So, we should not look upon 3BL as a rigid compartmentalized box, and use it as the "elastic tool" it can clearly be [20]. Most organizations could find an image of the concept and a system of measurement that will resonate with their circumstances and their stakeholders and use that in their internal assessment and their external communications with stakeholders.

Clearly, this raises a challenge of "Greenwash" creating the illusion of commitment to sustainability, but there are a large and increasing number of lobby groups and consultancies developing tools to control and measure sustainable activity that will expose the greater part of these illusions. Indeed, we will look at some of these in Chapter 10, as they are certainly a part of the future of our commitment to sustainability.

However, the key is that organizations and specifically in our view manufacturers of products must evaluate the performance of their products, their production methods, and their downstream materials distribution and upstream supply chains to optimize their sustainability performance, in terms of:

- embedded energy in their products
- efficiency of use of resources including optimized materials sourcing
- simplicity of design, including minimizing materials use, minimizing material removal-based manufacturing processes, and simplified assembly processes
- longevity and ease of life extension through prioritizing repair and reuse over recycling and waste
- minimizing or eliminating the impact on the local environment in terms of water use and pollution and air and ground pollution levels, without simply displacing these problems to earlier or later stages of the total supply and use chain
- optimizing packaging to ensure that it meets the needs of product protection and marketing without generating excessive waste

These issues can all be measured as indices using (possibly refined versions of) current modeling techniques and reported in the annual statement as part of the TBL reporting.

9.2 BENEFITS TO PRODUCERS AND BUYERS OF DESIGNED-IN SUSTAINABILITY

The benefit to the environment of designing in sustainability is quite clear. If we improve fundamentally our approach to sustainability within the field of engineered and manufactured product, we reduce our impact on the environment while simultaneously continuing to offer a maintenance of or even improvement in the standard of life or living of the inhabitants of the planet.

But what is in it for the buyers and consumers of these better-engineered and more consciously sustainable products? Will they be inferior in some way? Will they mark a step backward toward some imagined bucolic utopian past? Will they simply be another means of marketing ever-increasing consumption to those who already overconsume?

And what of the designers, manufacturers, and producers of these products, machines, and equipment? Will they trade in their hair shirts for the silk or satin garments of smugness and superiority? Will they use the illusion of a fundamental shift in the design paradigm to continue on down the path of unsustainable growth?

9.2.1 Benefit to buyers

The buyers and consumers of manufactured goods and product will potentially benefit both directly and indirectly from designed-in sustainability.

The direct benefits are described in the following subsections.

9.2.1.1 A lower running cost, due to improved energy efficiency, or the in-built use of sustainable energy sources

Exemplars of improved efficiency designs include domestic appliances such as dishwashers and washing machines, which are now designed to use less water and consume less power by washing at lower temperatures, and by improving insulation systems, etc. The benefits of these designed-in features are clearly demonstrated using the EU rating system for appliances, and using consumer buyers guides, such as Which and Energy Star. Indeed *Which*, for instance, offers not only a guide to energy efficient machines but also a guide on how to optimize their use for environmental impact and cost reduction. They calculated annual energy costs of 198 washing machines and found they vary between £9 and £47 per year for four loads a week at 40 °C [21].

The Bosh Machine below occupied 15th place in terms of a combination of cost/year, cost/kg, cost/cycle, and CO_2 emissions according to ratings by the UK web site Sust-It. Apart from having a good rating in terms of water and energy use, it allows the user to decide whether to operate the machine for fast results on a quick cycle, or to optimize energy use on an Eco-cycle (Figure 9.4) [23].

Built-in renewable energy sources for sustainability are increasingly being seen in areas such as street lighting, street signs, and GSM masts. For street lighting, for instance, the concept is that PV-generated energy is stored during the day and released at night to give a self-sustaining product that can be installed without the need for grid access. The lighting unit uses LED lighting to reduce

FIGURE 9.4

The Bosch Logixx WAS24461GB Washing Machine [22]. (For color version of this figure, the reader is referred to the online version of this chapter.)

power consumption. This solution is proving popular in Sub-Saharan Africa and the Middle East, but it has been used in the United Kingdom (Hackney borough among others) (Figure 9.5).

Typical features and benefits for PV solar/LED street lighting include:

- Over 10-year LED lamp life
- Immediate lighting without preheating
- Cool white light without flickering
- LED lamps retain luminosity for their full life span
- In case of a single LED breakdown all other LED modules continue to work
- LED lamps are nontoxic
- Safe 24 V dc circuit with reduced risk of electric shock
- Simple installation and low to no maintenance [22]

9.2.1.2 Lower overall cost due to improved longevity or reparability

The net cost-benefit argument will not always win customers, as some are working to limited budget and cannot afford to pay more now for savings over the longer term. However, as new and more efficient technologies become more widely adopted, the premium price charged by the early producers of these technologies is eroded, and they become available to a wider market.

FIGURE 9.5

Street Lighting with Built-In PV Solar Power Unit—Carey Glass Solar [24]. (For color version of this figure, the reader is referred to the online version of this chapter.)

From a long experience in developing and marketing internationally engineered products from advanced lubricants to mechanical handling components and equipment, the author can confirm that offering features that relate to an identified need clearly gives the client a benefit. The skill is in identifying the future needs of the client base, developing products that fulfill those needs, and finally making the client aware of the solutions offered.

Thus, for the example of the Dualit toasters mentioned in Chapters 3 and 10, it is clear that the initial cost is typically 30% above that of other comparable products. However, the extended life generated by the option to replace failed elements or the timer, etc., allows the unit to outlast its competitors by a factor of 3 or more. Thus, over its total life, and including the cost of replacement parts, such a toaster might offer a 50% saving.

In a similar vein, the rock drill design also found favor with clients in Sub-Saharan Africa, where simple robust design offering ease of maintenance and longevity is placed above sophistication and complexity of design offering what are often regarded as unnecessary labor-saving features. The rigs currently in use, whether sourced cheaply from China or at greater cost from Europe or USA, would succumb sooner or later (respectively) to the ravages of the hostile environment in which they were operating. Replacing work shafts was an expensive and difficult operation, involving high costs stocks or long procurement delays. Replacing a sacrificial sleeve, kept as an inexpensive stock item in the central depot is quicker, easier, and cheaper, and allows the working life of the drill to be extended by a factor of two to three times.

9.2.1.3 Intangible benefits

These direct benefits have hitherto largely translated ultimately to time, trouble, and money. There are other benefits to the user, which are more intangible. These include:

- *Status*—owning a "greener" motor vehicle or domestic appliance could enhance a consumer's statue among their peer (conspicuous consumption); similarly, a company operating with a low environmental impact may improve their status within their clientele.
- *Satisfaction*—the owner of a more sustainable product may derive direct satisfaction and pleasure simply form the knowledge that they are reducing their impact on the environment. This is often the motivation for the "early adopters" to make purchase decisions in favor of, for example, hybrid or electric cars, long before these are easy to use or indeed tried and tested in the market.
- *Aversion*—the possibility of legislation or taxation measures being imposed in the future may also influence the purchasing decisions of both domestic and industrial buyers. As an example, since vehicle road tax in the United Kingdom is now based on CO_2 emissions, many consumers have been influenced in their buying decisions by the emissions performance of a car, rather than just fuel consumption, style, or engine or road performance.
- *Altruism or enlightened self-interest*—those who place a great importance on sustainable living will seek out products and services that reflect that importance (Figure 9.6).

It is our view that these intangible benefits will play an important role in driving forward the sustainability agenda, but ultimately, many buyers and consumers will look first to cost and other direct benefits.

9.2.2 Benefit to producers

The prevailing view is that there are increasing financial incentives and benefits to the producers of demonstrably sustainable products and services. The evidence of this is limited, according to Blackman and Rivera [26]. Where the manufacturer is able to gain market share or to charge a premium price for a

FIGURE 9.6

Modern Environmental Message from "Green Is Sexy" Web Site [25]. (For color version of this figure, the reader is referred to the online version of this chapter.)

demonstrably sustainable product, the profit benefit is clear. However, this needs to be weighed up against any potential additional cost from sustainable materials sourcing, for instance.

It can be argued that sustainable sourcing and intelligent design, resulting in reduced materials usage, reduced metal-removal in manufacture, and optimized embedded energy will all contribute to the traditional bottom line in terms of materials costs and materials flow management benefits. Similarly, optimizing packaging will also lead to reduced cost in the downstream logistics chain, giving a competitive advantage in the market.

The market does not always respond positively toward sustainability as an argument: Perkins Diesel, for instance, found that the Chinese market was reluctant to pay a premium price for an engine with fuel-efficient performance, designed-in longevity, and ease of repair, but the markets of Europe and USA responded very well to the same product marketed with precisely that message. Evidently, those markets where there is sufficient available income for buyers and consumers to make choices based on the medium-term returns and long-term effects of a product will offer a more attractive market in the short term to the sustainable producer or manufacturer.

Nonetheless, for an SME manufacturer, the extra burden imposed by the increasing requirements to certify their sustainable credentials either for the product or the company as a whole may be seen by some to appear on the liability side of the balance sheet.

Clearly, there are also intangible benefits for manufacturers, distributors, and retailers of sustainable products in terms of their increasingly important obligations under corporate responsibility toward sustainable growth, and these can be clearly seen by the way many larger organizations are rushing to ensure their environmental credentials at a corporate level are seen by their customers.

Any benefit to the manufacturers of sustainably engineered products must be demonstrable, if those companies are to embrace the concept of designed-in sustainability.

At the product or component level, this means that we need to develop further the current tools and models that allow the designer to make decisions on materials sourcing, selection, and manufacturing method to include a cost-benefit calculation.

At the company or corporate level, it means using one or more of the many certification schemes to verify the company's commitment. Of course, some of these schemes also apply at the product level: indeed, the EU is currently consulting on standardizing measurement systems and measures in both categories.

Two methods have been put forward by the Commission to measure environmental performance throughout the product lifecycle. They are called Product Environmental Footprint (PEF) and the Organisation Environmental Footprint (OEF). The use of these methods would be voluntary for Member States, companies, private organizations, and the financial community [27].

9.3 THE SUSTAINABILITY MEASUREMENT AND CERTIFICATION INDUSTRY

As outlined earlier, in order to benefit commercially from a shift to engineered sustainability, manufacturers of machines, equipment, and components must be able to show the market that their products are offering the key feature of designed-in sustainability, and demonstrate the benefits this will bring to the client, user, or buyer.

Currently, companies wanting to highlight the environmental performance of their products face must choose between a multiplicity of methods, standards, and labels each promoted by governments, private consultancies, and lobby groups, leading to extra costs for multiple verification schemes and customer confusion as to the validity of any given scheme.

Much of the research carried out so far has been focused on food-based sustainability labels, such as Fair Trade, Rainforest Alliance, and Organic status. In Germany, pioneer country in the EU on such issues, the earliest labels such as *Der Grüne Punkt* were based on the recyclability of first and second tier packaging. In this book, we are more interested in those labels that can be applied to a broad spectrum of products, and which try to encapsulate the whole life cycle. This is not to say that there is no value in specific industry-based labeling, but rather that these labels impinge less on the world of the design engineer, especially where they refer to food and other natural products (Figure 9.7).

The US-based research carried out by the Hartman group [29] has indicated that many consumers do not currently place a high value on sustainability in the selection of supplier or manufacturer of household goods. Indeed many of the people surveyed, including those nominally interested in the subject had no clear concept of sustainability. Typically, it seems, those who have, to look for "green" labels to reassure them of a product's credentials. It is also apparent that there is a decline in the recognition of strict definitions of sustainability (Figure 9.8).

In their 2009 paper, Delmas *et al.* held that the increasing number of eco-labels could lead to information overload, consumer confusion, and skepticism. They give the example of the fact that there were currently more than six coffee eco-labels and that consumers apparently have difficulties recognizing the differences between them [30].

Their conclusion is that any successful and effective label must fulfill a number of criteria under the headings of increasing awareness and confidence in the consumer, and stimulation of a willingness to pay (Figure 9.9).

FIGURE 9.7

A Selection of Sustainability Labels [28]. (For color version of this figure, the reader is referred to the online version of this chapter.)

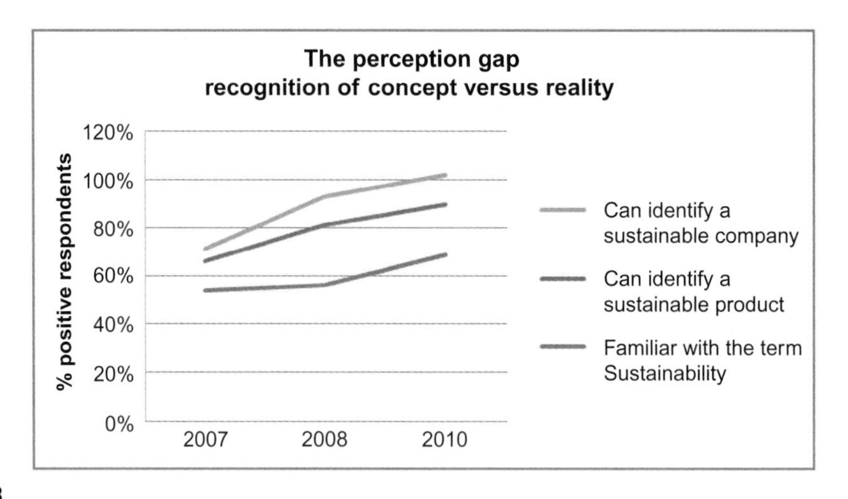

FIGURE 9.8

THE GAP! Consumers Do Not Yet Make the Connection Between Sustainability and Sustainable Companies and Products [29]. (For color version of this figure, the reader is referred to the online version of this chapter.)

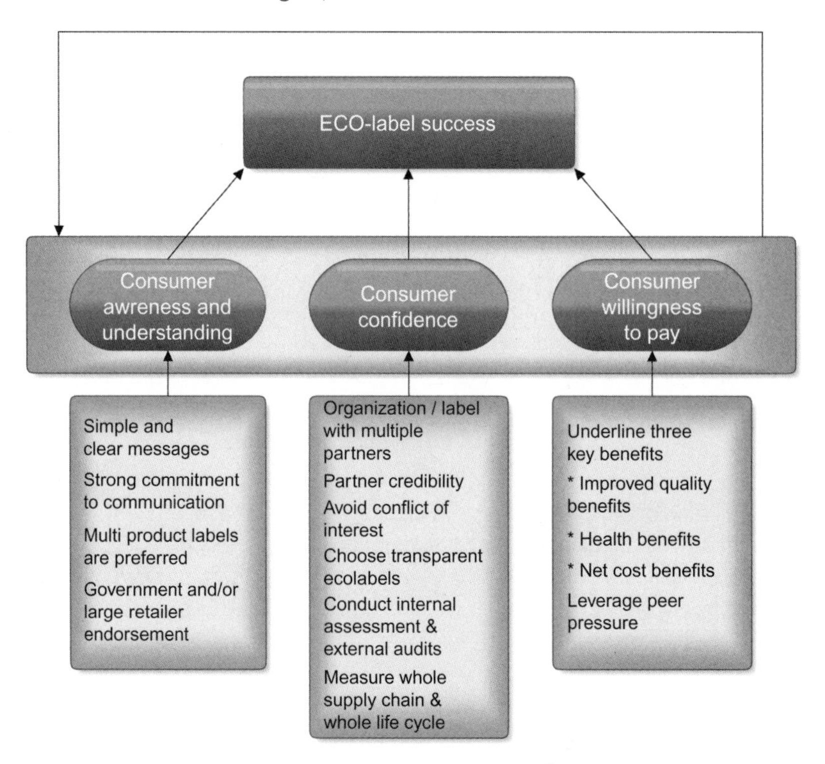

FIGURE 9.9

The criteria Affecting Success of a Sustainability Label. (For color version of this figure, the reader is referred to the online version of this chapter.)

Modified from Ref. [30].

For the manufactures too, there is risk associated with the choice of verification or audit entity and concomitant label, as a poor label may not carry recognition with the consumers, and indeed may mislead or confuse the customer, becoming a nonbeneficial cost.

The key then for producers is to align themselves with one or more of the recognized labels confirming the status of the company or its products as sustainable. There are currently a large number of such rating and grading systems across the developed world; so, it is important to have some understanding of the main ones and discuss some of the reviews of those systems. This will enable the designer to decide which systems, standards, and labels offer a real, concrete unit of measure, and which could be classified as potential "Greenwash."

These could be standards imposed by government, such as the EU requirement for energy rating under their D to AAA rating system or the US Governments Energy Star label. Or it could be a rating or seal of approval from one of the many so-called independent institutes. The consultancy and lobbying group Terrachoice produced reviews in 2009 and 2010 of the wealth of "green standards and labels" adopted by US companies, and concluded in principle that many of these labels were "Greenwash," but that "legitimate green standards help fight Greenwashing" [31].

9.3.1 ISO 14000 series

Building on the experience of the ISO 9000 systems, which fulfilled the need for quantifiable systems and audit procedures in quality management, the ISO 14000 series of standards were derived from the need for environmental protection. As with ISO 9000 series, the standards measure the creation and maintenance of audit systems and the systematic adoption of an organization of a consistent approach to sustainability and environmental protection,

They are not a guarantee of sustainability in the products designed or made by an organization. Indeed, when the author was leading the drive for ISO 9001 certification by one German company, a simple grammatical error in the forward indicated that it was the intention of the company to produce poor quality goods. The TüV Rheinland assessors queried the error, but remarked that had it been our genuine intention to produce rubbish, as long as we achieved this goal systematically and within the remit of the audited ISO system, then the system itself would still be awarded ISO 9001 certification!

Nonetheless, it does form a good basis for other organization-wide certification systems, as it prescribes best practice for minimizing pollution risk, and establishes that a company must start the process by creating an environmental policy to identify all potential risks and effects of their production activity [32].

It also covers the issue of waste management, seeking to minimize waste and maximize reuse and recycling. It has been estimated that an average manufacturing company produces waste to the value of approximately 4.5% of turnover. Reducing this waste by 25% would thus save over 1% on turnover, making a direct contribution to the improvement of the bottom line.

9.3.2 Carbon neutral

This is an example of the CO_2-based certification schemes. It was developed by the Carbon Neutral Company in 1998 to measure the carbon impact of organizations, with a view to improving their performance. Clients' greenhouse gas emissions are assessed using external assessors, and strategies are developed to reduce or offset these emissions.

The carbon neutral certification is then awarded on the basis of credible action to reduce emissions to zero [33].

Some of the key product or service-based systems and labels are outlined and reviewed below.

9.3.3 The Energy Star label

The Energy Star Programme is a voluntary partnership of the U.S. Department of Energy (DOE) and the U.S. Environmental Protection Agency (EPA). An agreement, signed in 2000, partners the European Union in the labeling scheme, placing the European Commission in charge of the program. The label is well-established in both the USA and Europe, with thousands of products for home and office use across more than 40 product categories. The logo typically appears on office equipment such as monitors, printers, etc. [34].

Recognition levels have grown in the USA from 30% of consumes in 1999 to over 60% in 2005. Although recognition is one issue, action is another. There are apparently a number of factors militating against an increased uptake in the purchase of products carrying this label:

(a) According to Ward *et al.* in 2011, the offer of a direct monetary saving over the life of an appliance is a strong positive motivator. However, as energy costs decrease relatively, this net cost-benefit is eroded and take up is smaller [35]

(b) The target audience is also critical: those in the lower income brackets will be less concerned with ecology in their purchasing decisions, and often not in a position to pay a premium investment price for a net cost-benefit [36]

(c) Public pressure to improve the environment: the energy star label explicitly promises benefits both to the consumer and the planet. “Money Isn’t All You’re Saving” or “Save Energy, Save Money, Protect the Environment” [35]

9.3.4 The Blue Angel (Blauer Engel)

The Blue Angel (Blauer Engel) is a German certification for products and services that have environmentally friendly aspects. The first and one of the most well-known eco-labels worldwide, Blue Angel promotes the concerns of both environmental protection and consumer protection [37]. The fact that it is owned by the Federal Ministry of the Environment means that consumer confidence levels are high.

The label specifies that the product or service focuses on one of four different protection goals:

- Health
- Climate
- Water
- Resources

According to their own web site, they have achieved 76% brand awareness in their home market of Germany, and 39% of buyers are influenced by the label when shopping. They have over 1750 licensees with over 11,000 approved products across the spectrum of consumable goods from abrasives through floor coverings to domestic equipment [37].

9.3.5 The EU Ecolabel

This is an EU-wide standard, which is administered by officially designated "competent bodies" in each EU country. As with the other labels considered above, the EU Ecolabel covers a broad range of products from household appliances, TV and home entertainment, furnishings, etc., through clothing, DIY, and gardening equipment and supplies to cleaning chemicals lubricants.

Products that carry the Ecolabel have been independently assessed to be less damaging to the environment than some equivalent brands over their whole life cycle. They must meet a set of published environmental criteria agreed upon by EU member states after consultation with relevant interested parties.

The assessment covers the full life cycle of the product under consideration, broken up into four sections:

- *Extraction*: The environmental toll caused during mining, crop cultivation, petroleum extraction, forest management, and other activities
- *Manufacturing, packaging, and distribution*: Pollution emitted and resources used to construct or produce the products and get them to market
- *Use by consumer*: Consumption of energy and water resources during the working life of the product
- *End-of-life*: The ease of recovery and recycling [38]

Since the EU Ecolabel was established, the number of companies receiving the label has increased each year. At the beginning of 2010, more than 1000 EU Ecolabel licenses were awarded in nine established product groups. The United Kingdom currently has 31 Ecolabel holders. However, Delmas *et al.* remain unconvinced by the label, stating that while the Flower label indicated a "general message of sustainability," it does not communicate well the criteria by which the products are judged, and thus has a lower impact and recognition score than one might expect [29].

9.3.6 The Carbon Reduction Label

The Carbon Label Company was set up by the Carbon Trust in 2007 to help meet the needs of both businesses and their customers. In 2009, the Company became the Carbon Trust Footprinting Company to reflect the breadth of services it provided.

A product bearing this label indicates that the carbon footprint of the product (or service) has been measured and certified, and that the producer or manufacturer of the product or service is committed to a further reduction of that carbon footprint over a 2-year period. The calculations made across the whole life cycle, and are based on the relevant British Standards PAS 2050, using the organization's own dataset. This gives a consistent basis of calculation and allows easy comparison between similar products and services.

The system is not without critics; however, as the United Kingdom's leading supermarket chain, Tesco, have dropped the process after committing considerable resource to the process, citing prohibitively high costs and poor consumer recognition as the reason for doing so [39].

9.3.7 The Energy Saving Trust Recommended

Energy Saving Trust Recommended is a best-in-class product certification and labeling scheme that covers 32 product categories on the UK market. Manufacturers, suppliers, and retailers can join the scheme to help their customers identify the most energy efficient products in their ranges. The Energy

Saving Trust, which manages the certification, is an independent, nonprofit organization backed by the Government.

Product categories range across seven sectors:

- Appliances
- Consumer electronics
- IT
- Lighting
- Heating
- Insulation
- Glazing

Under the Energy Saving Trust Recommended scheme, only products that meet strict criteria on energy efficiency can carry the label. For example, only fridges which are A+ can be certified, and the washing machines on the scheme must be AAA—that is A for energy, A for wash quality, and A for spin. The criteria are set by an independent panel and reviewed annually to ensure that they always reflect development and changes in the market and technology. Not all the products certified by the trust will already carry an EU label: glazing, boilers, and TV, for instance, are not certified under the EU scheme [40].

All certification criteria go through a planning, consultation, and peer-reviewing process before they are implemented. This means that to be Energy Saving Trust Recommended products must be top in their class at energy efficiency.

9.3.8 The reviewers—Quis custodiet ipsos custodies?

There have been a number of learned articles and studies into the nascent industry that eco-labels and standards are becoming fast. Some of those reviews and studies have been referred to in the short descriptions above. The majority of these studies have concentrated either on the technicalities and effectiveness of one or more labels or standards or on the apparent effect that these labels are having on buyer behavior and hence the strategy adopted by manufacturers and retailers in tapping that changing behavior profitable. Some, like Murray and Mills [36], mention in passing the net effect that this might have on reducing carbon emissions.

To date, we have found relatively little in the way of studies that look at these standards and labels from the perspective of the 3BL benefits that they ultimately need to generate in order to promote sustainability improvements though technological change—which itself, as we argue, is brought about via the hands and minds of the designer.

There is, naturally, a wealth of web-based criticism and review of the labels and standards so eagerly embraced by some manufacturers and retailers. Indeed, the term "Greenwash" was coined to describe the use of such labels to mislead and confuse. Typically, advertisements often claim that a product is made from "only natural ingredients." And this means what, precisely? In these circumstances, we can apply the famous "Ja und?" or so what test. An unsubstantiated claim such as the one above means little or nothing, and is merely part of advertising copy.

Much of the criticism and discussion of the subject is anecdotal and online, often in the form of blogs and public rating systems, such as the Greenwashing Index [41]. On this site, members of the public are encouraged to look out for and rate claims by companies and organizations as to their sustainability credentials or those of their products. On the web site of the Stop Greenwash web site of Greenpeace, they claim that "green is the new black," and point out that while some businesses are "genuinely committed to making the world a better place," some are cynically using a convenient slogan to hide the inconvenient truth [40]. While these sites are serious-minded organizations offering an interesting means of raising awareness, and possibly keeping check on the gap between PR and reality, they are hardly scientific or objective.

Since 2007, the TerraChoice organization has produced a series of reviews of "the sins of Greenwashing." In their latest review in 2010, they noted a general improvement in the way companies were approaching the sustainability agenda: it seems that those more practiced at reviewing their own products, or by extension using external bodies to review their products, became better at avoiding the pitfalls of making exaggerated or misleading claims [31]. Nonetheless, an overwhelming 95% commit one or more of their "sins" of Greenwashing. These are:

1. *Hidden trade-off*—using a narrow set of criteria to claim sustainability and ignoring other aspects of sourcing, production of consumption. An example given for this is paper, which may well be made from sustainably sourced wood, but using an inefficient high-energy process.
2. *No proof*—where claims cannot be easily substantiated by supporting data or third-party audit.
3. *Vagueness*—where generic claims are made. The all-natural example is referred to, pointing out that lead, mercury, and uranium are natural, but their use is possibly not sustainable.
4. *Irrelevance*—where the claim is true but unimportant. The example given is CFC-free. Since the 1970s, CFC use has been illegal, so we would hope that the claim were true!—but irrelevant nonetheless.
5. *Lesser of two evils*—where the product is a less-polluting product than others in the same class, but the class itself performs poorly.
6. *Fibbing*—claims that are based on misleading information or telling downright lies.
7. *Worshipping false labels*—claiming third-party endorsements that are untrue.

In our view, while this system has some merit, any review needs to be based on firm evidence, including embedded energy of the product, a review of the full supply chain and the life-cycle, and other tangible, clearly defined criteria. This view was shared by Joel Makower in 2010 [42]. In his view, TerraChoice were guilty of at least three of their own sins, in that proof was rarely advanced or made available, and high-level unsubstantiated findings are given. A more important issue is one of vagueness: is it "Greenwash" to cry "Greenwash" where a company has made a valid marketing claim but failed to back it up," when the term was originally defined as the deliberate dissemination of misleading or false information to make the organization appear more environmentally friendly than it is in reality. And while Makower admits to subjectivity, his basic point is that, like many other green watchers, TerraChoice reports are also largely subjective, and the company itself is seeking to market their consultancy services to potential clients seeking to avoid being branded as Greenwashers.

This was also the conclusion of M.A. Delmas and V.C. Bulbano, for example, writing in 2011, who concluded that a comprehensive analysis of the drivers that lead companies to engage in the practice of Greenwashing was lacking. Hitherto, most reviews had been empirical, and focused on identification of the culprits, rather than promotion of best practice. Their paper aimed to address that gap [43].

In their analysis, a company that engages simultaneously in poor sustainability performance and positive media communication about their performance is Greenwashing. This may also be a little simplistic. For example, where the case against LG Electronics allegedly committing the sin of Greenwashing by mislabeling some of their domestic appliances as meeting Energy Star ratings may be clear-cut, a similar company some of whose products reach the required standards where others should not be so labeled in our view. It may well be that the more sustainably engineered products are available at a premium while the less so are marketed by the same company under a budget label to access that part of the market who cannot afford the more sophisticated product. In such a case, we would seek to encourage improvement in the sustainability of the budget range by better design rather than castigate the manufacturer.

The "Divers of Greenwashing" paper posits the concept of "green" and "brown" firms, where environmental performance is the differentiator, and focuses on the motivations that lead poor performers to claim better than actual levels of performance. They identify a number of internal and external drivers, which relate well to our earlier discussions about the benefits of adopting a sustainable approach to the consumer, the manufacturer, and society as a whole. These drivers are well represented by their figure on page 5, reproduced here as Figure 9.10.

This chapter avoids a careful definition of good and poor performance. This was a deliberate choice, justified on the basis that it was easier to change the market communications strategy for a "brown" organization than to change their sustainability strategy. To that end, they produce a template for improved communications and advocate a top-down commitment to transparency and accuracy.

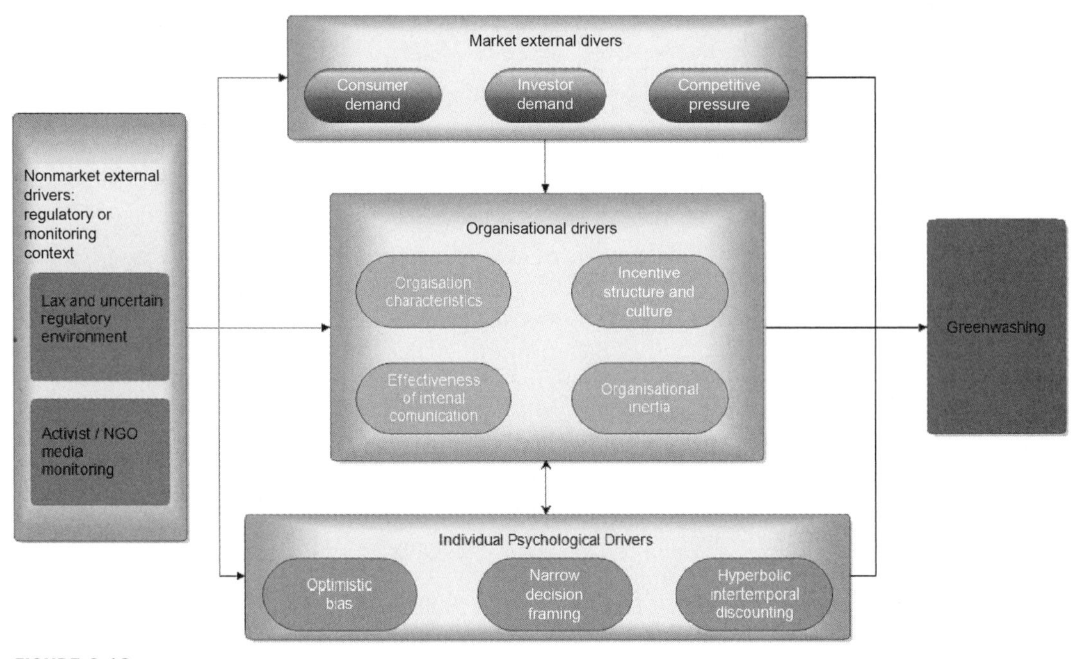

FIGURE 9.10

The Drivers of Greenwashing. (For color version of this figure, the reader is referred to the online version of this chapter.)

Modified from Delmas and Bulbano [43].

Our contention is that while this may represent a positive step forward, sustainable progress is made only when the whole company engages with a positive approach to sustainability, seeking to design it in to their products and services to give a long-term, sustainable future.

As regards, labels and quality management systems such as the Soil Association's Organic status, or Globalgap and Fair Trade, the author's own anecdotal experience can serve to indicate potential weaknesses in the system. When leading the renovation of a mango pack-house in Guinea, the small-scale producers lacked the financial means to invest in soil improvement products, herbicides, and insecticides. The product was therefore nominally organic. However, poverty, the very reason why these products were "organic," prevented them from buying in the skilled and seasoned consultants and auditors required to demonstrate the organic nature of the products.

It is clear that just as with quality management systems such as ISO 9000, people management systems such as *Investors In People*, the "green" or sustainability industry, including the ISO standard 14000 series has created a "monster" in that some of these systems and labels become an end in themselves, giving rise to a lucrative consultancy industry, rather than a means to an end.

That end should be to allow the designers and manufacturers of product to assess objectively their products against meaningful standards, and to allow the buyer, user, or consumer to evaluate the relative merits of individual products within a given class in terms of their impact on global ecology or lifetime sustainability.

In the words attributed to Juval, the Roman poet, quis custodiet ipsos custodies?

References and information sources

[1] J. Elkington, Cannibals with Forks: The Triple Bottom Line of 21st Century Business, The Conscientious Commerce Series, New Society Publishers, Gabriola Island, BC, Canada, September 1, 1998, ISBN: 0865713928.

[2] J. Elkington, Enter the Triple Bottom Line. On-line publishing by KMH Associates ES_TBL_7/1, August 17, 2004.

[3] A. Savitz, K. Weber, The Triple Bottom Line: How Today's Best-Run Companies Are Achieving Economic, Social and Environmental Success—and How You Can Too, E-Book, 2006, ISBN:0470534664.

[4] W. Norman, C. MacDonald, Getting to the bottom of "triple bottom line", Bus. Ethics Q. 14 (2) (2003) 243–262, 2004, ISSN: 1052-150X.

[5] M.L. Pava, A Response to "Getting to the Bottom of 'Triple Bottom Line", Paper to Harvard I-Site, 2007.

[6] Segelocum Ltd training materials, Retford, UK, 2012.

[7] Directive 2006/46/EC of the European Parliament and of the Council of 14 June 2006. http://eur-lex.europa.eu/LexUriServ/LexUriServ.do?uri=OJ:L:2006:224:0001:01:EN:HTML Available from eur.lex.europa.

[8] R. Isaksson, R. Garvare, Measuring sustainable development using process models, Manage. Audit. J. 18 (8) (2003) 649–656.

[9] T. Wiedmann, M. Lenzen, Triple-bottom-line accounting of social, economic and environmental indicators—a new life-cycle software tool for UK businesses, in: Paper Presented to Third Annual International Sustainable Development Conference Sustainability—Creating the Culture, 15–16 November, Perth, Scotland, 2006.

[10] B. Foran, M. Lenzen, C. Dey, M. Bilek, Integrating sutainable chain management with triple bottom line reporting, Ecol. Econ. 52 (2) (2005) 143–157, ISSN:0921-8009.

[11] M. Munasinghe, Environmental Economics and Sustainable Development, World Bank Environment Paper, World Bank, Washington, ISBN: 0821323520, 1993.

[12] R. Parker, Technology Investment and Shareholder Value Lecture at the Imperial Business Insights Series, February 9, 2013, see http://www3.imperial.ac.uk/business-school/programmes/innovation-and-entrepreneurship/.
[13] A. Bickmeyer, M. Hansch, EON 209 sustainability report, May 2013.
[14] J. Stacey, KPMG global sustainability advisor engineers are way down the 'responsible list', J Professional Engineer Journal 18 (9) (2005) 8.
[15] Cascade Engineering Triple Bottom Line Report 2012, Cascade Engineering, Grand Rapids Michigan.
[16] A. Skouloudis, K. Evangelinos, F. Kourmousis, Development of an evaluation methodology for triple bottom line reports using international standards on reporting, April 5, 2009, Environmental Management, Springer Science + Business Media.
[17] Incentives for greening UK business, Bus. Environ. ISO 14000 Updates 16(9) (2005) 7–8.
[18] J. Tullberg, Triple bottom line—a vaulting ambition? Bus. Ethics 21 (3) (2012) 310–324, Article first published online: June 18, 2012.
[19] G. Hubbard, Measuring organizational performance: beyond the triple bottom line, Bus. Strat. Environ. (18) (2009) 177–191, ISSN:09644733.
[20] B. Vivian, Blog, 2009. http://www.vivianpartnership.co.uk/is-triple-bottom-line-a-misguided-principle.
[21] Which report on washing machines: washing energy costs, 2009. http://www.which.co.uk/home-and-garden/laundry-and-cleaning/guides/washing-machine-energy-costs/published.
[22] Bosh Web-site. http://www.bosch-home.co.uk/our-products/washers-and-dryers/washing-machines/ (accessed August 2013).
[23] Sust-It. http://www.sust-it.net/energy-saving.php, published on-line by Sust-It Trurnaround Lts, of Cheltenham UK (accessed August 2013).
[24] E. Carey, Carey Glass Solar Power-Point Presentation 2011, Carey Glass Solar, Dublin, Ireland, 2011.
[25] Green is Sexy Website, blogs published online by Canadian actress Rachel Macadams, 2007.
[26] A. Blackman, J. Rivera, Producer-level benefits of sustainability certification, Conserv. Biol. 25 (6) (2011) 1176–1185.
[27] Initiative for Green Products, announcement published on-line by Enterprise Europe Yorkshire, 2013. http://www.ee-yorkshire.com/yf/news/newsid=773 (accessed June 2013).
[28] A selection of "Green" labels—open sourced via Google Images, June 2013.
[29] L. Demeritt, Understanding the sustainable consumer: lessons learned and upcoming research, Presented by The Hartman Group on-line sustainability webinar notes, 2013. www.hartman-group.com/pdf/sustainability-webinar-2013.pdf (accessed August 2013).
[30] M. Delmas, N. Nairn-Birch, M. Balzarova, Lost in a Sea of Green: navigating the eco-label labyrinth, Abstract 209 paper, published UCLA, January 11, 2009.
[31] The sins of Greenwashing, Home and Family Addition, A report on environmental claims made in the North American consumer market, Terrachoice, www.terrachoice.ca, www.sinsofgreenwashing.org/ (accessed July 2013).
[32] A.S. Morris, ISO 14000 Environmental Management Standards: Engineering and Financial Aspects, Wiley & Sons, Chichester, UK, 2003 Oct, ISBN: 978-0-470-85128-9.
[33] Carbon Neutral Certification, the Website of The Carbon Neutral Company, London and New York. http://www.carbonneutral.com/our-services/carbon-neutral/ (accessed August 2013).
[34] L. Bateman, The Green Label Guide, The Sixty Mile Publishing Company, London, 2013. http:www.greenwisebusiness.co.uk/.
[35] D.O. Ward, C.D. Clark, K.L. Jensen, S.T. Yen, C.S. Russell, Factors influencing willingness to pay for the energy star label, Energy Policy 39 (3) (2011) 1450–1458.

[36] A.G. Murray, B.F. Mills, Read the Label! Energy Star appliance awareness and uptake amongst US consumers, in: Paper Presented to Annual Meeting, Agricultural and Applied Economics Association July 24–26, Pittsburgh, Pennsylvania, 2011.
[37] Der Blaue Engel Umwelzeichen mit Markenwirkung web-site of RAL GmbH, D53757 Sankt Augustin, http://www.blauer-engel.de/ (accessed August 2013).
[38] EU Ecolabel, The European Commission, published by Office for Official Publications of the European Communities, 2006. www.eco-label.com (accessed August 2013).
[39] I. Quinn, Frustrated Tesco ditches eco-labels, The Grocer (2009). http://www.thegrocer.co.uk/companies/supermarkets/tesco/frustrated-tesco-ditches-ecolabels/225502.article. (accessed July 2009).
[40] About Energy Savings Trust, website. http://www.energysavingtrust.org.uk/Take-action/accessed (accessed August 2013).
[41] Greenwashing Index: help keep advertising honest promoted by Enviromedia Social marketing, supported by University of Oregon School of Journalism and Communication. www.greenwashingindex.com/ (accessed August 2013).
[42] J. Makower, Is TerraChoice Greenwashing? Greenbiz Website Blog November 1, 2010. www.greenbiz.com/blog/2010/11/01/terrachoice-greenwashing (accessed July 2010).
[43] M.A. Delmas, V.C. Burbano, The drivers of Greenwashing, Calif. Manage. Rev. 54 (1) (2011) 64–87.

CHAPTER

Predicting the Future

10

10.1 UNSUSTAINABLE FUTURES

10.1.1 Challenging views

When the author was studying mechanical engineering and German in the early 1970s as an undergraduate at Bath University, our German applied mechanics tutor, Gustav Winkler, was a champion of the simple over the complex and the mechanical over the mechanized. Following the presentation by a visiting lecturer on the nascent subject of "alternative technology," his challenging view was that what we needed were alternatives to technology not alternative technology.

In essence, this view has been expounded generally by "naysayers" for whom the solution does not lie within the development of new and better technological solutions, but rather within the return to some real or imagined bucolic golden era, where people interacted with their environment on a more sustainable level, using mechanical rather than electromechanical devices, and remaining in harmony with their resources: living sustainably.

Incidentally, we must credit the same now Prof. Dr. Winkler (Turbo Gustav) with the design of a wind-powered car—see succeeding text—so it would seem the view was expressed as a devil's advocate (Plate 10.1)!

PLATE 10.1

Prof. Dr. Winkler and a Student Demonstrating the Wind-Powered Vehicles [1]. (For color version of this figure, the reader is referred to the online version of this chapter.)

Sustainability in Engineering Design. http://dx.doi.org/10.1016/B978-0-08-099369-0.00010-8

The *contra-wind* car is a demonstration that the power of the wind can be harnessed to travel into the wind, without tacking. The Flensburg-built machine was the winner at the Danish wind-powered competition held at Stauning airstrip in 2010, reaching between 35% and 50% of the oncoming wind speed powered only by the wind [2].

10.1.2 The limits to growth

Thus, since the early discussions prompted by the publication of *The Limits to Growth* [3], there has been a polarization of views.

Those supporting Meadows are generally held to be in the minority. They argue that there are limits to growth set by the finite resources contained within our planet and that current global problems such as ecological, equity, conflict, and social cohesion problems and the differences between the developing and developed worlds cannot be resolved without radical change in the fundamental social and market models and principles. The key to this is the abandonment of high material living standards maintained by a largely unfettered market economy. The contention is that advances in technology cannot sufficiently reduce the use of resources required to maintain the current levels of consumption or the economic growth necessary to sustain the current prevailing economic model. Thus, extending these levels of growth or consumption to the developing world would prove fatal to the world's ecological balance, ensure the final consumption of a number of critical mineral and other resources, and cause the collapse of society.

What is required, according to this view, is an alternative approach that will change the dynamic of consumption and growth and ensure a more equitable distribution of the access to resources.

There are some well-presented arguments in favor of this view. In his paper [4], Ted Trainer supports the challenges the "technical fix" solution espoused by the positivists, taking issue, for example, with the prevailing view that modifications to the market and resource scarcity and a combination of legislation and technological development will bring about a sustainable future. He is justifiably critical of the arguments advanced by Julian Simon in *The Ultimate Resource* [5] regarding the apparently inexhaustible nature of resources, which will be controlled purely by the classical elastic pricing model, or that we will never run out of energy. Other reviewers have also challenged the simple or even simplistic logic of this book, often seen as an attack on the limits to growth espoused by the neo-Malthusians, including H.E. Daly, who exposed the illogicality of the infinity-based arguments, or at least how Zeno's classical paradox of Achilles and the tortoise may not be applied to unknown resource levels [6]. Nonetheless, although Daly's assertion that dismissing Simon's argument as flawed indicates that the neo-Malthusian first principles are unscathed is equally unconvincing, his conclusions are somewhat more moderate than the rhetoric would suggest, in that while sensibly objecting to "Further prolongation of the current compulsive quest for infinite growth, power, and control," he is actually seeking a steady-state future rather than a dismantling of current structures and activities.

10.1.3 The natural market-driving sustainability

The alternative viewpoint is that political and socioeconomic pressures will force change on commerce and industry and that those changes will substantially alter the consumption patterns and hence demand, while the advancements of science will be translated by engineers, microbiologists, geneticists, and the like into products and systems that help satisfy the need to material advancement while reducing or eventually eliminating the burden on the planet and its inhabitants. Thus, for proponents of this argument, ways will be and are being found to improve cultivation rates and yields to ensure that an ever-growing world population will be fed. Within this strain of thought, it is also assumed that elastic market forces will price out of reach increasingly rare resources and thus reduce consumption, without the need for a radical rethink or reorganization of the basic tenets and principles of westernized "civilized" society.

This could be styled the technical fix solution.

Certainly, the contributions to the 2010 REWAS symposium on sustainable metal production in 2010 suggested that new ways are being sought to treat current waste products to source key materials for future use and that recycling of base materials, such as metals, was becoming increasingly resource-efficient. In their review of the event, Anne Kvithyld and Christina Meskers acknowledged the importance of detail and the contribution of the material scientist and the metallurgist. They also observed that the detail is not enough: "Understanding sustainability requires also a bigger system or life cycle perspective." Their argument shows how sustainability is bigger than the engineering sciences and should involve the social sciences for context. They also argue that clear communication from the scientific to the lay communities is an essential ingredient for success in developing sustainability strategies [7].

Erickson and Gowdy conducted a historically based study, *Resource Use, Institutions, and Sustainability: A Tale of Two Pacific Island Cultures*. The study concluded that extending the historical results of two differing approaches to managing resources and population growth "would imply that the amelioration of natural resource scarcity through trade and emigration might artificially raise carrying capacity in the short-run, but would be limited in the long-run due to the destruction of essential natural resources" [8]. The natural market solution is, then, implicitly flawed.

This chimes well with the authors' observations in travel within West Africa, where subsistence existence within tropical rainforests is largely unchanged by industrial development, but continues nonetheless at the expense of resource depletion. Life expectancy is relatively short, and by all typical measures of adult literacy, infant mortality, access to clean water, etc., these communities are on the margins of what we regard as developed society. Since the vast majority are "equally poor," perception of poverty only really comes into play when these same people are displaced or migrate into the cities, where their condition and hence status can and will be measured against their neighbors. The key is, however, that resource depletion, such as denuding forests of edible wildlife or even deforestation for charcoal burning and manufacturing of domestic products or implements, does not seem to limit population growth.

10.1.4 The third way

There is a third strain of thought that recognizes the current starting position and proposes a mix of the two opposing views:

- Advancing technology will indeed help to reduce the burden of growth on the planet's resources.
- At the same time, political and philosophical pressures will modify the current pure market economy, replacing it with a more collaborative and inclusive model

Indications of this sea change in approach are found in

- collaboration projects such as Wikipedia and open-source software such as Linux;

- crowdfunding of projects and product development, where the innovator appeals directly to the user for funding and support;
- the huge increase in awareness of sustainability issues as witnessed by the wealth of informal discussion, articles, and blogs.

In this vein, the World Economic Forum report in 2009 concluded that innovation was the only way forward: there is a long-term need to dematerialize the economy and focus on satisfying the consumers' rather than the producers' needs. This will also demand an improvement of collaboration along the whole value chain, based around improvements in resource efficiency, product takeback, and sharing [9]. They also comment on the temporary drop in some material resource prices, driven in their view by the world downturn, and due to recover before the middle of the decade as demand recovers. They do emphasize the roles played by both producer and consumer in the way ahead. The producers must seek to diversify material sources and reduce waste while becoming fundamentally more energy efficient in the production process. To make this an effective strategy, companies will need to share resources and collaborate. The consumers will react to the inevitable price rises in energy, food, and water resources by seeking out more efficient ways to consume these resources.

Jacqueline Novogratz in her talks about patient capitalism posits the theory that sustainable jobs, goods, and services add dignity to the world's poorest and that we need to change the way of working by listening to the demands of the consumers, including low-income consumers to improve their access to goods and services in a sustainable manner. The key from her viewpoint is to fuse the interests of commerce, society, and ecology. Effectively, she is also advocating the triple bottom line, both at the bottom of the business pyramid and at the top [10].

The more utopian ideals of happiness index from Nic Marks' claim that measuring the success of an organization or country by the happiness index as an alternative to production will lead to a sustainable future: the "Happy Planet Index" (HPI). The index uses life expectancy experienced well-being and ecological footprint. The results are displayed in tabular form and as a colored world map. Countries such as Argentina, Vietnam, and Mexico score well under this system; India and Brazil are ahead of the United Kingdom, most of Europe, and Canada; while the USA ranks alongside Denmark, just above most of sub-Saharan Africa at the bottom. While this may be at the more extreme end of the systems of measurement, it does indicate that we can and should be thinking along new lines to encompass the full impact of sustainability on the world [11].

Adnan Khan in *The Khalifa* 08/10/2008 article proposed Islam as an alternative and more sustainable way forward, but while his arguments may carry some weight as regards Islamic banking as an alternative to the current Western banking model, there was no real evidence of a commitment to sustainable use of resources in his article [12].

Beers and colleagues of Wageningen University in the Netherlands reported on a *Transforum* study of the way public perceptions and attitudes shaped the direction of sustainable development. In their paper [10], they accept that ambitions for sustainable development require a structural change of existing societal (sub)systems. These changes will result from experimentation and innovation, and the images associated with these transitions can have a decisive influence on decisions whether to adopt, support, or reject particular ideas and innovations. An example cited is that of biofuels, where their initial positive image as a sustainable fuel drove a specific response by many governments, particularly in emerging economies. The consequential changes in cropland usage and unsure net outcome in terms of net greenhouse gas emissions could be said to be a result of an initial overpositive image.

On the more positive side, Hartley reported in 2010 for the Energy Savings Trust that sustainable goods and services represent both a growth opportunity in themselves, valued at £200 billion in the United Kingdom alone, but that adopting sustainable policies, practices, and standards could produce annual productivity savings of 100 billion. This suggests that it is already making commercial sense for companies to adopt a positive approach to sustainability and develop standard procedures and designs to embed their strategy.

Supermarket group Asda (part of the Wal-Mart family) is working with the University of Leeds to set up a large study into public attitudes to sustainability, cost, and value. By looking at "mainstream" (rather than the committed "deep green") consumer, we can start to influence opinion and generate positive action on waste management (including food, energy, and rubbish waste) and buying habits and thereby help the move to a low-carbon society. Furthermore, the survey intends to help clarify communications with the consumer to improve trust in quality and sustainability symbols and measurements [14].

In the words of a rival supermarket's advertising campaign, "every little helps."

10.1.5 How the viewpoints collide: example, the milk bottle

Consider the issue of the milk bottle. Our childhood memories of milk delivered in truly recycled (i.e., steam-sterilized) bottles have been replaced by blow-molded HDPE (high-density polythene) containers purchased from the local convenience store or supermarket.

The culture of re-using glass bottles continues in Germany, where the collection, sterilisation and re-use of glass beer bottles has become standard practice, being both commercially successful and widely accepted and understood by the consumer, and is encouraged by a returnable deposit system, as we had many years ago in the United Kingdom.

This older model apparently offers a more sustainable alternative: over 90% of the bottles are reused typically up to 10 times, with each reuse being subject to the added embedded energy of return transport, condition check, and cleaning. According to Van Doorsselaer and Lox [15], the break-even point is around 5% breakages, below which multiple-use glass bottles tend to win out over single-use ones, based on energy needs over the life cycle.

The newer solution of HDPE offers reduced mass and lower embedded energy of production and recycling, but it is a single-use product, with the possibility of recycling, currently in the United Kingdom at a rate of around 76% in 2010 [16]. A study in 2011 by Singh, Krasowski, and Singh suggested that modern HDPE containers, even packed in reusable crates, still offered an advantage in terms of overall energy used in transportation [17].

So, which of the two is actually the more sustainable and is the traditional reuse model better overall, as many activists would believe? It appears that most bloggers writing between 2010 and 2013 opted for the traditional route of reusable glass bottles. Although some contributors accept that they have limited access to facts on the relative embedded energy or greenhouse gas emissions, they nonetheless hold opinions based on the "feel" of one solution over the other or based on single against multiple use: a single aspect of the overall whole life cycle issues.

The truth is that each system needs to be examined as a whole. A calculation of the embedded energy or ecological footprint of both systems would be required, based on properly gathered evidence and using one of the currently accepted calculation models or better still using the algorithm in development by the authors, taking into account the embedded energy across the whole life cycle of the product.

The conclusions of the existing study in India, referenced earlier in the text, must be correct for that specific distribution mode, whereas the sustainability merits of recycling (largely locally) beer bottles in Germany need to be examined in that context and may well differ from the situation regarding milk, produced increasingly in larger, more efficient dairy centers and distributed regionally or nationally.

10.2 THE ENGINEERS' VIEW

In ASME's third annual survey published in November 2011, Brown indicates that while many of the 2000 engineers and students surveyed have apparently embraced the concept of sustainability in their work, some among them see the term as a "trendy name for what used to be called good engineering." And while some dismiss the concept as "flavor of the month," most see that sustainable practices are ultimately cost-saving.

As an indicator, the responses to the question "Over the past year, approximately what portion of all of your projects included specifications that were based on sustainable and/or green design principles beyond those mandated by regulations?" are given in Table 10.1.

So what are these engineers and their organizations actively doing to promote sustainability? 62% of the respondents cited the need to design for reduced energy use, 35% were seeking to reduce material waste in manufacturing, and 27% were looking to improve manufacturing efficiency in both energy and material usage terms [18].

The importance of sustainability to the organizations for whom these engineers were working is also used as an indicator of the level to which sustainability is embedded into engineering practice at an organizational level. The result for the question "How involved is your organization with sustainability or sustainable design practices?" gives a moderately positive reading, with over 40% indicating that their organization is "somewhat involved" (see Table 10.2). It will be interesting to see how this

Table 10.1 ASME Members' Responses to "Proportion of Projects Involving Sustainability Question" [18]

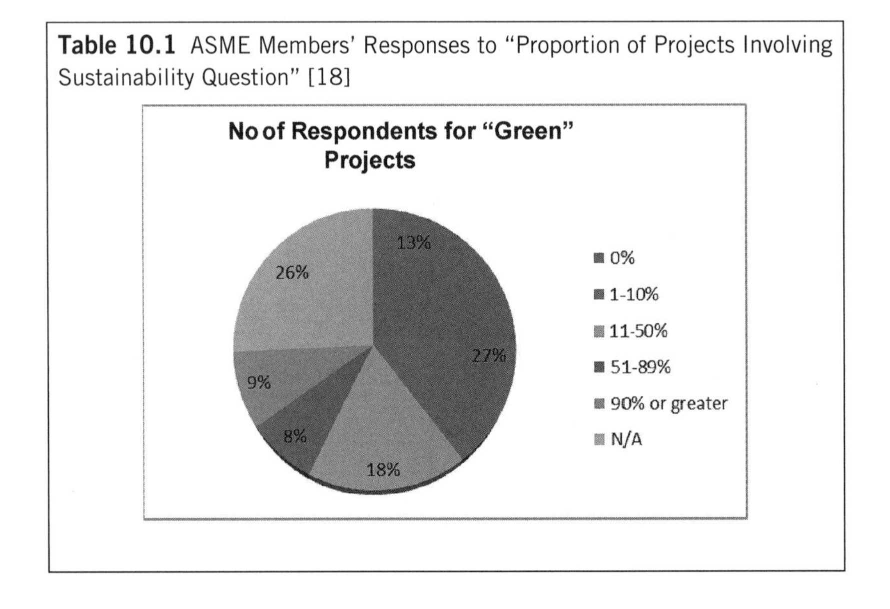

Table 10.2 ASME Members' Responses to "Proportion Organizations Involved in Sustainable Practices Question" [18]

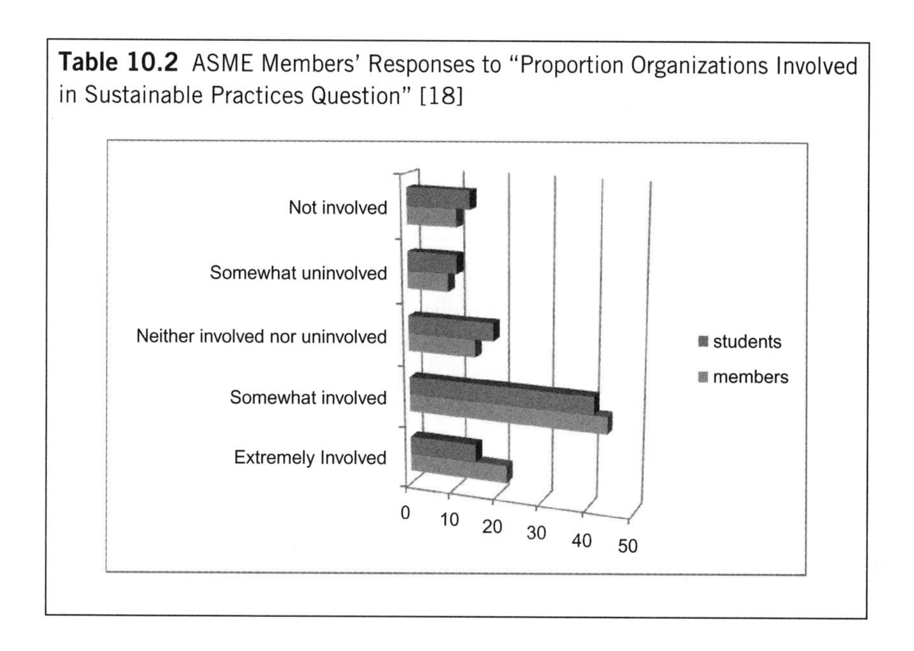

changes as the pressures of market and cost come to bear more heavily in the future and as smaller organization begin to embrace the 3BL principle.

Neither the Institute of Mechanical Engineers nor the UK Engineering Council annual membership surveys have to date addressed this subject, although the Engineering Council does issue sustainability guidance to its members, including a six-point plan:

- Contribute to building a sustainable society, present and future.
- Apply professional and responsible judgment and take a leadership role.
- Do more than just comply with legislation and codes.
- Use resources efficiently and effectively.
- Seek multiple views to solve sustainability challenges.
- Manage risk to minimize adverse impact to people or the environment [19].

The German Association of Engineers (VDI) is also keen to promote sustainable development and focus its membership on these issues, with a number of studies conducted, and an active lobbying voice in Europe and domestically for improved legislation in this regard [20].

10.3 CONCLUSION

In conclusion, it seems that in many articles and publications concerning these issues, several key factors seem to emerge:

(a) The number of articles, blogs, and papers attacking the current system actually outweighs those supporting it; thus, the status of this as a minority view is uncertain: it is becoming the popular view that the current system as a whole is not sustainable.

(b) The observation and listing of failures of the current system and the impending doom for human society or indeed survival of the planet are rarely accompanied by a credible argument as to the alternative.

(c) There is far more published opinion in online sources, blogs, articles, and social networking site entries than there are learned papers, and there is far more from social scientists and economists than from physical scientists and engineers.

(d) Arguments from all sides are often based on a limited view of the effects of specific products, packaging, or energy sources. Few consider the whole life cycle.

(e) Can the differentiation between the sustainability of two alternative products, materials, or processes be measured entirely in terms of greenhouse gas emissions or embedded energy? How then do we account for the means of generation of the energy so embedded? Or factor in the effect of pollution of the ground or watercourses?

10.4 IF I WERE YOU, I WOULDN'T BE STARTING FROM HERE!

10.4.1 We can only move forward

In our view, sustainable development is not an issue for articles of faith or belief systems alone, nor is it a discussion that should provoke intolerance of opposing viewpoints at an almost fanatical level.

What is critically important is that the general public, opinion formers, and indeed politicians understand and debate the concepts associated with sustainability based on observed fact rather than wild speculation and that we seek to develop a solution from where we are, not where we would like to be.

The famously apocryphal advice, genuinely given to the author on seeking directions in Ireland, "If I was you I wouldn't be starting from here," is not applicable in this case.

We are where we are and need to move from here, not from some imagined ideal starting point.

10.4.2 The benefit to the environment

The benefit to the environment of designing in sustainability is quite clear. If we improve fundamentally our approach to sustainability within the field of engineered and manufactured product, we reduce our impact on the environment while simultaneously continuing to offer a maintenance of or even improvement in the standard of life or living of the inhabitants of the planet.

Even if the pessimistic position is that we delay the inevitable overexploitation of the planet and the resulting destructive trigger points by two or three hundred years, the optimist, and indeed the engineer, in us suggests that developing technology will continue to push that envelope forward, allowing a little longer for the much slower process of evolutionary social and economic change to take place. It is in our view clearly better to drive forward the pace of technological change while being mindful of the need to report openly and honestly on that change and its sustainability impact.

10.4.3 The benefit to the consumer

The benefit for buyers and consumers of these better-engineered and more consciously sustainable products is less immediately clear. Will they be more expensive? Will they be inferior in some way? Will they mark a step backward toward some imagined bucolic utopian past? Will they simply be another means of marketing ever-increasing consumption to those who already overconsume?

The answers to these questions are by no means certain, but the indicators are currently that consumers in the developed world are already seeking to change their buying habits with the deliberate intent of improving or securing their environment. Clearly, this is a change of small steps—evolution not revolution—and some believe it is either too little too late or simply a veneer of sophistication on a basically flawed consumption model.

Our view, based on the current evidence and trends, is that the increasingly well-educated younger generation is genuinely seeking ways that they can positively influence a sustainable future for themselves and their offspring while striving to maintain a certain standard of living for themselves and their nearest and dearest.

10.4.4 The challenge of the designer

And what of the designers, manufacturers, and producers of these products, machines, and equipment? Will they trade in their hair shirts of penance for the silk or satin garments of smugness and superiority? Will they use the illusion of a fundamental shift in the design paradigm to continue on the path of constant growth necessary to stoke the fires of traditional capitalism?

It is clear that the command economy model has failed. It is equally clear that unfettered (natural) capitalism and unchecked growth have also failed. So, if we are to be ruled by neither the politics of envy nor those of greed, can we devise a way that the natural human ingenuity, that which has marked us out for a stellar path in evolutionary terms, will become a force for good rather than a force for ill?

All around us, we can see evidence of how standing still is not an option: our own curiosity, greed, and desire for constant "improvement" will ensure that this is not so.

From history, we can learn that the scientist and the engineer must take moral responsibility for the effect and use of their systems, machines, and creations. We also see that if public reaction can promote or stifle alternative ways forward, then the public must become better informed, allowing an open and clear debate based on issues and facts rather than dogma, opinion, and self-interest.

This involves giving designers the tools by which they can make a positive contribution: educating other decision-makers as to the whys, wherefores, and possibilities offered by this approach and communicating the resulting successes and failures to the public.

10.5 THE WAY FORWARD

Do we anticipate an immediate response to or recognition of the fundamental principle espoused by us in the book?

Taking the historical view shows that many now cannot see what the "fuss" was about when Taguchi suggested that the design engineer should be fundamentally responsible for quality: this is now accepted as common sense. We anticipate that in 10 years time, the idea that the designer is responsible for the sustainability or otherwise of their products, designs, and creations will be similarly thought of as common sense. In the meantime, we need to ensure that

(a) designers the world over take this message on board;
(b) the companies and organizations within which they work accept that this is so and that it forms part of the companies responsibility under 3BL;

(c) the buyers and consumers of engineered products, white goods, etc., expect to be able to evaluate with some confidence the sustainability of any given product and compare it to others of similar vein, just as we can for the nutritional value of foodstuffs at present;
(d) that the bases on which these evaluations take place are factual, well researched, and straightforward in application;
(e) that the generation of designers, engineers, managers, social scientists, politicians, and consumers we are now producing will have received sufficient education on the critical issues surrounding sustainability that they too will become the solution and not the problem.

References and information sources

[1] Kurioses Rad. Mit Gegenwind immer Rückenwind, Spiegel Online,11.08.2008. http://www.spiegel.de/unispiegel/wunderbar/tueftelei-eines-professors-turbo-gustav-und-sein-gegenwindfahrrad-a-570961.html (accessed May 2010).
[2] Aktuelle Meldung Gegenwind für "Turbo-Gustav" 10.09.10 FH Flensburg web site, Kategorie: Pressestelle Autor: Pressestelle. http://www.fh-flensburg.de/fhfl/aktuelle_meldungen.html?&cHash=b3f5e8ef18&tx_ttnews%5BbackPid%5D=230&tx_ttnews%5Btt_news%5D=336 (accessed August 2010).
[3] D.H. Meadows, D.L. Meadows, et al., "The Limits to Growth": A report for the Club of Rome's project on the predicament of mankind, (1979).
[4] Natural Capitalism Cannot Overcome Resource Limits, Ted Trainer, Social Work, University of NSW, Australia. https://socialsciences.arts.unsw.edu.au/tsw/D50NatCapCannotOvercome.html (accessed July 2010).
[5] J.L. Simon, The Ultimate Resource, Princeton University Press, 1981, ISBN 978-0-85520-583-6.
[6] Julian Simon, Herman E. Daly. Review of The Ultimate Resource, Originally published in the Bulletin of the Atomic Scientists, January 1982.
[7] Anne Kvithyld, Christina Meskers, Different perspectives in enabling materials resource sustainability, J. Miner. Metals Mater. Soc. 65 (8) (2010) 982–983.
[8] Jon D. Erickson, John M. Gowdy, Resource use, institutions, and sustainability: a tale of two Pacific Island Cultures, Land Econ. 76 (3) (2000) 345–354.
[9] Sustainability for Tomorrow's Consumer, The Business Case for Sustainability, Report was prepared in collaboration between Deloitte Touche Tohmatsu and the World Economic Forum, January 2009.
[10] Jaqueline Novogratz. Patient capitalism. Filmed June 2007, Posted August 2007, TEDGlobal 2007. http://www.ted.com/talks/jacqueline_novogratz_on_patient_capitalism.html (accessed August 2010).
[11] Nic Marks, The Happy Planet Index. http://www.happyplanetindex.org/ (accessed August 2010).
[12] There are Alternatives to Free market Capitalism Article in *The Khalifa* 08/10/2008 Adnan Khan. http://www.khilafah.com/index.php/the-khilafah/economy/3919-there-are-alternatives-to-free-market-capitalism (accessed August 2010).
[13] D.F. van Apeldoorn, K. Kok, J.M. Vervoort, F.L.P. Hermans, P.J. Beers, Future sustainability and images, Futures 0016-328742 (7) (2010) 723–732.
[14] Garry Hartley, The Green Economy: delivering the goods, Energy Savings Trust blog 14/08/2010 (accessed August 2010).
[15] K. Van Doorsselaer, F. Lox, Estimation of the energy needs in life cycle analysis of one-way and returnable glass packaging, Packag. Technol. Sci. 12 (2000) 235–239.
[16] B. Boukley. HDPE milk bottle report shows recycling rates up. November 8, 2011, William Reed Business media. http://www.dairyreporter.com (accessed August 2010).

[17] J. Singh, A. Krasowski, S.P. Singh, Life cycle inventory of HDPE bottle-based liquid milk packaging systems, Packag. Technol. Sci. 24 (2011) 49–60.
[18] Alan S. Brown, ASME's third annual survey finds that engineers are still trying to understand how sustainability fits into their workflow, Mech. Eng. 103 (11) (2011) 36–41.
[19] Engineering Council web site. http://www.engc.org.uk/about-us/sustainability (accessed August 2010).
[20] VDI web site. http://www.vdi.eu/about-us/ (accessed August 2010).

Index

Note: Page numbers followed by *f* indicate figures, *t* indicate tables, and *p* indicate plates.

D

E

F

G

H

I

J

K

L

M

N

O

P

T